Galvani's Spark

Galvani's Spark

The Story of the Nerve Impulse

ALAN J. McCOMAS, MBBS
Emeritus Professor of Medicine
McMaster University
Hamilton, ON
Canada

Oxford University Press, Inc., publishes works that further Oxford University's objective of excellence in research, scholarship, and education.

Oxford New York
Auckland Cape Town Dar es Salaam Hong Kong Karachi Kuala Lumpur Madrid Melbourne
Mexico City Nairobi New Delhi Shanghai Taipei Toronto

With offices in
Argentina Austria Brazil Chile Czech Republic France Greece Guatemala Hungary Italy
Japan Poland Portugal Singapore South Korea Switzerland Thailand Turkey Ukraine
Vietnam

Published by Oxford University Press, Inc.
198 Madison Avenue, New York, New York 10016

www.oup.com
Oxford is a registered trademark of Oxford University Press

Library of Congress Cataloging-in-Publication Data

McComas, Alan J.
Galvani's spark : the story of the nerve impulse / Alan J. McComas.
p. ; cm.
Includes bibliographical references and index.
ISBN 978-0-19-975175-4
1. Neural conduction—Research—History. 2. Neural transmission—Research—History. 3. Action potentials (Electrophysiology)—Research—History. I. Title.
[DNLM: 1. Hodgkin, A. L. (Alan L.) 2. Huxley, Andrew, Sir. 3. Neurophysiology—history. 4. Action Potentials. 5. History, 20th Century. 6. Nervous System. 7. Neurophysiology—instrumentation. WL 11.1]
QP363.M385 2011
616.8'047547—dc22 2010049113

Dedicated to Sir Andrew Huxley, O.M., F.R.S.
Nobel Laureate, 1963

UNIVERSITY COLLEGE LONDON

Though so many years have passed, I am still able, if I close my eyes, to see some of their faces again. I picture them at a meeting of the Physiological Society, sitting on the wooden benches in the old lecture theater at University College London. There, halfway back, is Alan Hodgkin, younger colleagues on either side. As he turns his head to one of them, there is that half-smile and the glint in the eye that speaks of his confidence, of his great enjoyment of life. Everything seemed to come so effortlessly to him. On the other side of the lecture theater, in the second row from the front, a dark-haired man in a gray three-piece suit gets to his feet and looks intently at the person who has just finished his communication to the Society. In his beautifully modulated voice, Andrew Huxley will ask whether the speaker had considered an alternative explanation for his findings. And, of course, the answer will be no. Who else but Andrew would have thought so deeply, and would be so familiar with such a range of problems? Sitting towards the back of the theater, a slightly older man nods his head in appreciation of the question: Bernard Katz. Then comes a distraction: the door at the rear of the lecture theater opens and heads turn to watch a fine-looking elderly man, ramrod straight, walking carefully down the steps to the front: A. V. Hill.

Four Nobel Laureates in the same room.

Though he was sometimes there too, Eccles is more difficult to place in the same lecture theater. I see him instead attending one of the many symposia to which he continued to be invited after his retirement. Here he comes now, in his navy blue suit, with those intimidating dark-framed spectacles, walking to the platform where he will chair the next session.

Kenneth Cole I saw but once, and then in the dimmed light of the large Los Angeles convention hall, addressing hundreds of neuroscientists with an unmistakable bitterness in his voice. . . .

Six men, all of them giants in their field, and all of them interested in the nature of the nerve impulse—the basis of all our thoughts and actions. True, there were others, some of whom had gone before, and there would be a select few who would provide remarkable detail afterwards. In their own time, however, it was the work of these men that captured attention and was to prove especially important. Much of it had been done without technical help and with little or no financial assistance. Often they had built their own equipment and typed their own manuscripts. All of them had been driven by a natural curiosity, without thought of reward. In different ways, their lives had been extraordinary.

I had half-hoped that someone else would tell their story, and the events that led up to it, but the years have passed and soon there will be none to remember it. For those interested in science and in human nature, it is a story worth telling.

PREFACE

In writing a book of this kind there is the perennial problem of deciding where to start. Since the motivation for writing it was, in large part, derived from my respect and admiration for Sir Andrew Huxley, one possibility was to start with the summer of 1939, when Hodgkin and Huxley, respectively research fellow and student, had set off from Cambridge to do their research in Plymouth at the Marine Biological Laboratory. However, such a beginning would have neglected to place their experimental work, and the theoretical insights that stemmed from it, within a proper historical context. There was also the point that the nerve impulse was not the only form of "animal electricity" (Galvani's term) to have attracted the attention of previous workers.

There were other considerations, too. While the understanding of the nature of the nerve impulse was to be the theme of the book, there was the question as to what happened when the impulse reached its destination at the end of the nerve fiber. And would it not be appropriate, whenever possible, to bring in the work of other Nobel Laureates who had received their awards for research on the nervous system? Finally, there was a debt to the Physiological Society that had to be repaid. For more than 30 years I had the privilege of membership of the Society, and the meetings of the Society gave me the opportunity to see some of the greatest names in physiology, to hear them present their work, and, very often, to see them demonstrating their experiments. And, apart from that, there were opportunities to stay at a variety of universities, including some venerable colleges, and, in the tradition of the Society, to enjoy dinners in the company of old and new friends. With this in my mind, the book came to include the founding of the Society and the early days of British physiology.

As I began writing, I could not help reflecting that someone else might have done it better, either a professional scientific historian or perhaps a membrane physiologist with first-hand experience of the kinds of problems that Hodgkin and Huxley had encountered. The year 2002 would have been an appropriate time for such a book to appear, for it marked the 50th anniversary of the publication of the all-important Hodgkin–Huxley papers on the nerve impulse in the squid giant axon. Though there were a few journal articles, including an account of the mathematical treatment of the impulse by Andrew Huxley himself,[1] it was not to be. There were, however, two qualifications. A few years earlier, there had been an authoritative historical review of research into "animal electricity" and the nerve impulse by Marco Piccolino of the University of Ferrara.[2] And then, in 2000, Stanley Finger's history of some of the pioneering studies in neuroscience had appeared. Entitled *Minds Behind the Brain: A History of the Pioneers and their Discoveries*,[3] this excellent book, the work of a professional historian, did cover some of the research on the nerve impulse, notably that of Galvani and Volta, and much later, that of Adrian, Erlanger, and Gasser. However, Finger's book was, and is, broad in its scope, with much attention given to the localization of function in the different areas of the cerebral cortex. Although the first draft of my own book had been completed before I became aware of Dr Finger's, it would be misleading to pretend that *Minds Behind the Brain* had no influence on the final version, and my notes will make this obvious. As one example, I had not known that one of the homemade cathode ray tubes used by the St. Louis group to display nerve impulses for the first time was still in existence among the archives in Washington University. I hope that *Minds Behind the Brain* and *Galvani's Spark* will be seen as complementary to each other, even if the prize for erudition is awarded to the former.

Though hesitant, I was not entirely unprepared for my new task. Other than an acquaintance with Andrew Huxley, one I viewed as between apprentice and master, I had carried out some simple experiments on peripheral nerves as a student, employing the same techniques that Lorente de Nó had published eight years before.[4] Later, I had used microelectrodes to record from the muscle fibers of patients with hereditary disorders causing stiffness or paralysis of their muscles. Later still, there had been research on other neurophysiological topics—muscle fatigue, spinal reflexes, and a method for estimating the numbers of motor nerve cells in the human spinal cord during life. It was the kind of work that caused small ripples rather than a big splash, but it had been enormously enjoyable all the same. Then there had been the scientific meetings, especially those of the Physiological Society, at which it would be possible to see Alan Hodgkin and Andrew Huxley in action, not to mention many other fine neurophysiologists, among them Bernard Katz and John Eccles. From the point of writing a book, the combination of laboratory work and attendance at meetings had, as it were, kept me in the game. Even now, half a century after the first faltering attempts, I still spend part of the week looking at nerve impulses on an oscilloscope screen and I find them just as fascinating now as I did before.

As the book began to take shape, I had to reject a number of preconceived ideas. There were some discoveries as well, such as the reason for George Bishop not receiving a share of the Nobel Prize in 1944, and the cause of Kenneth Cole's displeasure. But even if there were character blemishes, there was usually much more to admire, to the point that I was obliged to shed some longstanding prejudices. Indeed, all of those about whom I have written were remarkable persons, accomplishing much and often doing so with very little in the way of resources. As Eccles has recounted,[5] in all Sherrington's time at Oxford, not only was the laboratory equipment primitive and sparse, but there was very little money for research and no secretarial assistance in preparing the papers. It also became clear to me that, in our desire to press on, to devise and use new techniques for examining the nervous system, we were in danger of overlooking some of the older observations, and, more importantly, the deductions from those observations—observations and deductions that are still valid. Who, studying consciousness today, will have taken the trouble to read Helmholtz's *The Facts of Perception*? Yet this magnificent essay,[6] the written version of an address made at Berlin University in 1878, contains much that is relevant to the beginnings of consciousness in the human brain, and surpasses many of the studies that have appeared since.

I also made discoveries of a different kind, one being that the founding chair of surgery at my university, now an emeritus professor in his nineties, had been taught physiology by Joseph Erlanger and, still a student, had assisted George Bishop in his research.[7] Another discovery, quite unexpected, was that all Sherrington's regalia and awards were not in Oxford, as one might have anticipated, but in the Woodward Library on the University of British Columbia campus. It was there that I was able to hold and inspect one of Sherrington's laboratory notebooks, a school exercise book with the experiments written up in pen and ink. Through the kindness of the Waynflete Professor of Physiology at Oxford, at that time Dr. Colin Blakemore, my sense of communion with Sherrington was raised still further by being able to spend a summer month working in his old office and sitting at his huge, double-sided desk. In the same year, while visiting my sister in England, I chanced upon the modest terrace house in Ipswich where Sherrington had lived after retirement, in the back garden of which he had been photographed with Adrian and Forbes.[8] Yet another surprise, a pleasant one that provided encouragement during the long literary searches, was that some of the journals in the McMaster University medical library had once belonged to John Fulton, the eminent Yale neurophysiologist and medical historian, and a friend and former research associate of Sherrington.

The writing of the book was also helped by the availability of so much published material, not only the scientific papers but also the autobiographical essays and, later, the obituary notices of the main characters. Especially helpful were the *Biographical Memoirs of Fellows of the Royal Society* and the *Nobel Lectures* (both available on the Internet). Though somewhat averse to electronic communication, I have every reason to be grateful to the young men who invented Google and I remain astonished at the ever-increasing wealth of information that can be so easily summoned in the comfort of one's home through this magnificent search engine.

There are, however, two cautionary notes that combine to make the situation for future historians less favorable. The first is the reproduction of copyrighted material. Formerly, permission for this was freely and often graciously given. Unfortunately the process has become a business and, while some publishers remain generous, others are not averse to charging fees—in some cases, hundreds of dollars for single illustrations! The second concern is that the custodians of our reference libraries continue to replace shelves of bound journal volumes with computer terminals. It is a trend that those of a similar age and background to mine can only view with dismay. Indeed, how can one compare the sight of a "hard copy" emerging from the computer printer with the pleasure of turning the pages of a book printed almost a century before, and in another continent? And who might have read that same book in the intervening years? It was thoughts of this nature that consumed me as I looked through a copy of Keith Lucas's 1917 book *The Conduction of the Nervous Impulse*, and discovered a series of tram tickets from pre-WWII Vienna, and penciled margin notes written in English. How had such a book, perhaps once the property of an aspiring neurophysiologist, ended up in an American secondhand bookstore 80 years later?

As the work progressed, it became apparent that certain symmetries had been observed in the studies on the nerve impulse, in the sense that an early event had been balanced by a later one in the same location. For example, it was T. H. Huxley who had helped to bring about the first appointment in physiology at Cambridge, and it had been Huxley's grandson, Andrew, working in the same university with Alan Hodgkin, who had made the important discoveries before and after the 1939–45 war. Then, too, there was the fact that the Rockefeller Institute (now the Rockefeller University) had been the site of both Lorente de Nó's monumental but largely unsuccessful study of peripheral nerve in the 1940s and of Roderick MacKinnon's spectacular triumphs on excitable membranes 50 years later. Failure followed by redemption.

It was also interesting to compare the contributions of two very different "schools" of physiology—the American one, composed of inventive and energetic young men, largely self-taught in science, and the British one, in which distinguished academic names followed from one generation to another. It should have been no contest, with the vitality and resources of the New World winning easily. As it turned out, for a long time it was the British neurophysiologists who were supreme.

I have already mentioned some of the problems in writing a historical survey. There is another one in having to decide whether to keep the account of events in a strict chronological order, or to follow the work of an individual scientist over a period of time, or to deal with a single topic from its beginning to its present state. In the end, I chose to follow the chronological route, with a bit of hunting around each time-point. This policy inevitably presents some challenges for both author and reader. For example, it was difficult to write about the period between Galvani's work, in the late 1700s, and that of Bernstein a century later, without adopting a formal style. The events that followed, however, lend themselves to a more flexible approach, partly because of the greater amount of testamentary material available and partly because of personal recollection.

The book includes several vignettes, written in the present tense, in which a real event has been surrounded by a fictional penumbra. In creating these, my intention was to give new life to a particular happening, usually one that had occurred at least half a century ago, instead of allowing it to languish within the text of the main account. Even so, it would be dishonest to deny that the characters and the occasions involved had become so vivid in my mind that there seemed no alternative but to portray them in this way. I accept the criticism that must surely follow. Less blameworthy, perhaps, have been passages that describe the political, social, and economic environments of the time. Although there have been exceptions, research can be pursued only when the conditions are favorable. Had it not been for the two world wars, those of 1914–18 and 1939–45, and the intervening Depression, progress on the nature of the nerve impulse would have been swifter, while the contribution from continental Europe would have been greater had there not been an exodus of Jewish physiologists from Germany and the neighboring countries threatened by Hitler.

Another issue was how much science to include in a historical survey. If I have not succeeded in getting the balance right, I can only refer those who feel short-changed to the original sources, and ask those for whom there is too much detail to skip the offending pages. Then there is the matter of the subtitle. The nature of the nerve impulse, and the nature of chemical transmission at the nerve endings, are only parts of the whole story concerning the way that the nervous system works—though, to be sure, they are very important parts. Yet within this restricted territory there are as many deserving names left out as there are names that have been included. To have gone into these matters more deeply, however, would have been to obscure the main path of the story, and to have produced more of a *Who's Who of Neurophysiology*. To those whose work has been omitted, to their colleagues and families, I can only apologize. And finally, in a book of this complexity, there may well be some errors—if so, they are unintentional and will doubtless be corrected by those keen enough to spot them.

Life, alas, is full of compromise.

ACKNOWLEDGMENTS

There are so many people to thank that it is difficult to decide where to begin. Perhaps one should start in Newcastle upon Tyne (UK) with the late Professor Alfred Harper, whose well-organized and up-to-date lectures engendered, in at least one of his medical students, a lifetime's fascination with the nervous system. Dr. Bill Catton, the only neurophysiologist in the university, looked kindly on my early misadventures in his laboratory, while Dr. John Walton, now Lord Walton of Detchant, proved the best patron that an aspiring neuroscientist and neurologist could have wished for. It was Sir John Gray at University College London, however, who gave me the chance to begin serious experimental studies on the nervous system, and whose scientific rigor provided important lessons to a beginner. Many years later, it was Professor Jack Diamond and the late Professor Moran Campbell who incurred my everlasting gratitude for inviting me to join them at McMaster University.

As to the book itself, I am indebted to a host of librarians. At Trinity College, Cambridge, I was able to consult the Lord Adrian papers, while at the Rockefeller University, René Mastrocco allowed me access to the Gasser archive, and also showed me the rooms that he, Gasser, and his fellow neurophysiologists had used in the 1940s. In the Woodward Library at the University of British Columbia it was possible to examine Sherrington's notebooks and to admire his many scrolls, medals, and academic regalia. The opportunity to actually hold and look through a copy of Galvani's *Commentarius*, published in 1791, came through the kindness of Estela Durkan at the Library of the Royal College of Physicians of Edinburgh; Ms. Durkan also made other scarce publications available to me.

Among the librarians who went out of their way to find photographs of neurophysiologists for me, I must salute Jack Eckert (Francis A. Countway Library of Medicine, Boston), Florence Gillich (Cushing/Whitney Medical Historical Library, Yale), Lee Hiltzik and Margaret Hogan (Rockefeller Archive Center), Joanna Hopkins (Royal Society of London), Venita Paul (Wellcome Library Images), Ulrike Polnitzky (Austrian National Library, Vienna), Philip Skroska (Becker Medical Library, Washington University School of Medicine), and Jocelyn Wilk

(Columbia University Archives, New York). Then there were Heads of university departments, including Professor Bill Harris (Department of Physiology, Development and Neuroscience, Cambridge, UK) and Professor Jack McMahan (Texas A & M University). Of the Nobel Laureates who made my life easier, I would like to mention David Hubel (Harvard University), Bert Sakmann and his assistant Filomina von Hofmann (Max Planck Institute of Neurobiology, Martinsried, Germany), and Erwin Neher (Max Planck Institute of Biophysical Chemistry, Gottingen, Germany), while I am especially indebted to Professor Roderick MacKinnon, who allowed me to visit him in his laboratory at Rockefeller University. Other present or retired university persons include Professor Tony Angel (Sheffield, UK), Professor Ian Fleming (Cambridge), Professor Hugh Huxley (Brandeis University, Waltham, Massachusetts), Professor Martin Rosenberg and Professor Tilli Tansey (both of Queen Mary and Westfield College, London), and Professor Susan Schwartz-Giblin (New York University).

In a different category are the family members of those famous neuroscientists no longer with us. Here I would like to acknowledge Professor Deborah Hodgkin and Lady Marion Hodgkin, Lady Lucy Adrian, Ruth and Simon Weidmann, and especially Professor Simon Keynes and Professor Peter Matthews, both of whom enriched my life with their lively correspondence. And it was a great pleasure to get in touch with Dr. Rose Mason (née Eccles), briefly my mentor at University College London, after an interval of 50 years; had it not been for Sharon Carleton, of the Australian Broadcasting Corporation, this valuable contact could not have been made.

Nearer home, I owe much to Jane Butler, Claudia Castellanos, and Aurelia Shaw, who helped with the manuscript, and to the patient and forgiving artist, Steve Janzen. My friends and retired McMaster professors, Norman Jones, Karl Freeman, Lud Prevec, and George Sweeney, kindly read drafts of the book and made valuable comments, often while we were out walking together.

I am also indebted to publishers, especially the Rockefeller University Press, Cambridge University Press, Elsevier, and Macmillan, each of whom allowed me to reproduce several illustrations without charge. If John Wiley & Sons had imposed fees, I should have been in penury because of the large number of figures taken from the *Journal of Physiology*—a big thank you therefore to Mr. Duncan James. And I cannot resist crying "Shame on you" to the publisher who not only asked $600 for using a single photograph (of myotonic goats) but threatened to employ a debt collection agency unless it received a response to their offer! Fortunately, an enterprising University of Michigan law student, Mr. James Knapp, Jr., whose family owns several such animals, came to the rescue and made the request no longer necessary.

I come now to those to whom I owe most. My wife, Kate, gracefully accepted the long absences while I was word-processing elsewhere in the house. At a time when authors are having difficulty finding publishers, Oxford University Press accepted the manuscript without any quibbles, and for that I am obliged to Mr. Craig Panner. I was very fortunate to have Mr. David D'Addona as my editor; it was David who patiently guided me through the different steps in creating the book, often making valuable suggestions and always adding his encouragement and concern for my well-being in his e-mails.

Finally, there is the inspiration for the book in the person of Sir Andrew Huxley, who kindly accepted the dedication. Though I never had the privilege of working with Sir Andrew (who would have noted my shortcomings immediately), I had ample opportunity to observe him and marvel at his genius, and, later, to enjoy the gift of his friendship. Though T. H. Huxley declared himself an agnostic in religion, those who know his grandson would surely agree that Sir Andrew is a close approximation to a scientific deity.

CONTENTS

GLOSSARY OF TERMS AND ABBREVIATIONS

Å = Ångstrom, one ten-millionth of a millimeter

Acetylcholine = chemical transmitter released from nerve endings

Actin = muscle protein involved in contraction process

Action current (impulse) = membrane currents flowing during nerve or muscle fiber impulse

Action potential (impulse) = change in membrane potential during nerve or muscle fiber impulse

Adrenaline = chemical transmitter released from nerve endings, also a hormone

ATP = adenosine triphosphate, the main energy compound of a cell

Autonomic nervous system = nerve cells and fibers concerned with regulating automatic body functions (e.g., heart rate, blood pressure, sweating, pupil size)

Axon = slender, thread-like process conveying impulses from body of nerve cell to other nerve cells or tissues

Capillary electrometer = device used for measuring small changes in voltage

Cerebellum = posterior part of brain below the cerebral hemispheres

Channel = opening in cell membranes for ions or other molecules to pass through

Conductance = the ease with which ions pass through a membrane

Curare = naturally occurring plant poison able to paralyze muscle fibers by blocking acetylcholine receptors

Cytoplasm = contents of cell

Demarcation potential = difference in electrical potential between an intact region of nerve or muscle fiber membrane and a region that has been injured or otherwise modified

Dendrites = branching processes of nerve cell that receive information, in the form of chemical transmitters, from endings of other nerve cells

Depolarization = reduction in membrane potential of a nerve or muscle fiber membrane

DNA = deoxyribonucleic acid, the substance that makes up genes

Dopamine = chemical transmitter released from nerve endings

EEG = encephalogram, encephalography; the recording of electrical activity from the surface of the brain
EMG = electromyogram, electromyography; the recording of impulse activity in nerve and muscle fibers
End-plate potential = decrease in muscle fiber membrane potential following impulse in (motor) nerve fiber
Eserine = physostigmine, a plant compound that inhibits the degradation of the transmitter acetylcholine at the junctions between nerve and muscle fibers
Facilitation = enlargement of response when a second stimulus is given
GABA = gamma-aminobutyric acid, an inhibitory chemical transmitter released from nerve endings
Galvanometer = an instrument for measuring current or voltage
Gate = an opening or closing mechanism in an ion channel
Impulse = brief electrical event transmitted along nerve or muscle fibers
Inhibition = reduction in response due to application of a second stimulus
Ion = electrically charged atom or group of atoms
Kymograph = motorized rotating drum, with biological and other events recorded on surface
m = meter
m.e.p.p.'s = miniature end-plate potentials in muscle fibers following spontaneous release of packets of acetylcholine from nerve terminals
Migraine = severe episodic headache due to cortical hyperexcitability
millisecond = one thousandth of a second
mm = millimeter
Motoneuron = nerve cell in brain stem or spinal cord innervating (skeletal) muscle fibers
Motor unit = a single nerve fiber and the colony of muscle fibers that it supplies
ms, msec = millisecond, one thousandth of a second
Myelin (sheath) = lipid coating the larger nerve fibers
Myelinated fiber = a nerve fiber with a myelin sheath
Myofibril = longitudinal contractile element in a muscle fiber
Myograph = device for recording muscle tension
Myosin = large contractile protein in muscle fibers
Myotonia = episodic stiffness due to abnormal impulse activity in muscle fibers
Neuron = nerve cell
Neurophysiology = the study of how the nervous system works
Node (of Ranvier) = regular discontinuities in the myelin sheath surrounding a nerve fiber
Noradrenaline = transmitter released from nerve endings
Oscilloscope (cathode ray) = instrument with special screen for instantaneously displaying electrical potentials
Patch clamp = electrical isolation of small area of cell membrane

Permeability = the ease with which molecules, including ions, diffuse across membranes
Physiology = the study of the workings of the body (or organism)
Receptor = molecular structure on cell membrane that combines with chemical transmitter
Reflex = automatic response to a stimulus
Resting potential = potential difference across cell membrane in absence of impulse activity
Retina = nerve cell layers at back of eye
RNA = ribonucleic acid, involved in expression of DNA
Saltatory conduction = conduction of the nerve impulse from one node of Ranvier to the next
Saxitonin = toxin produced by marine algae
Subliminal fringe = neighboring neurons excited but not discharging impulses
Sympathetic ganglion = collection of neurons belonging to the sympathetic part of the autonomic nervous system
Synapses = meeting points between nerve cells, or between nerve and muscle fibers
Synaptic vesicles = rounded structures containing transmitter in nerve terminals
Tetrodotoxin = toxin contained in Japanese puffer fish
Threshold = the smallest current or potential required to excite a cell
Transmitter = chemical substance, released at nerve endings, for exciting or inhibiting another nerve cell (or muscle fiber)
µm = micron, a thousandth of a millimeter
µV = microvolt, a millionth of a volt
Voltage clamp = electronic circuit for holding the cell membrane potential constant

What is now proved was once only imagin'd.

William Blake

I am the splendid impulse
That comes before the thought,
The joy and exaltation
Wherein the life is caught.

Bliss Carman (Earth Voices)

1

Introduction

One should begin by explaining to those who do not already know that a *physiologist* is a person who examines the ways that the body works, and that a *neurophysiologist* is a physiologist who specializes in the study of the nervous system. It was, in fact, during a physiology lecture more than 50 years ago, that I, together with the other first-year Newcastle medical students, first heard the names of Hodgkin and Huxley. Even then, in 1953, it had been 14 years since Alan Hodgkin and Andrew Huxley had made the first of their dramatic breakthroughs in the understanding of the nerve impulse. Seven years later, qualified in medicine but eager for some research experience in the workings of the nervous system, I found myself in the Department of Physiology at University College London, the department to which Andrew Huxley had just been appointed Jodrell Professor. Like the other young people in the department, I was surprised to discover that, despite his many achievements, the newly appointed professor was still in his early forties, and that his formidable intellect was accompanied by considerable athleticism. He could, as he demonstrated at the annual departmental retreat, more than hold his own at tennis and, with a sharp eye and strong wrists, hit a baseball—a new challenge for him—soaring into the outfield. It was the intellect that intimidated, however. Not being mathematically inclined, it took me longer than most to appreciate the significance of the Hodgkin–Huxley equations, their solutions, and the predictions that arose from them. The experiments on which the equations had been based had been equally remarkable and, like the theoretical treatment, had set new standards. As a result of the Hodgkin–Huxley work, it had become possible to understand details of the way in which an electrical impulse was transmitted along a nerve fiber.

Someone other than a neurophysiologist might immediately ask why an understanding of the nerve impulse, work for which Hodgkin and Huxley received a Nobel Prize in 1963, should be considered so important. The best answer is that the nerve impulse is the mechanism by which the brain conducts its affairs, the currency for all its transactions. A sense of the role of the impulse is given in the vivid description of an awakening brain by the eminent physiologist Sir Charles Sherrington. The description, the once-famous "enchanted loom" passage, comes in Sherrington's monograph *Man on his Nature*, published in 1940.[1] The author was an accomplished poet and his lyricism, the use of metaphor, and the unusual sentence structure help to capture the wonder of the brain. Reflecting on a lifetime's study of the nervous system, and now well into his eighties, Sherrington likened the impulses to points of light, while the nervous system, with its billions of fibers, was:

> "A scheme of lines and nodal points, gathered together at one end into a great revelled knot, the brain, and at the other trailing off to a sort of stalk, the spinal cord. Imagine activity in this shown by little points of light. Of these some stationary flash rhythmically, faster or slower. Others are travelling points, streaming in serial trains at various speeds, and junctions whence diverge, the lines of travelling lights. The lines and nodes where the lights are, do not remain, taken together, the same even a single moment. There are at any time nodes and lines where lights are not.
>
> Suppose we choose the hour of deep sleep. Then only in some sparse and out of the way places are nodes flashing and trains of light-points running. Should we continue to watch the scheme we should observe after a time an impressive change which suddenly accrues. In the great head-end which has been mostly darkness spring up myriads of twinkling stationary lights and myriads of trains of moving lights of many different directions. It is as though activity from one of those local places which continued restless in the darkened main-mass suddenly spread far and wide and invaded all. The great topmost sheet of the mass, that where hardly a light had twinkled or moved, becomes now a sparkling field of rhythmic points with trains of travelling sparks hurrying hither and thither. The brain is waking and with it the mind is returning. It is as if the Milky Way entered upon some cosmic dance. Swiftly the head-mass becomes an enchanted loom where millions of flashing shuttles weave a dissolving pattern, always a meaningful pattern though never an abiding one; a shifting harmony of subpatterns. Now as the waking body rouses, subpatterns of this great harmony of activity stretch down into the unlit tracks of the stalk-piece of the scheme. Strings of flashing and travelling sparks engage the lengths of it. This means that the body is up and rises to meet its waking day."

In Sherrington's description is the concept of an impulse as the conveyor of information from one part of the nervous system to another. It is impulses that, for example, provide the summons for finger muscles to contract so that a pencil may be grasped, and bring messages from the skin telling the brain that the side of the nose is being touched. But there is more to it than this, for, regardless of their destinations, it is the impulses that bring the brain to life. In Sherrington's words, "The brain is waking and with it the mind is returning."

Thus, not only is the nerve impulse the means by which we experience the outside world and move around within it, but it is also the basis of all our thoughts and feelings, our personalities, and our dreams and ambitions. It was through patterns of impulses in certain parts of the brain that Shakespeare wrote his sonnets and Beethoven his symphonies, and Einstein discovered relativity. It is through the nerve impulse that, for better or for worse, we express ourselves.

The impulse was likened to a currency. Like a good currency, it had to be reliable, to hold its value, to be rapidly transferable, and to be accepted wherever it goes. The nerve impulse has all of these qualities. Lasting only 1 millisecond (a thousandth of a second), it can, in the largest fibers, be transmitted from one part of our body to another at velocities of up to 70 meters a second (approximately 160 mph) and at rates of up to 500 a second. At the lower end of the range of sizes are fibers with diameters of no more than 1 μm (a thousandth of a millimeter), invisible to the naked eye, yet each capable of transmitting impulses independently of its neighbors. Also remarkable is the fact that, regardless of the size of the nerve fiber, and of the speed and distance traveled, the nerve impulse maintains its amplitude throughout its journey.

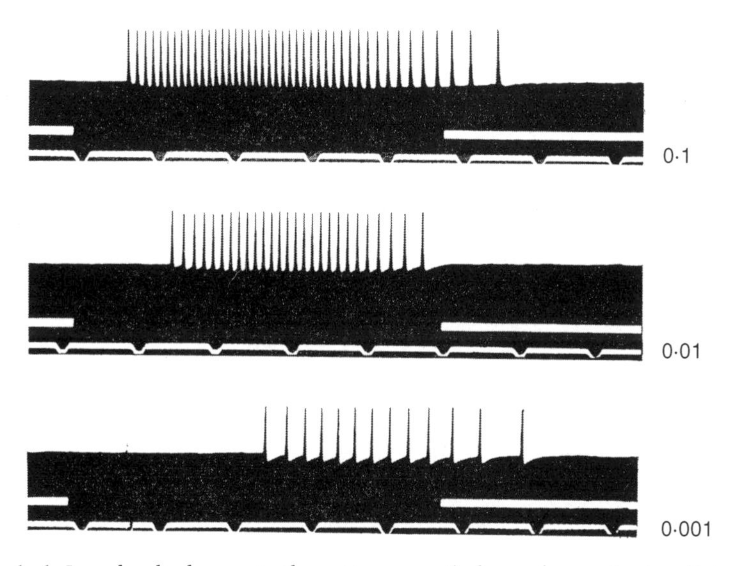

Figure 1–1 Impulse discharges in the optic nerve of a horseshoe crab when lights of different intensities are shone on the eye. In each case the delay before the start of the discharge, the maximum firing rate, and the duration of the response depend on the brightness of the stimulus, the relative intensities being shown at the right. The interruptions in the lower white lines indicate 0.2 seconds, while that in the upper white line gives the duration of the stimulus. (Modified from Hartline HK. Intensity and duration in the excitation of single photoreceptor units. *Journal of Cellular and Comparative Physiology*; **5**: 229–247, 1934. Courtesy of John Wiley & Sons.)

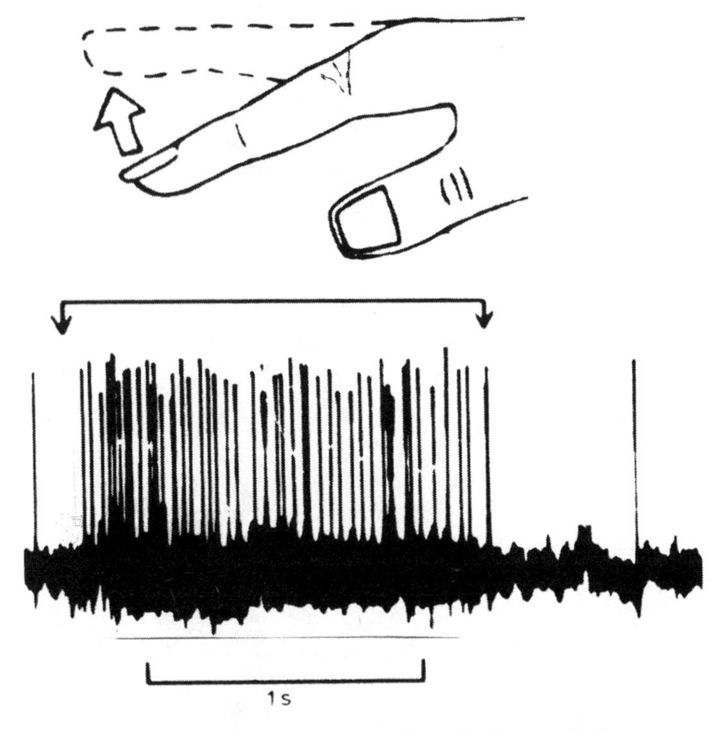

Figure 1–2 Responses of two or more nerve cells in the human thalamus to passive straightening of the index finger on the opposite side of the body.

Figures 1–1 and 1–2 give two very different examples of nerve impulse activity. In Figure 1–1, three lights of different intensity were shone onto the eye of the horseshoe crab, *Limulus*, which, in each case, responded by firing a burst of impulses down its optic nerve. The recordings from a single nerve fiber show that although the length of the burst, and the initial rate of impulse discharge, depended on the brightness of the stimulus, the amplitudes of the impulses were perfectly maintained in all three instances. The recording was made by the American neurophysiologist Kieffer Hartline in 1934, the year I was born.[2] The second example of impulse activity (Fig. 1–2) is one that I made myself many years later, at a time when neurophysiologists were assisting neurosurgeons in their operations on patients with parkinsonism and intractable pain. Since people's brains, like their heads, vary so much in their anatomy, it was necessary to map out the different regions of the thalamus, a large collection of nerve cells at the base of the cerebral hemisphere. In one area, close to the neurosurgeon's target zone, it was possible to record impulses in cells responding to touch or to joint position. In this case, it was the straightening of the index finger of the opposite hand that caused the thalamic cells to discharge. It is unlikely that these were the actual impulses felt as movement of the finger by the patient, but they were certainly part of the impulse pattern that, most likely in the cerebral cortex, was responsible for that sensation. However, the essential point of the figure is to show that, just as in the eye of the horseshoe crab, the conscious human brain makes use of impulse discharges in carrying out its business.

It is enough that the nerve impulse is the basis of all our thoughts, words, and deeds, but in recent years the study of the nerve impulse has acquired a new significance. It has become apparent that a wide variety of diseases and disorders, ranging from paralytic migraine to fatal ventricular fibrillation, may be caused by disturbances of the same membrane mechanisms that are responsible for impulse activity. More than this, nearly all the drugs that we use achieve their effects by acting on receptors, the same receptors that combine with the chemical transmitters released by impulses at the ends of the nerve fibers. And then we can hardly forget that, by blocking nerve impulses with local anesthetic, the dentist and the surgeon can carry out their various procedures.

Given the importance of the nerve impulse, why was it that, outside the neurophysiology laboratories, the pioneering studies of Hodgkin and Huxley attracted little attention? Most probably, the answer lay in the timing. The five papers in which Hodgkin and Huxley gave a complete account of their work were published in 1952.[3] In the following March, when we students were learning about the physiology of the nervous system, I went to a public lecture in the Chemistry Building. The speaker was Sir Lawrence Bragg, the Head of the Cavendish Laboratory in Cambridge, who, many years previously, at the precociously young age of 25, had been awarded the Nobel Prize for his work on the structure of inorganic crystals. At the end of the lecture, which presumably had been on some aspect of X-ray crystallography, Sir Lawrence had made a tantalizing remark. It was to the effect that there was great excitement in the Cambridge Laboratory over the work that "two young men" had just completed. In the following month, on April 25, 1953, a two-page letter appeared in *Nature* announcing the discovery of the DNA double helix.[4] The two young men had been James Watson and Francis Crick.

Important though the research on the nerve impulse was, the discovery of the double helix was undeniably the greater event. Indeed, the recognition of evolution as the basis of inter-species differences, and the discovery of the double helix, may well remain, perhaps for all time, the two greatest intellectual advances in the life sciences.

That being said, it is instructive to compare the two pieces of research, the one on the nerve impulse and the other on the DNA double helix. The solution to the structure of DNA had come surprisingly easily, and only 9 years after the demonstration by Oswald Avery and his colleagues that DNA, rather than protein, was the cell's hereditary material.[5] Further, the double helix work had not required any experimentation on the part of Watson and Crick, unless Watson's model building could be counted as such. The only laboratory work had been that of Rosalind Franklin, who had grown the DNA crystal and taken the X-ray diffraction photograph that Watson had been able to see during one of his visits to London. A quick look had been enough to tell him that the structure of DNA must be helical.[6] In contrast, Hodgkin and Huxley had done all their own experiments, devising—or else exploiting—new techniques as they went along. Again, although there had been heavy mathematics in the DNA work, both in the analysis of the diffraction pattern and in the constraints that had to be observed in the model building, the Hodgkin–Huxley mathematics was more original and certainly more arduous. In contrast to the 9 years between Avery's work and that of Watson and Crick, it had taken almost two centuries to get from the first intimation of a nerve impulse, by Galvani in 1770, to the differential membrane conductances of Hodgkin and Huxley. As Crick himself would discover in the latter part of his career, the nervous system does not give up its secrets easily.

I have said that Andrew Huxley's was an intimidating presence at University College London. The intimidation was as unintentional as it was inevitable. As we junior members of the department walked along the corridor and passed the professorial laboratory, the door would sometimes be sufficiently ajar for us to glimpse the lathe in the far corner, a reminder that here was a scientist who liked to design and build his own equipment. Andrew Huxley had already invented, or, as it turned out, reinvented, the interference microscope, and there had also been a novel ultramicrotome—for cutting the very thin sections needed for electron microscopy—and a micromanipulator to assist him in his very fine dissections of single muscle or nerve fibers. There was another fact of which we junior staff could hardly be unaware. In addition to having such extraordinary biological, mathematical, and engineering expertise, Andrew Huxley was the grandson of T. H. Huxley, the noted Victorian scientist and educator, and the person who had defended Charles Darwin so ably in the famous debate on evolution at the Oxford meeting of the British Association in 1860. An event 150 years ago, yet still only two generations removed from the present! Yet, despite Andrew Huxley's distinguished lineage and his multiple talents, there was never a hint of arrogance in his dealings with those less gifted. If a member of the department needed help, it would be willingly given, and given in a very unassuming, matter-of-fact way, with just a hint of surprise that a solution so obvious to Andrew Huxley had not occurred to the other person.

In 2003 there was an important development in the nerve impulse field. Roderick MacKinnon was awarded the Nobel Prize in Chemistry for elucidating the three-dimensional structure of a voltage-gated membrane channel, the type of adjustable opening in the nerve fiber membrane that makes the impulse possible. Like the 1953 studies on the DNA double helix, it had been based on X-ray crystallography, but, because of the markedly more complex structure of the channel molecule, the task had been much greater, and the solution had taken considerably longer—years rather than a few months.[7] MacKinnon's approach to the problem had been ingenious, and, though his research was chemical rather than neurophysiological, there was the same kind of beauty in his solutions as there had been in Hodgkin and Huxley's 50 years before.

The publication of MacKinnon's papers brought two centuries of scientific endeavor to a close. It was at last possible to envisage, in astonishing detail, how the nerve impulses, once initiated, undertook their travels and—as Sherrington so eloquently described—brought the brain to life.

But how had this scientific quest started?

2

The Spark

One can start with the nerves or with what came to be called "animal electricity." That there were nerves that connected the eyes and ears to the brain and others that ran from the spinal cord to the muscles and skin was well known to the Roman physician Galen (129–199), who had performed dissections in a number of animal species as well as in human cadavers.[1] However, a description of the finer structure of nerve had to await the studies of Antonie van Leeuwenhoek (1632–1723) in Delft. Using a simple single-lens microscope of his own making to study cross-sections of peripheral nerve and spinal cord in parts of cows and sheep brought to his home, Leeuwenhoek found that nerves were composed of very fine "vessels" running lengthwise (Fig. 2–1, *top*).[2] Rather later, Felice Fontana (1730–1805), working as Director of the Museum of Physics and Natural Sciences in Florence, went a step further by cutting a slit in the sheath of a nerve and exposing the very fine fibers (Leeuwenhoek's "vessels"). In his words, "a nerve is formed of a large number of transparent, homogenous, uniform and simple cylinders" (Fig. 2–1, *bottom*).[3]

As to how the nerves conveyed instructions to and from the brain, Galen favored the concept of "spirits" (*pneumata*), which were secreted and stored in the ventricles of the brain before entering the nerves and traveling to their destinations. Much later—14 centuries later—came the idea that a nerve caused a muscle to contract by secreting a drop of fluid onto its surface. A third idea, held by Isaac Newton among others, was that a nerve produced its effect by vibrating, like the string of a violin.

The correct concept, that electricity was the method of signaling, was the conclusion of Luigi Galvani (1737–1798; Fig. 2–2), and it was based on observation, deduction, and experiment. Though the term was not coined until the 17th century,[4] "electricity" had long been known to exist in the tissues of certain fish. The ancient Egyptians were familiar with the electric catfish, *Malapterurus*, and Galen himself employed powerful shocks from the electric ray, *Torpedo*, to treat headache and epilepsy. In this he followed in the footsteps of another Roman physician, Scribonius Largus. It was Scribonius who, a century before, had instructed his patients to stand on the ray, though sometimes the fish would be applied to the forehead instead.[5] It is not difficult to imagine how impressive, perhaps even magical, the treatment must have seemed to the recipients and

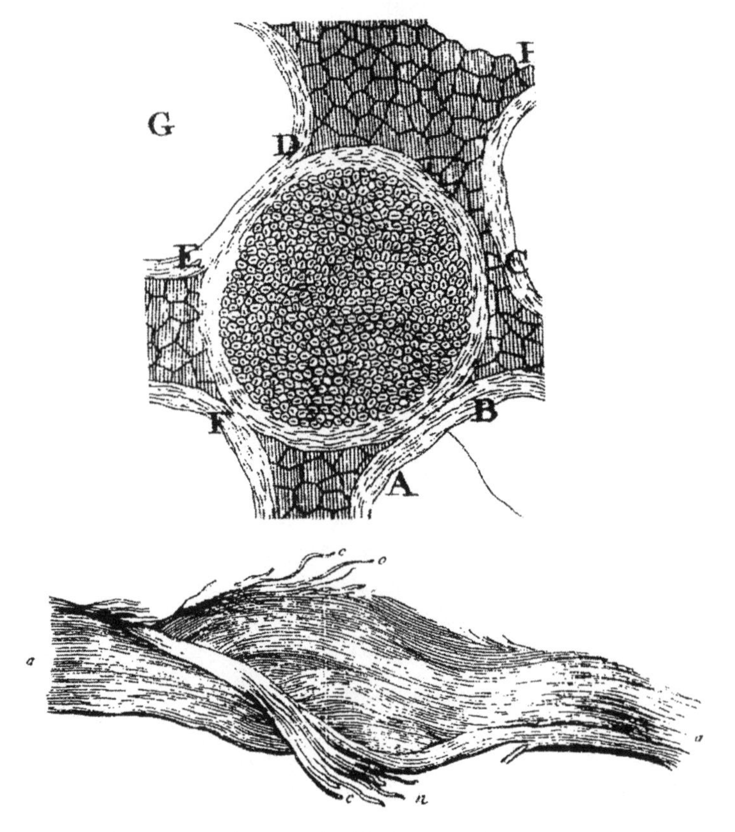

Figure 2–1 *Top*: Antonie van Leeuwenhoek's drawing of a cross-section through a small nerve (*A, B, C, D, E, F*). The many small fibers (Leeuwenhoek's "vessels") are enclosed by an "external coat" of fibrous tissue. The larger cells surrounding the nerve are fat cells. (From Leeuwenhoek, 1719) *Bottom*: Fontana's drawing of nerve fibers (Fontana's "primitive cylinders") exposed by slitting open the fibrous sheath. *c, n, o* depict several of the broken nerve fibers. (From Fontana, 1781)

onlookers, especially since it was likely to have been effective in the majority of cases.[6] Though this form of therapy fell from favor, or was perhaps forgotten, the electric organs of the ray and eel would come to capture the interest and imagination of generations of neuroscientists, and would eventually contribute to an understanding of the mechanism of the nerve impulse. It was Galvani's achievement to show, clearly and for the first time, that animals other than fish were also able to generate electricity, and that the electricity in their nerves and muscles was probably important for their functioning.

Luigi Galvani[7] was born in Bologna in 1737, and it was in that Italian city, then one of the many properties of the Pope beyond Rome, that he was to carry out his life's work. As a student Galvani's inclination had been to study theology, with the intention of entering one of the monastic orders. However, the strong papal influence extended beyond the Church into the arts and sciences, and there was

Figure 2–2 Luigi Galvani (1737–1798). Though Galvani is shown in the finery of a Bolognese gentleman, the artist has included dissections of frog legs at the bottom of the portrait.

also pressure from Galvani's father for his son to study medicine. In the end, the son studied both medicine and philosophy, obtaining degrees in the two subjects on the same day. After graduation, it was medicine that he chose for his career and, following a successful dissertation on the formation and development of bones, Galvani was appointed lecturer in anatomy at the University of Bologna. Even today, 240 years on, it is possible to gain an impression of the city as it was in Galvani's time. The large central square with its cobbled stones, the Piazza Maggiore, remains the focal point of the city. It was a square that Galvani would have crossed many times on his way to and from the university. In the original university building, on one side of the square, was the anatomy theater where he was likely to have performed dissections and given lectures and demonstrations.[8]

It is commonly thought that Galvani's famous experiment, the experiment of "the spark" that provided the foundation for his later research and that led to the recognition of animal electricity, was no more that a lucky observation. In the sense that something quite unexpected happened one day in the laboratory, this is true. It is also true, as Pasteur would remark many years later, that, in the fields of observation, chance favors only the prepared mind,[9] and Galvani's mind was very well prepared. As a newly appointed lecturer, he had studied the comparative anatomy of the renal tubules, nasal mucosa, and middle ear, looking for similarities and differences between one species and another. It was the sort of work that was being carried out in anatomy departments elsewhere, as well as by amateur scientists, invariably gentlemen sufficiently wealthy to have their own microscopes and to have the time to indulge their curiosity by looking through them in the comfort of their homes. It was, for anatomists, professional and amateur, a golden era, a time when the fine structure of the various tissues and organs would be described for the first time. Inevitably, as a structure was revealed, there were the questions "What was its function?" and "How did it work?," questions that formed the basis of the fledgling science of physiology.

Increasingly, it was to physiology that Galvani leaned. There was another scientific interest too, and this was electricity—not the animal type found in the different species of electric fish, but the electricity of the external, physical, world. It had become a subject suitable for study, and indeed for home and public entertainment, once it had become possible to capture and store it. The breakthrough had come with the Leiden jar, named after the Dutch city where the invention had been made. With a covering of tinfoil, the narrow-necked glass jar contained a piece of metal or a solution as its other conductor and would then become a capacitor, holding its electrical charge until a connection was made between the inside and outside of the jar. That was the *discharge*. The *charging* of the jar would be done with some form of friction machine, usually with a hand-cranked wheel turning a belt, a process equivalent, in the modern world, to walking across a nylon carpet, or pushing a shopping cart over a waxed floor, in a dry atmosphere. It was during the operation of one such friction machine that Galvani had made the crucial observation that would ultimately lead to the discovery of the nerve impulse.

At some time before the discovery, probably in 1770, his scientific interest had moved to nerves and muscles, with those of the frog becoming his favored preparation. It was after such a dissection had been made, with the nerves of the leg left attached to the spinal cord, and with a wire inserted through the divided spinal column, that the crucial observation of the spark was made. It is not known who, Galvani or his wife, the beautiful Lucia Galeazzi, had performed the dissection and was now touching the nerves with a lancet, or who was operating the friction machine in another part of the room.[10] Nor is it known what the original purpose of the experiment had been. The fact remains that, at the very moment a spark was given off, either from the machine or from one of the Leiden jars on the table (Fig. 2–3), the leg muscles twitched. The response was totally unexpected. Overcome with curiosity, Galvani attempted to repeat the effect and was able to convince himself that the production of the spark was the cause of the muscle contraction. Further experiment showed that there was a poor correlation between the size of the contraction and that of the discharge, an observation that led him to suppose, correctly, that the spark, rather than acting directly on the muscle, had released some internal force in the muscle:

> "The electrical atmosphere hit, and pushed, and vibrated by the spark is that, which brought to the nerve, and similarly pushing and commoting

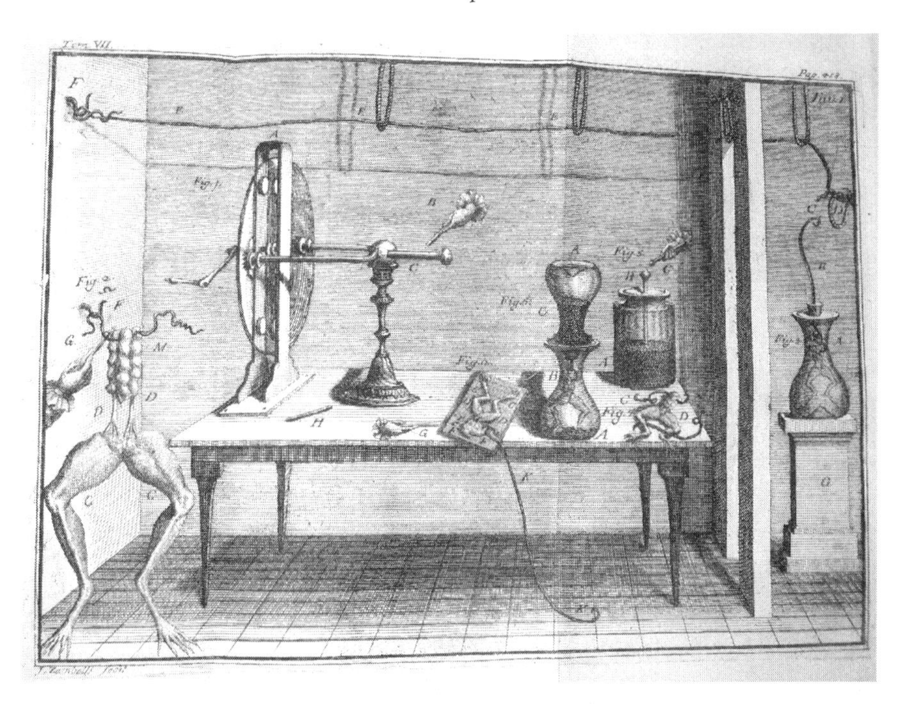

Figure 2–3 Galvani's laboratory in Bologna. This beautiful and much-reproduced engraving is from Galvani's *Commentarius* and contains several smaller figures. On the left of the table (labeled *Fig 1*) is a hand-cranked friction machine, used for charging a Leiden jar at the right of the illustration (*Fig 3*) through an overhead wire (*E, E, E*). There is another Leiden jar on the right end of the table (*Fig 5*). Also shown are three frog leg dissections.

> some extremely mobile principle existing in nerves excites the action of the nerveo-muscular force."[11]

Galvani's next step was one that would, paradoxically, lead both to his ultimate scientific triumph and his personal downfall. It was influenced by Benjamin Franklin's well-known experiment a few years earlier, in which a kite, flown during a thunderstorm, plucked electricity from the atmosphere. Galvani wished to see whether atmospheric electricity, like the friction machine, was capable of causing muscles to twitch. It was—after running wires through the spinal cords of frogs and placing the frogs at the top of his house, he observed that, during a thunderstorm, each lightning flash was associated with a contraction. On a calm day, in contrast, the legs, hung upon an iron railing, remained motionless. As he attempted to take the legs down, however, there was a twitch as the metal hook in the spinal cord touched the railing. It was another chance observation, and once again Galvani pursued it in the laboratory, showing that contractions could be produced whenever a metallic connection was made between the leg muscles and the spinal cord. His interpretation of the phenomenon was that the contractions resulted from a flow of internal, "animal," electricity from the muscles, made possible by the presence of the metal conductor. Galvani further suggested that the electricity in the muscles was stored with its two forms (i.e., opposing charges) distributed on the internal and external surfaces of the fibers respectively, much like the conductors on the inside and outside of a Leiden jar. It was a remarkable speculation, and one that anticipated the measurement of the resting membrane potential a full century and a half later. Galvani summarized his years of experiment by publishing *De Viribus Electricitatis in Motu Musculari. Commentarius* in 1791.

By all accounts the *Commentarius* was well received in the academic circles of the time, and especially by the Professor of Physics in the University at Pavia, another of the historic Italian cities and lying some 200 km west of Bologna. At the time he received his copy of the *Commentarius*, Alessandro Volta (1745–1827; Fig. 2–4) was 46 and already an internationally known scientist of distinction.[12] Born of an aristocratic family in the lakeside town of Como, in a mountainous region of northern Italy, Volta did not speak until the age of 4, but after that he progressed rapidly. Like Galvani, Volta might have entered the priesthood, but instead he had become fascinated by electricity, conducting his own experiments, and had determined to become a physicist. By 18, he was already reporting his

Figure 2–4 Count Alessandro Volta (1745–1827). There are several portraits of Volta, in some of which he is holding one of his "piles" (batteries). In this portrait the pose presumably indicates approval of the Napoleonic regime. (Wellcome Library, London)

experiments and setting out his ideas in correspondence with leading scientists. Three years later, still in Como, came the announcement of the first of many inventions, a "perpetual electrophorus"; this resembled the Leiden jar in its ability to hold an electrical charge, but differed from it by being able to donate some of its electricity repeatedly before needing to be recharged. Later Volta performed experiments on the inflammability of marsh gas, which he collected from around Lake Como. There were also studies on the composition of the atmosphere, and on the surface tension of water. And still more inventions—a gas lamp, sensitive devices for detecting and measuring small amounts of electricity, and a condenser (capacitor). By 1791, the year that Galvani published the *Commentarius*, Volta's fame as a scientist was well established, aided in part by extensive travels in Europe, travels during which he had met other scientists and been elected to learned societies, including the Royal Society of London.

On reading Galvani's treatise on animal electricity, Volta was initially enthusiastic. Indeed, his curiosity led him to repeat some of Galvani's experiments. Then came doubts, and after the doubts, the beginning of a major controversy. Pavia, the city in which Volta reigned as an esteemed university professor, was not unused to battles. There had been a particularly important one in 1525 in which the ancient Lombard city had witnessed the capture of the French king, Francis I, by the emperor, Charles V, of Spain. The battle between Galvani and Volta, though an academic one, was hardly less momentous. Volta began by acknowledging that, as Galvani had claimed, the contact of metal with tissue could provoke a muscle contraction in a frog leg. Though neither Galvani nor Volta would have been aware of it, a Dutchman, Jan Swammerdam, had made the same observation over a century earlier, and had been prepared to demonstrate the phenomenon to visiting dignitaries. It was Volta, however, who took the matter further by showing that the muscle twitch was elicited more reliably if the metal arc connecting the muscles and nerves consisted of two dissimilar metals, brass and steel for example, rather than a single one. Further, the energy for the contraction did not seem to come from the muscle, as Galvani had supposed, because a contraction still occurred when both points of the bimetallic arc were touching nerve. Volta's interpretation of his results was that any electricity was more likely to have come from the contact between the two dissimilar metals than from the muscles. That the dissimilar metals were capable of stimulating sensory nerves, without any muscle being involved, was shown by applying the tips of the arc to the tongue and noting the metallic taste that resulted. Electricity, to Volta, was "external," being generated outside the animal, rather than by the tissues.

Galvani did not capitulate, however, and devised a clever riposte, showing that a muscle twitch occurred when the cut end of the sciatic nerve was brought into contact with the leg muscles (Fig. 2–5, *left*). Although no metal had been involved, Volta was not convinced, arguing that muscle and nerve, like metals, were dissimilar materials. Galvani now conceived the ultimate proof. Having prepared two frog legs, he carefully positioned the sciatic nerve of the first leg so that its cut end

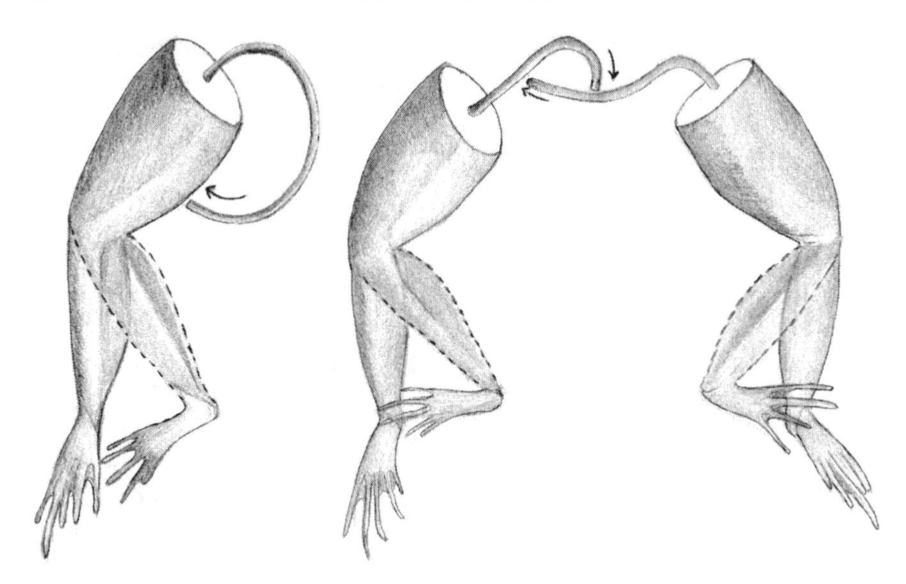

Figure 2–5 Galvani's two critical experiments, proving that nerve and muscle alone (without any contact with metal) could produce muscle twitches. *Left*: The cut end of a sciatic nerve touches the surface of a muscle. *Right*: The cut end of each nerve is made to touch the intact trunk of the other nerve.

touched the sciatic nerve of the second leg, with another contact between the two nerves being made elsewhere (Fig. 2–5, *right*). He found that the contacts between the two nerves were sufficient to elicit a muscle contraction in the first leg, and sometimes in the second leg also. Since no metals had been involved in the circuit, the electricity could only have come from the animals and from the nerves themselves. It was a conclusive experiment in favor of the existence of "animal electricity."

Unfortunately for Galvani, Volta was also right, this time for thinking that dissimilar metals in contact were a source of electricity. Spurred on by his animal experiments, he used his electroscope to detect the charge generated by such contacts. Then in 1800 came his greatest achievement. By placing alternate discs of zinc and copper on top of each other, with a brine-soaked cloth between each pair, he was able to produce a steady source of electricity. The "Voltaic pile" had become a battery, and, as Volta recognized, the pattern of alternating metal discs resembled the stack of muscle plates in the electric organ of the ray. Not only did Volta write of his invention to the Royal Society of London, of which he was already a member and the recipient of a medal, but he also demonstrated it to Napoleon Bonaparte in Paris. The demonstration proved to be a rewarding experience. Another medal materialized, this time a gold one, and, a few years later, a pension and titles. Moreover, it was the success of the battery that, quite unfairly, swung scientific opinion behind Volta in the dispute over the existence or absence of animal electricity. Just as the invention of firearms had won the battle of Pavia for the Spaniards in 1525, so the creation of the battery had enabled Volta to win the battle for public opinion in his dispute with Galvani. How could the inventor of such a remarkable device be wrong?

By 1800, however, Galvani had already been dead for 2 years. The final years had not been kind to him. Not only had Volta instigated widespread disbelief in his work and conclusions, but Galvani's wife and companion in the laboratory, the beautiful and talented Lucia, had died the year before the publication of the *Commentarius*. Then, after Napoleon had invaded northern Italy in 1796 and had taken Bologna, Galvani had refused, on religious grounds, to swear the oath of allegiance to the new Cisalpine Republic. Scientific and personal martyrdom had followed. Banned from teaching, he had lost his position as a university professor and, with it, his income. Destitute, childless, largely discredited, and without any professional honors, Galvani died in 1798. Volta, in contrast, remained in favor, even after the French had been ousted from northern Italy and replaced by the Austrians. Retiring to the country, famous and respected, he died at the age of 82. Later, his life and work were commemorated by the introduction of the term "volt" to quantify the electromotive force, Volta's "tension," driving an electric current. History treated Galvani kindly, too. Galvanometers measured current, and a steady current was referred to as a galvanic one. Iron became galvanized by the electrolytic deposition of zinc, and even humans could be galvanized into activity. And, many years later, in Bologna, the city where he spent his life, there would be a Piazza Galvani, with a statue of the scientist holding a dissecting board on which the spinal cord and two frog legs lay exposed.[13]

For the 30 years immediately following Galvani's death, there was little work on animal electricity. One person who was involved, however, was Galvani's nephew, Giovanni Aldini, who had been his uncle's confidant, prior to becoming a professor at the University of Bologna. It was Aldini, who, in extensive European travels that included lecturing at medical schools in London, reaffirmed his uncle's observations. Importantly, he demonstrated that frog legs would twitch when their nerves touched muscle tissue and when no metal contacts were involved. It was this important experiment of Galvani's that had proved the existence of "animal electricity."

When the next advance came, it was once more in the Italian universities. Galvani, in his experiments, had always used the contraction of a frog leg to detect animal electricity. In Florence, some 36 years later, Leopoldo Nobili became the first to detect the current flowing from the cut surface of a frog muscle with a physical device, an "astatic" galvanometer of his own making. Unfortunately, the influence of the deceased Volta was still powerful and, rather than recognizing the current as a biological one, Nobili attributed it to a thermal effect as moisture evaporated from the exposed tissues. Nevertheless, his measurement was an important advance, in that it had removed much of the mysticism associated with the experiments on animal electricity.

Figure 2–6 Carlo Matteucci (1811–1868) in 1853. By this time he had completed his study of muscle and nerve. (Wellcome Library, London)

It was Carlo Matteucci (1811–1868; Fig. 2–6)[14] who would make the greatest advances, however. A graduate of Galvani's university, Bologna, Matteucci in 1840 became a professor of physics in Pisa, yet another of the old university cities in the north of Italy and situated close to the Ligurian Sea on the west coast. Four years previously, while directing a hospital pharmacy in Ravenna, Matteucci had had his research on the electric organ of the ray presented in Paris as a communication to the highly prestigious Academy of Sciences (*Académie des Sciences*) by one of the Academy members. It was part of the most comprehensive study yet of the organ that Scribonius and Galen, so long ago, had used to treat their patients during the days of the Roman Empire. Matteucci had shown that the discharge of the electric organ was controlled by a special structure, the "electric lobe," in the brain stem of the fish, and he had dissected the nerves connecting the two (Fig. 2–7). He had also shown the existence of sensory pathways to the electric organ from different regions of the head and body. In the following year, 1837, as a young man of 26, he presented more of his work to the Academy, this time in person. As with the first communication, it was well received. Scientists, like other mortals, have their prejudices and jealousies, however, for, when the work was subsequently assessed for publication by the Academy, there was a dissenting voice. In a letter to a friend, Matteucci described what happened:

> "I will tell you about the scene at yesterday's session. What a shame my dear friend! The only Italian of the Academy, an old friend of mine, Guglielmo Libri, the miserable, the person who ten minutes before reading the Report had shaken my hand and spoken with me about the praise

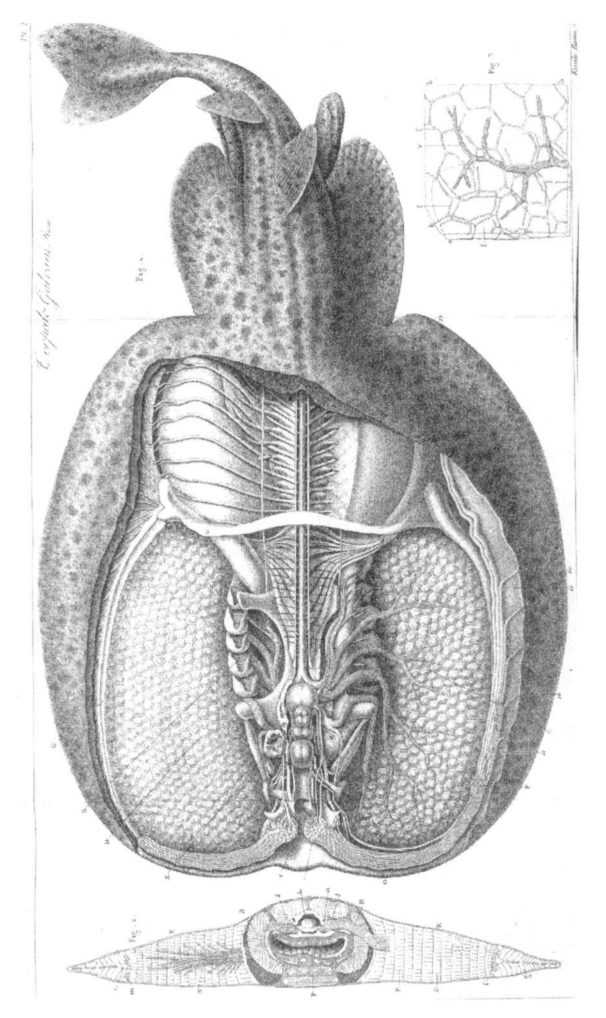

Figure 2–7 Dissection of an electric ray (*Torpedo*) to show large flat electric organs on each side of the head, together with the nerves supplying them. This lovely dissection and illustration is that of Paolo Savi, an associate of Matteucci. (See Matteucci, 1844.)

> of my work recently expressed by von Humboldt, declared against me. It is impossible for you to imagine the sense of disgust felt by those present. You know how they normally vote. Well, this time all the members stood up, he alone remained seated."[15]

The work was published.

At about the time that he was presenting and publishing his work on the electric ray, Matteucci also became interested in the animal electricity of the frog. Familiar with the research of Galvani and of Nobili, he recognized that the current that Nobili had measured was a true biological one, and driven by a difference in electrical potential between the cut and intact surfaces of nerves and muscles. This conclusion would probably not have surprised Galvani, who had proposed that there was a difference in electrical charge between the interior and exterior of a muscle fiber and that, like a Leiden jar, could provide current when a connection was made between them. Matteucci went on to show that this "proper" current, as he called it—or *demarcation current*, as it came to be known—was not peculiar to the frog but could be detected in warm-blooded animals, such as pigeons and rabbits, as well. Ingeniously, he went on to build a biological "pile" (battery) out of muscles, much as Volta had done with his stack of dissimilar metal discs. In Matteucci's pile, a number of thighs were cut across and placed end to end, with the cut surface of one thigh touching the intact surface of the next. Matteucci found that the amplitude of the demarcation current increased in proportion to the number of thighs in the series. Matteucci then made two other extremely important experimental observations. One in particular must have impressed all those to whom Matteucci demonstrated it. He placed the cut end of a frog sciatic nerve on the intact thigh muscles of a second frog. When the thigh muscles were made to twitch, so, too, did the muscles supplied by the sciatic nerve of the other frog. The explanation for the phenomenon was given by one of the distinguished members of the *Académie des Sciences de Paris*:

> "A l'instant où la grénouille se contracte, il y a une décharge électrique qui passe dans l'extrémité du nerf de la jambe, quand cette extrémité pose sur le muscle. . ." (As soon as the frog contracts, there is an electric discharge which passes to the extremity of the leg nerve, when this extremity rests on the muscle).[16]

This was the first clear statement that there was an electrical discharge, or "action current," associated with the muscle contraction. Matteucci's other observation, that the muscle contraction momentarily abolished the demarcation current, recorded with the galvanometer when the muscle had been at rest, provided further evidence on the nature of the action current. Unfortunately, it was at this point that Matteucci began to encounter difficulties, both scientific and political. Inside the laboratory, he tried various types of experiment but was no longer able to show that muscle contraction abolished the demarcation potential; instead the demarcation potential appeared to increase. Outside the laboratory, he had already encountered the opposition of Libri at the Paris Academy, and then two other of his countrymen had claimed priority for some of his observations. Neither of these attacks, however, could match that of a rising young physiologist at the University of Berlin.

Emil du Bois-Reymond (1818–1896; Fig. 2–8)[17] had learned his subject from Johannes Müller, the person regarded as the founder of the discipline of physiology in Germany, and the widely respected author of the definitive *Handbuch der*

Figure 2–8 Emil du Bois-Reymond (1818–1896). Despite his French name and ancestry, du Bois-Reymond was born in Germany and made his career in that country. A neuroscientist of remarkable insights, he was an influential teacher, a formidable foe in argument, and a considerable wit (see Note 29). (Wellcome Library, London)

Physiologie des Menschen. Having trained under Müller, it was not long before du Bois-Reymond's special aptitude gained him the position of assistant to Müller in Berlin. Eight years younger than Matteucci, du Bois-Reymond had also entered the animal electricity field through experiments on electric fish, and had then moved on to the nerves and muscles of the frog leg, using a galvanometer for the latter studies. While the designs of galvanometers vary, all models depend on the same physical principle—a current flowing along a wire induces a magnetic field around the wire, and the magnetic field can produce movement of a pointer. To make the galvanometer responsive to very small currents, such as those in a nerve or muscle, the wire is wound into a coil, the sensitivity increasing with the number of turns. Whereas Matteucci's galvanometer coil had 3,000 turns of fine, silk-covered, copper wire, du Bois-Reymond's coil contained over 5 km of wire; du Bois-Reymond had made each of the 24,000 turns himself (Fig. 2–9, *left*).[18] Du Bois-Reymond made two other innovations, the use of an "inductorium" for stimulating nerves, and non-polarizable electrodes for recording the demarcation current.

The inductorium consisted of two overlapping wire coils (Fig. 2–9, *right*), of which the inner, the primary, created a magnetic field as current passed through it. This magnetic field, in turn, induced a brief current in the secondary, outer, coil, both when the current in the primary coil began and when it was terminated. Because of their short durations, the two currents in the secondary coil (the *make* and the rather larger *break*) were very suitable for stimulating nerve; further, the strength of the stimulus could be varied by adjusting the overlap between the primary and secondary coils.

The non-polarizable electrodes were a necessity if the recordings from nerve and muscle were not to be contaminated by much larger "junction" (electrochemical) potentials, such as those existing between dissimilar metals (Volta's discovery) or between a metal electrode and conducting fluid or biological tissue. Du Bois-Reymond's solution to the problem was to contact the nerve or muscle with a salt-impregnated clay electrode, which was connected to a zinc sulfate solution containing a zinc rod. If the recording circuit was completed by an identical, reference, electrode in the bathing fluid, then the junction potentials balanced each other out, leaving the demarcation potential.

With his sensitive galvanometer du Bois-Reymond was able to confirm Matteucci's earlier observation that the demarcation current in a nerve disappeared when its muscle was made to contract. He also confirmed that a muscle twitch could somehow stimulate the cut end of a nerve placed on its surface. Quite properly, du Bois-Reymond gave credit to the Italian for both findings. It was du Bois-Reymond's interpretation of them, however, that irritated and then, as the correspondence grew, angered the older man.

Du Bois-Reymond contended that the two observations reflected the presence of an oscillation, the *negative Schwankung*, which he had detected in the

Figure 2–9 *Left*: The very sensitive galvanometer designed by du Bois-Reymond. The many turns of wire (24,160) were coiled round two parallel plates (*e*, *e*). The vertical thread supported two magnetized needles, of which only the upper can be seen. The lower needle, hidden from view between the plates, rotated in response to the magnetic field generated by the biological current flowing in the two coils of wire; a slit in the plates made this movement possible. The poles of the upper and lower needles faced in opposite directions so as to nullify the effect of the earth's gravitational field. *Right*: Du Bois-Reymond's inductorium.

demarcation current, and that this oscillation was, in fact, the action current responsible for excitation. The controversy between the two men had none of the gentlemanly restraint that had characterized the scientific argument of Galvani and Volta, Matteucci hanging on to his claims of priority for observations and du Bois-Reymond providing the most plausible explanations for them, explanations delivered at length in a monumental treatise published in 1848. During the great debate, Matteucci had received encouragement and sympathy from an unexpected source, Michael Faraday, the President of the Royal Society of London. Faraday, who had performed the definitive experiments on the relationship between electricity and magnetism, would have met Matteucci when the latter had presented his work to the Society. Without taking sides in the scientific argument, Faraday concluded a letter written in 1853:

> "How earnestly I wish in such cases that the two champions were friends yet I suppose I may not hope that you and du Bois-Reymond may some day become so. Well, let me be your friend at all events and with the kindest remembrance to Madame Matteucci and yourself believe me my dear Matteucci ever truly yours,"
>
> M. Faraday[19]

Whether it was Michael Faraday's letter, or the mellowness of age, or the reluctant but inevitable acceptance of scientific truth, the fact is that Matteucci did come to terms with du Bois-Reymond. In the preface to his *Course of Electrophysiology*, published in 1856, Matteucci wrote:

> "The controversy having arisen, the subsequent publications of du Bois-Reymond have given me the desired opportunity to recognise and to confess my error, and I am glad to say that among the reasons which encouraged me to publish this Course, not the least was to have an opportunity to put in the true light the services rendered to Electro-physiology by that person whom I would like henceforth to call my old adversary."[20]

By then, however, Matteucci was secure, still a professor in Pisa, but now head of the best-equipped institute of physics in the country, famous internationally for his earlier research, the recipient of prizes, and a future senator. All his physiological studies, on the electric ray and on nerve and muscle, had been carried out in his youth and had taken him no more than 8 years. Before the story moves any further, however, it is well to recapitulate.

To Galvani belongs the credit for discovering a source of electricity in nerves and muscles, a source somehow responsible for the contraction of muscles. The crucial experiment was his demonstration of a muscle twitch when the cut end of one sciatic nerve was placed on the intact region of another nerve. Galvani thought that, under normal circumstances, electricity flowed from the ventricles of the brain, through the nerves and out to the muscles.

Volta showed that the muscle contractions in some of Galvani's earlier experiments were probably due to excitation of the nerve fibers by electric currents induced by contacts between dissimilar metals.

Matteucci demonstrated that an injury current, the "demarcation current," flowed between the cut and intact surfaces of a muscle or nerve, that this current momentarily disappeared when the muscle contracted, and that a cut nerve could be excited when an underlying muscle twitched (the "induced twitch phenomenon").

Du Bois-Reymond recognized that Matteucci's findings could be explained by an action current, the "negative Schwankung," which was an oscillation of the demarcation current and which reflected the excitation of the nerve fibers. It was the first intimation of the nerve impulse and it was from this observation of du Bois-Reymond that the all subsequent work on the nerve impulse evolved. The understanding of the nature of the nerve impulse would become the most important task—a holy grail, as it were—for later neuroscientists.

In reviewing the early work, the only uncertainty today, one that could easily be resolved by a simple experiment with modern equipment, is the detailed basis for the observations of Galvani and Matteucci. On the one hand, it is true that a current flows between the injured and intact regions of a muscle or nerve. On the other hand, it is doubtful if such a current in a nerve would be large enough to excite the fibers in a second nerve placed in contact with it. The alternative explanation is that contact with a second nerve, or with a muscle belly, immediately lowers the resistance of the electrical circuit, causing a surge in the demarcation current and excitation in the segment of nerve next to its cut end. There is a similar concern about the explanation of the induced twitch.[21]

After Matteucci's work, the leadership in electrophysiology moved away from Italy. At the Collège de France, in Paris, was Claude Bernard (1813–1878), a contemporary of Matteucci and a man whose many contributions to physiology included the observation that the South American poison, curare, produced its paralysis not by any direct action on the muscle fibers themselves, but rather, he thought, by affecting the nerve twigs supplying the muscle. This powerful observation proved to be the forerunner of a host of studies on the neuromuscular junction (see later). Etienne-Jules Marey (1830–1904), 17 years younger than Bernard and one of his successors at the Collège de France, investigated the electric ray, *Torpedo*, and measured the strength of its powerful discharges.[22] The electric organ, Marey realized, was composed of highly specialized muscle fibers, a conclusion that had been hinted at by the Italian scientist Stefano Lorenzini two centuries earlier.[23,24] Also in Paris, but this time on the wards of the Salpêtrière Hospital, the solitary figure of Guillaume Duchenne (1806–1875) might be observed, using electricity not just for treatment but also for diagnosing the causes of the muscle weakness that affected many of his patients. Also in the wards and lecture theater of the Salpêtrière Hospital would be Jean-Martin Charcot (1825–1893) and his coterie of neurologists who, by dint of careful observation, were describing and defining some of the disorders of the nervous system for the first time.

So far as the nerve impulse was concerned, however, it was in Germany[25] that the next main developments took place. The movement had already started with Johannes Müller and du Bois-Reymond in Berlin and was consolidated by du Bois-Reymond's students, but in between was the contribution of a man of outstanding genius.

Hermann Helmholtz (1821–1894; Fig. 2–10) was the son of a schoolteacher.[26] Like Volta, he was a slow developer but, once started, made rapid progress, helped in part by discussions with his father. At age 17 he left Potsdam to study medicine in Berlin and came under the influence of Johannes Müller, the "father" of German physiology. One of his classmates was du Bois-Reymond, whose accomplishments have already been described. Obliged to serve in the army for 8 years, Helmholtz somehow found time to experiment and to investigate the chemical changes and heat production accompanying muscle contraction. After leaving the army, Helmholtz obtained academic positions, first in Berlin and then in Königsberg. In that Baltic city, in 1848 at the age of 27, Helmholtz set out to study in more detail the processes involved in a muscle contraction. To do so, he designed and built a

Figure 2–10 Hermann Helmholtz (1821–1894). This photograph was taken in 1848 while Helmholtz was in Königsberg and beginning his experiments on muscle and nerve. The many other photos of Helmholtz show him considerably older, having achieved fame in multiple fields. (Wellcome Library, London)

device in which the shortening of the muscle belly drew a trace on a thin sheet of mica. He found that:

> "the energy of the muscle does not develop completely at the moment of an instantaneous stimulus. Rather, in most cases, after the stimulus has already ceased, it increases gradually, reaches a maximum, and again subsides. The force of the muscle was not strongest directly after the stimulation, but rather increases for a time and then falls."[27]

Helmholtz further noticed that the delay between the stimulus and the muscle response was shorter when the nerve was stimulated close to the belly of the muscle—a finding that could be explained only if the nerve impulse traveled towards the muscle with a finite conduction velocity. Helmholtz then set out to measure this velocity. It was an ambitious undertaking, and one that his teacher, Johannes Müller, had deemed impossible. Müller had assumed that an impulse, being an electrical event, would have a velocity similar to that of light, and that the time differences along a nerve would be far too short to be measured:

> "Probably we will never have the means to measure the speed of nervous action since we lack, in order to establish comparisons, those immense distances whereby we can calculate the speed of light, which, under this respect, has some relation with it."[28]

Amusingly, there had been an earlier attempt, by Albrecht von Haller—a contemporary of Galvani—to measure impulse propagation, based on the speed with which a passage from the *Aeneid* could be read aloud. The calculation depended on the number of times, in one minute, that a tongue muscle would have to contract.[29] Helmholtz, undeterred by Müller's skepticism and any results based on the *Aeneid*, set about the problem by using apparatus constructed to his own ingenious design (Fig. 2–11). As the muscle contracted and lifted the weight pan, it also interrupted a current that had started to flow when the stimulus was delivered.[30] While the current flowed, however, it deflected the magnetic needle of a galvanometer, the extent of the deflection being proportional to the duration of the current. To magnify the deflection, a beam of light was reflected from a mirror attached to the galvanometer needle. Having stimulated the nerve at its cut end, Helmholtz then moved the stimulating electrodes as close to the muscle belly as possible and repeated the measurements. Since the impulses now had less distance to travel, the current through the galvanometer was briefer and the excursion of the needle smaller. By comparing the needle deflections, Helmholtz was able to determine the time taken for the impulses to travel the 40-mm length of the nerve, and hence their velocity. The critical experiment was performed on

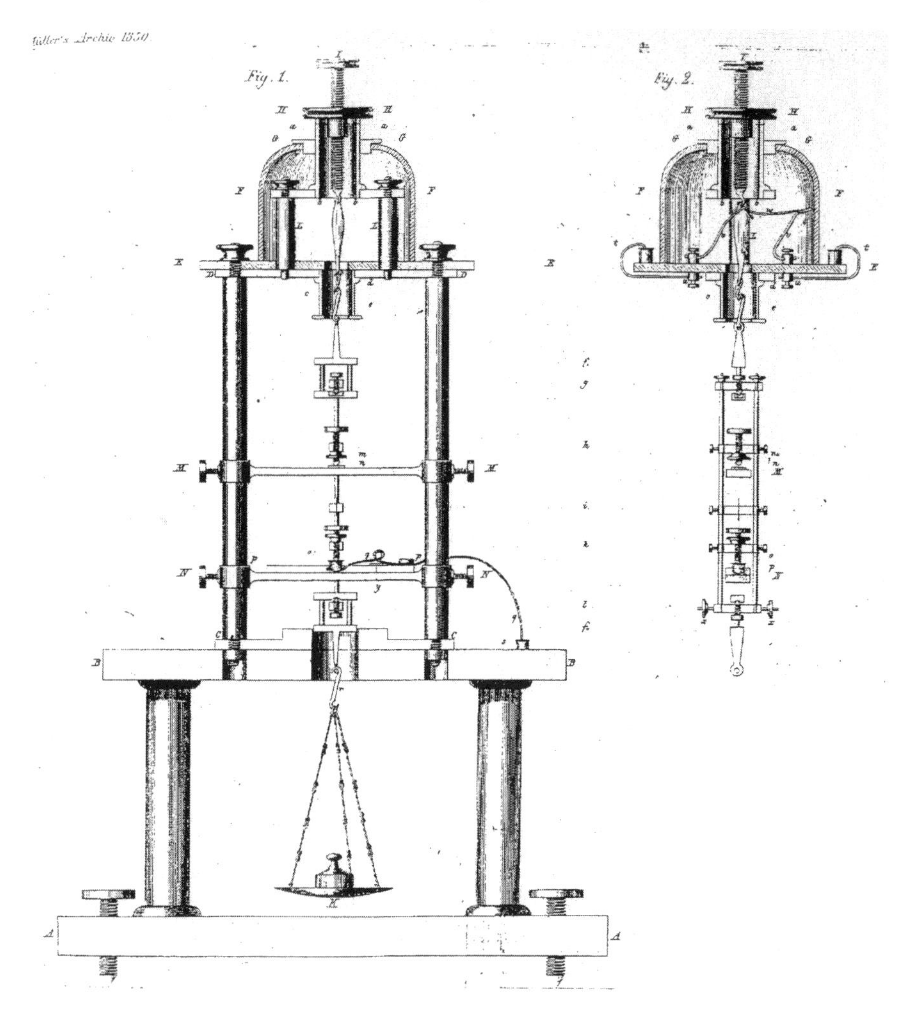

Figure 2–11 Helmholtz's equipment for measuring the conduction velocity of the nerve impulse. The muscle and nerve are mounted in a humidified chamber, shown in more detail on the right; the nerve can be stimulated at either of two points. As the muscle contracts it lifts the weight at the bottom (*K*), at the same time breaking an electrical contact and interrupting the current to a galvanometer. Although much of the lettering is too small to be deciphered, the beauty in the design is evident. (From Helmholtz, 1850)

December 29, 1848, with the aid of Helmholtz's wife, Olga. Helmholtz estimated the conduction velocity of the fastest-traveling impulses to be 30.8 $m.s^{-1}$ (meters per second), a very reasonable value, though one that would have depended on the temperature of the laboratory. Ironically, it was a value close to that derived from reading the *Aeneid* aloud, though every step in von Haller's reasoning had been at fault.

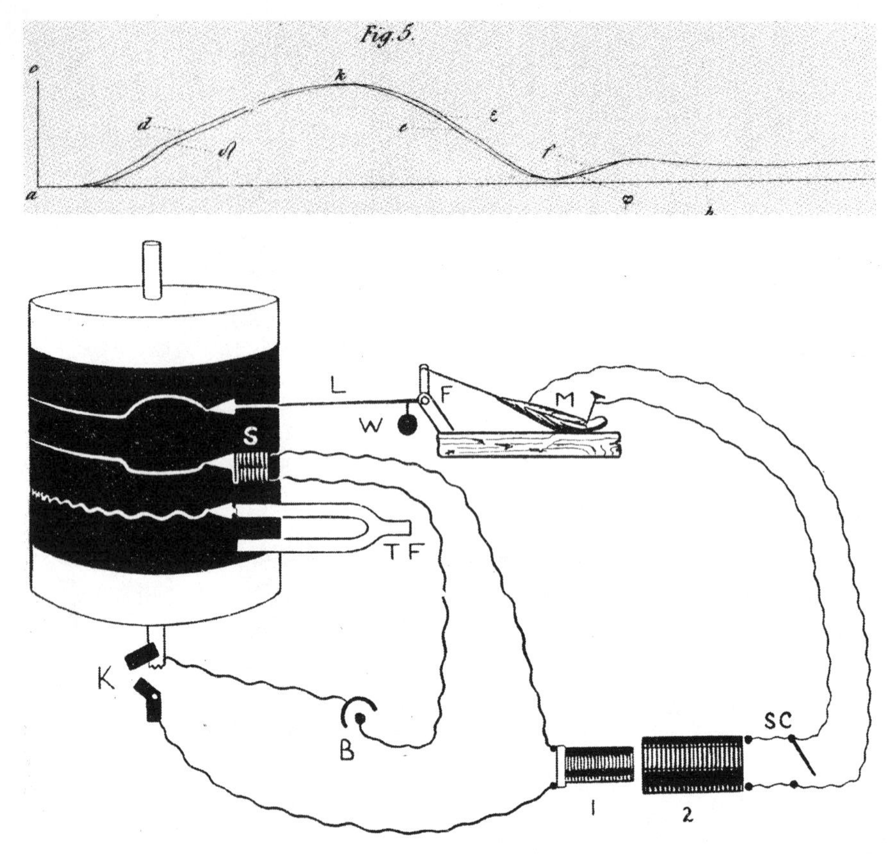

Figure 2–12 *Top*: Tracings of two muscle twitches on a kymograph, the nerve having been stimulated at two sites. The separation of the twitches is evident and was used by Helmholtz to calculate the maximum conduction velocity of the nerve impulses. (From Helmholtz, 1852) *Bottom*: A kymograph for recording muscle contractions. The point of the lever writes on paper covered with soot, while a tuning fork provides a record of time. Physiologists and medical students continued to use this type of equipment for more than a century, trying not to cover their hands in soot and varnish as they did so.

Not withstanding the simple logic of Helmholtz's experimental design, the critical measurements could not have been easily made, since the time difference for the calculation of velocity had been only 1.3 milliseconds and the excursion of the galvanometer needle would have been non-linear. It was the sort of measurement that would nowadays be made on an oscilloscope with a very fast trace. There was also the possibility that the muscle contractions, evoked by stimulating the nerve at two sites, were not equally forceful. Helmholtz, however, was very much aware of the potential errors in his system and took steps to minimize them—including the use of averages from repeated measurements, and the application of statistics to express the variability.

The success of the experiment, in showing that nerve impulses traveled with finite velocities, was a first step in removing some of the mysticism from the workings of the brain. It also established Helmholtz as a leading experimental physiologist, and this despite his age—he had published his nerve studies while still in his twenties. Later, Helmholtz would repeat his velocity measurements, this time by using the tracing of the muscle contraction on a rotating smoked drum, a "kymograph" (Fig. 2–12, *bottom*), the invention of Carl Ludwig (1816–1895).[31] Though the separation of the traces on the rapidly revolving drum might be no more than 1 mm, when the nerve was stimulated at two points, there was the satisfaction of seeing the curves produced by the contraction and relaxation of the muscle, rather than the movement of a galvanometer needle. Later still, Helmholtz, with Baxt, was able to apply the same method to the measurement of impulse conduction velocities in nerves of the human forearm, obtaining values slightly higher than those in the frog.

At the suggestion of Johannes Müller, Helmholtz went on to measure the time delay incurred in responding to a stimulus as quickly as possible (the *reaction time*). By applying weak electric stimuli to various parts of the body and asking the subject to immediately move his or her hand on feeling the shock, Helmholtz obtained values of between 0.12 and 0.20 seconds. The results were a surprise, not only to Helmholtz's mentor, Johannes Müller, but also to the scientist's father, August. The elder Helmholtz, in congratulating his son, expressed his astonishment that there should be a delay between a mental event, in this case the intention to move, and the body's reaction in the form of the contraction of the muscles. August's attitude was understandable. The mind is unaware of delays, either in the generation of a movement or in the development of a sensation. Everything, it seems, is instantaneous. The truth, however, is that our minds live a fraction of a second in the past, the interval being occupied by the generation and conduction of impulses from one part of the nervous system to another.

The determinations of impulse conduction velocity and reaction time were not Helmholtz's only contributions to physiology and medicine. At about the same time that he was investigating nerve and muscle, he also became interested in the eye, an interest sparked by his requirement to lecture to the medical students in Königsberg. Aware of the need to visualize the retina, he quickly hit on the idea of using a half-silvered mirror, and built the first ophthalmoscope himself in only two days. Continuing his fascination with the eye, he proposed a theory of color vision, one in which all the colors could be appreciated by the varying excitation of three types of receptor in the retina—a theory shown to be correct a century later. In addition, there was work on hearing and on the properties of sound involved in musical appreciation. Outside the biological sciences, Helmholtz had earlier published an extremely important paper on the conservation of energy, arguing that in those situations in which energy appears to be lost—as in a moving object coming to a standstill, for example—the energy is simply converted into another form, in this case heat. Later would come another influential paper, this time a theoretical treatment of the motion of a perfect fluid. Through much of his work ran a powerful mathematics, a subject he had largely taught himself, and which would lead to papers on non-Euclidean geometry. Given Helmholtz's many contributions to science, it was not surprising that he should have ended his career as Professor of Physics at the University of Berlin and as the founding President of the Imperial Physical-Technical Institute in the same city.

One of Helmholtz's friends from his student days was Emil du Bois-Reymond, and it was two of the latter's pupils who would complete the transformation of the nerve impulse from a metaphysical concept to a physiological event capable of measurement. Julius Bernstein (1839–1917; Fig. 2–13) and Ludimar Hermann (1838–1914) had grown up in Berlin as boyhood friends, the sons of prominent Jewish intellectuals.[32] While working in du Bois-Reymond's laboratory, Bernstein had devised a differential rheotome, an instrument that enabled electrical connections to be made for extremely short, but nevertheless precise, time intervals (Fig. 2–14, *left*). By stimulating at a high frequency, the small galvanometer deflections derived from the nerve impulses summed and became measurable. The time

Figure 2–13 Julius Bernstein (1839–1917). Following his success with the differential rheotome, Bernstein was appointed to the Chair of Physiology at the University of Halle at the age of 34. This photograph, the original of which is in the archives of the University of Halle, was taken in 1890 at the time he was Rector of the university.

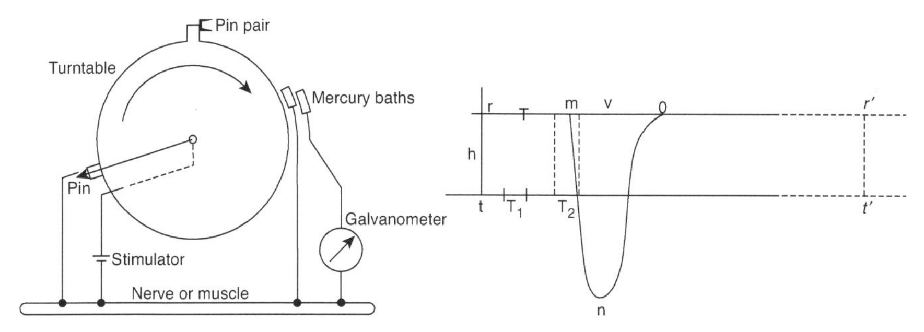

Figure 2–14 *Left*: The principle of Bernstein's differential rheotome. A pin on a rotating turntable completes a circuit, causing the nerve or muscle to be stimulated. At another point on the turntable a pair of pins momentarily complete a recording circuit as they contact two overlapping baths of mercury. The duration of the recording can be adjusted by altering the overlap of the baths, and the nerve or muscle current flowing during this short period deflects the galvanometer needle. The paired pins can then be moved nearer or further away from the "stimulating" pin, permitting a different part of the impulse to be sampled. *Right*: The nerve impulse (action potential), as deduced from the rheotome experiments. The impulse lasts less than a millisecond and momentarily exceeds the demarcation potential (i.e., the potential across the membranes of the fibers in their "resting" states). This was the first picture of the impulse, albeit a reconstruction. (From Bernstein, 1868)

window was then moved and the process repeated until the entire form of the nerve impulse had been mapped out. It was an ingenious method and one that showed the nerve impulse to have a rapid rise and a rather slower decay, the total duration being less than a millisecond (Fig. 2–14, *right*). Further, the experiment showed that, at its peak, the impulse not only abolished the demarcation current, as Matteucci and du Bois-Reymond had previously demonstrated, but also exceeded it.[33] It was a crucial finding and one that, had it been followed up, might have led to an earlier understanding of the membrane mechanisms underlying the nerve impulse. Using the same methodology as before, but stimulating the nerve at two points rather than one, Bernstein was also able to measure the impulse conduction velocity, obtaining values similar to those of Helmholtz.

Not to be outdone, Hermann repeated his friend's observations, eventually confirming them. Hermann's greater contribution, perhaps, was to describe the electrical properties of the nerve fiber in terms of resistances and capacitors.[34] Thus the membrane of the nerve fiber could be represented by a number of resistors and capacitors in parallel (Fig. 2–15, *top*). There was another resistance, in the form of the cytoplasm that made up the core of the nerve fiber, and yet another resistance, this time a much smaller one, in the fluid that surrounded the nerve fiber. Hermann's way of thinking, and his use of an analogue model, would prove helpful, if not essential, to future generations of neurophysiologists. It explained, for example, how the demarcation ("injury") potential was always less than the

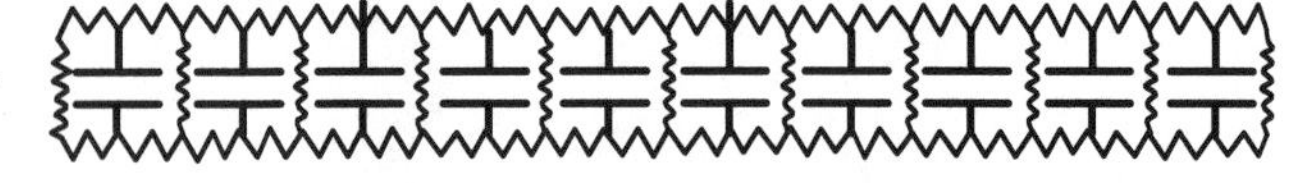

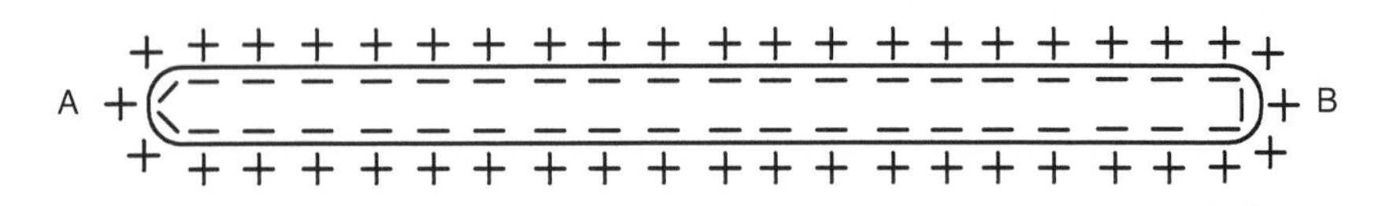

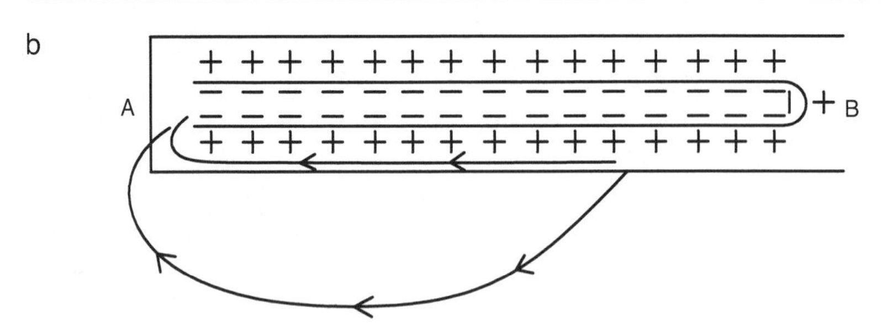

Figure 2–15 *Top*: Hermann's electrical analogue of the cell membrane, cell interior and exterior, as depicted by Gasser, using conventional symbols for resistance and capacitance. (From Erlanger J, Gasser HS. *Electrical Signs of Nervous Activity*, 1936, p. 139. Reprinted with permission of the University of Pennsylvania Press.) *Bottom*: Bernstein's concept of the membrane. (*a*). An excess of negative charges is present in the interior of the cell. (*b*). If the end of the fiber is cut, as shown, a current (of positive and negative charges) will flow between the intact portion of the fiber and the cut end. As this *demarcation (injury)* current passes through the intact membrane, it reduces the potential across it (i.e., *depolarizes* it). (From Bernstein, 1912)

potential across the membrane of an inactive, "resting" fiber. Similarly, the electrical analogue explained how the nerve impulse, when recorded with two electrodes resting on the nerve, was smaller than the actual change in potential across the nerve fiber membrane. And, because of these inadequacies in recording, it did not really matter whether one referred to the impulse in a nerve or muscle fiber as an *action current* or as an *action potential.* In fact, the two terms were often used interchangeably, though *action current* was favored by most of the early neurophysiologists.

Hermann's analogue model of the fiber membrane and its surrounding fluids was accepted by Bernstein, among others. Where the two differed, at least initially, was over the basis of the demarcation potential. To Hermann, the potential was something that a nerve or muscle developed at the site of injury, whereas Bernstein believed, correctly, that the injury, or the application of potassium, diminished a potential that normally existed across the membrane. In 1902 Bernstein published a lengthy thermodynamic analysis of this potential difference. His proposal was that, in the resting condition—when the fiber was not conducting impulses—the membrane was selectively permeable to potassium ions, and that, because ions have different mobilities, there would be concentration differences as well as a potential across the membrane.[35]

One can think of the potential developing because of the tendency of the potassium ions to diffuse out of the fiber down their concentration gradient. Being positively charged, the outward movement of potassium would leave a surplus of negative charges inside the fiber. The inside of the fiber would thus be negative with respect to the outside. The amplitude of the potential across the fiber membrane would be proportional to the difference in the concentrations of potassium ions on the two sides of the membrane and could be calculated using an equation derived by the German physical chemist, Walther Nernst (1864–1941), some years before:

$$E = \frac{RT}{F} \ln \frac{K_o}{K_i}$$

where R, T, and F are, respectively, the universal gas constant, the absolute temperature, and the Faraday, K_o and K_i are the concentrations of potassium ions outside and inside the fiber, and *ln* is the natural logarithm.

The equipment that Bernstein used to measure the potential across the membrane was very simple and is shown for muscle fibers in Figure 2–16. The excised muscle belly was immersed in oil, with one electrode applied to its cut end and the other to its intact surface. To avoid the spurious potentials that would occur with metal electrodes, Bernstein used saline-soaked clay electrodes instead. The two electrodes were connected to a galvanometer. The potential recorded—the *demarcation potential*—would have been rather smaller that the true potential across the muscle fiber membranes because of the tendency of the cut ends of the fibers to seal off and produce an electrical resistance, thereby diminishing the flow of the injury current. With this equipment Bernstein was able to demonstrate that the demarcation potential became larger as the temperature of the muscle was gently raised, as the Nernst equation predicted. A more convincing validation of the equation came a little later, from the work of Höber, who was able to show that the demarcation potential fell as the concentration of potassium in a solution bathing the muscle was increased.[36]

Both Bernstein and Hermann recognized that, just as an injury to one part of a nerve or muscle fiber produced a (demarcation) current (see Fig. 2–15, *bottom*), so an impulse in one segment of a fiber would induce a current across the membrane of an adjacent part, causing the latter to depolarize—a *local response*—and to become excited in turn. In this way the impulse would move along the fiber.

Ten years later Bernstein summarized his thinking in a monograph, *Electrobiologie.*[37] With excellent illustrations and a little over 200 pages in length, the book was by far the most complete account of electrical phenomena in living cells. Not only did it deal with the potential of the membrane in its resting condition, but it also proposed that the nerve impulse was the result of the membrane briefly

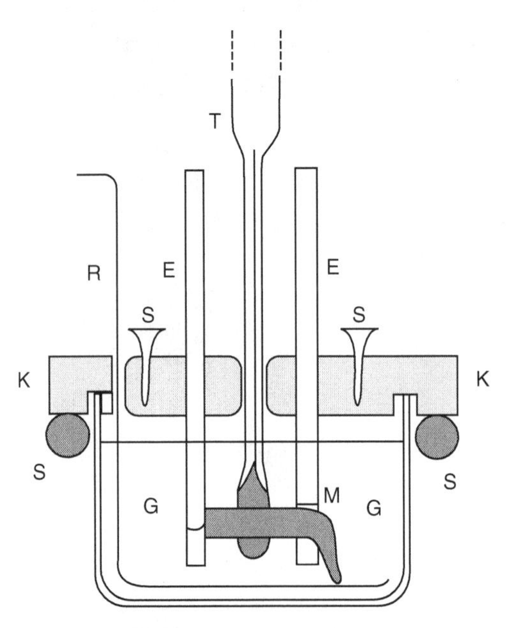

Figure 2–16 Bernstein's apparatus for measuring the potential difference between intact and cut portions of muscle fibers (*demarcation potential*). Saline-soaked clay electrodes (*E, E*) contact the injured and intact regions of the muscle (*M*). The chamber is filled with oil and covered with a cork lid (*K*), through which a thermometer (*T*) and stirrer (*R*) intrude. (From Bernstein, 1902)

losing its selective permeability for potassium by becoming permeable to all the ions around it. In one respect, however, the book fell short. In making the proposal of a non-selective increase in membrane permeability, Bernstein had discounted the significance of his earlier finding, that the impulse (the *action potential*) briefly exceeded the demarcation potential. It was a serious omission and the resulting conceptual error was one that would not be corrected for many years.

Despite its modest size, Bernstein's was a very influential book and came to dominate thinking on the nerve impulse for the next 40 years. The book did something else. On the very last page Bernstein predicted that the true form of the nerve impulse would be obtained by using a cathode ray tube, the recent invention of his countryman, the Nobel Prize winner Karl Braun. Bernstein did not live to see his prediction fulfilled, but within his lifetime the potential across the resting nerve membrane had been measured, admittedly crudely, in the form of the demarcation potential, the conduction velocity of the nerve impulse had been determined, and he himself had not only mapped out the shape of the impulse but had given a physico-chemical explanation for the potential across the nerve fiber membrane.

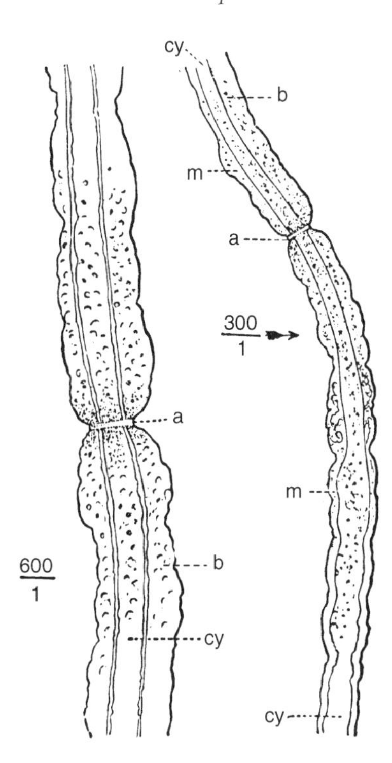

Figure 2–17 Two nerve fibers, drawn at different magnifications, showing the narrow gaps (*a*) in their myelin sheaths (*b*). Ranvier termed these gaps "constrictions," but they are today known as "nodes of Ranvier." (From Ranvier, 1872)

While the neurophysiologists were making their discoveries, there was an important contribution to knowledge about the nerve fiber from the histologists. Earlier, using a single-lens microscope, Leeuwenhoek had envisaged the nerve fibers as very fine vessels, while, rather later, Fontana had simply described them as translucent cylinders (see above). In 1878, however, with the aid of the more powerful compound microscope, Louis-Antoine Ranvier (1835–1922) recognized that the interior of a fiber was surrounded by a myelin sheath and that the sheath was interrupted every 1 to 2 mm; these narrow gaps were subsequently referred to as the *nodes of Ranvier* (Fig. 2–17).[38] At the time of his discovery, Ranvier was the Professor of Anatomy at the Collège de France, in Paris, serving under Claude Bernard. Much later it was shown that the presence of the myelin sheath and the nodes of Ranvier made nerve fibers much more efficient in conducting impulses.

More than a hundred years had elapsed since the first observations of Galvani, and it had been the Italians and the Germans, with some assistance from the French, who had carried on with the story of the impulse—Galvani's "extremely mobile principle existing in nerves." There had been one powerful country, always heavily involved in European affairs, that had been almost silent in this new venture: Britain.

49 Queen Anne Street, London, possibly February 15, 1890[1]

The coal fire had warmed the dining room nicely and the silver cutlery gleamed under the candelabra. Two places had been set for John and herself at the ends of the table, and four along each side for the guests. Lady Ghetal Sanderson adjusted the position of one of the wine glasses, stole a quick glance at herself in the mirror above the sideboard, and patted her hair. Everything was as it should be. She called for Kitty to start bringing in the tureens of vegetables, the gravy boat, and the large side of roast beef. Then she rang the bell to summon John and his guests down from the study.

Lady Sanderson enjoyed entertaining, especially in their London home. How lucky they had been to find the house—so close to the theaters and to the shops along Regent Street and Oxford Street, and within easy reach of Hyde Park. John, of course, had liked it because he could walk to University College, and because there was always the chance of an interesting encounter with one of the physicians and surgeons in neighboring Harley and Wimpole Streets. There had been no question of their selling the house when John had been appointed to Oxford. The university city was interesting, quaint even, with its centuries-old customs, but London was always the place to be. But what a peculiar bunch these physiologists were, so immersed in their work, and forever talking about their unappealing experiments on frogs and cats. It was difficult to take them seriously, though John certainly did. Indeed, John was one of them. She thought back to the first meeting, when the idea of forming a society had been discussed. Had it really been John's idea, or had it been, as she suspected, Michael Foster's? Anyway, it had been in this house that they had met, and she had been the one who had written out each invitation for John to sign. She remembered the wording:

> *"It is proposed to hold a Meeting at my house at 5.30 p.m. on Friday next (31st) of a preliminary character, for the purpose of considering whether any, or, what steps ought to be taken with reference to the Recommendations of Lord Cardwell's Commission. It will probably also be proposed at the Meeting to form an Association of Physiologists for mutual benefit and protection.*
>
> *Sharpey, Huxley, Foster, Lewes and others, have promised to attend. I shall be glad if you can come also."*

They had arrived like conspirators, in their cloaks and top hats. Huxley had been the first, with those long mutton-chop whiskers on either side of the dark eyes. Though everyone seemed terrified of that great intellect, it was a pity the old man no longer came to their meetings.

Lady Sanderson stepped into the hall and listened. She could just make out their voices from upstairs but there were no sounds of movement. How typical. This time she called out in her loudest, most authoritative voice, before going back into the dining room and seating herself at the far end of the table.

At last. The sounds of footsteps on the stairs. Still talking. Gaskell was the first to enter, holding forth in his resonant voice, followed by John and their newest guest, young Charles Sherrington. Lady Sanderson watched in disbelief as Gaskell went to the other end of the table, indicating that John—her John, the host of the dinner—should sit at his right. Before she could say anything, Gaskell, still talking, had picked up the carving knife and fork and had set to work on the roast.

No one else spoke. Gaskell, the great authority on the mechanism of the heartbeat, at last aware that something was amiss, looked up from his carving. A flush spread over the handsome face. Flustered, he looked apologetically at his hostess at the other end of the table:

"Pardon me, I thought I was in my own house."[1]

3

Catching Up

As already noted, it had been the Italians, Galvani and Volta, and then Matteucci, who had carried out the first experiments on nerve and muscle excitation, and there would be other countrymen, notably Angelo Mosso and Camillo Golgi, who would come after them and make major contributions to neuroscience. Similarly the French would produce Claude Bernard, François Magendie, Charles-Edouard Brown-Sequard, and Étienne-Jules Marey. Despite the appearance of these major figures, however, by far the greatest concentration of physiologists in the mid-1800s was in Germany, where it seemed that every major city had its university and medical school, with excellent teaching and research in anatomy and histology, and in the newly emerging science of physiology. In Berlin du Bois-Reymond had detected the action current for the first time, Helmholtz had determined its speed of propagation, and Bernstein had described its form. There was also Johannes Müller, Helmholtz's teacher, who among many other achievements, had formulated the doctrine of specific nerve energies, a doctrine that stated, in effect, that a particular sensation depended on the nature of the sensory receptors that had been stimulated, and that it was the function of a receptor to convert the energy of the stimulus into impulses (action currents) in the nerve fiber. Until then it had been supposed, for example, that the eye was somehow capable of generating light itself and that it was this internal light that was the perception. In Leipzig there was Carl Ludwig, as famous as Müller, with his own school of physiologists, and elsewhere in Germany could be found a score of physiologists of almost equal distinction.

Nowhere was the contrast with the situation in Germany greater than in Britain, where physiology was regarded as being of minor importance, certainly below anatomy, and far removed from medicine and surgery. Indeed, it was usually a young physician or surgeon who would be coerced into giving lectures in physiology to medical students; then, as his clinical practice prospered, he would discontinue his teaching and cause someone else to take over the unwanted responsibility. On the other hand, a small number of British clinicians, scattered throughout the country, were making, or would shortly make, important physiological observations. In the unlikely premises of the West Riding Lunatic Asylum in Yorkshire, David Ferrier was investigating the cerebral cortex in a number of mammalian species, including monkeys, by observing the effects of either stimulating its exposed surface or removing a small part.[1] Similarly, in London's St. Mary's Hospital, Augustus Waller was combining clinical practice with the first recordings of the human electrocardiogram, employing a toy train to move his photographic plates.[2] Then there was Richard Caton, in Liverpool, like Waller a forgotten name today, who was making, in animals, the earliest tracings of spontaneous electrical activity in the brain. And there were a few others, for the most part lonely men struggling to pursue a particular interest among their other affairs. But, so far as is known, only one person in the country was prepared to commit his career completely to the teaching of physiology.

William Sharpey (Fig. 3–1) was born in Scotland in 1802, at the time that Volta was at the height of his fame. After graduating in medicine from Edinburgh, Sharpey studied surgery in Paris but then decided to make his career in anatomy and physiology.[3] To gain further exposure to these subjects, he spent five years wandering around Europe, starting in Paris and finishing his studies in Berlin, and sometimes walking from one city to another with his belongings in a rucksack. Eventually he accepted the newly founded Chair of Anatomy and Physiology at University College London and embarked on a career that combined the teaching of medical students with research into the fine structure of bone. Sharpey was an excellent lecturer but, other than his microscope, had no apparatus for demonstrations to students. True, he could repeat Galvani's simple experiments, but at the time that Helmholtz was using the kymograph to measure the time-course of muscle contraction and the speed of nerve impulse conduction, Sharpey could

Figure 3–1 William Sharpey. (Wellcome Library, London)

Figure 3–2 Thomas Henry Huxley, 1876.

only ask his students to imagine a kymograph, spinning his top hat on the table in front of them.

Sharpey was not the only person aware of the shortcomings of medical education in physiology. The Professor of Anatomy at Cambridge, George Humphry, had agitated for the creation of a Chair in Physiology, and no less a person than T. H. Huxley[4] (Fig. 3–2) began to speak out about the dearth of informed instruction and research in physiology in Britain. Though Huxley himself had wanted to be a physiologist, the opportunity had never presented itself, though this did not prevent him from writing *Lessons in Elementary Physiology*[5] and lecturing on this subject to schoolteachers and workingmen in London. To a considerable extent, he was self-taught as a scientist, making his reputation on the basis of several fine papers describing the detailed structure and phylogenetic status of several marine organisms; he performed this work and wrote it up during his service as assistant surgeon on H.M.S. *Rattlesnake* in 1846–50.[6] The four-year voyage had taken the ship to Australia for an exploration of the east coast and Great Barrier Reef, and of the islands to the north that bordered the Torres Strait. A fine microscopist and comparative anatomist, Huxley was attracted by the relation between structure and function, and by the adaptations exhibited by different species. For one who had wanted to be an engineer but had been diverted into medicine, there was much in the architecture and mechanics of the body to occupy his interest.

As to the state of physiology in the British universities, Huxley was only too aware of the shortcomings in teaching and research, especially in comparison with the situation on the continent. Speaking to the graduating medical students at University College London in 1870, he remarked:

> "I get every year those very elaborate reports of Henle and Meissner—volumes of, I suppose, 400 pages altogether—and they consist merely of abstracts of the memoirs and works which have been written on Anatomy and Physiology—only abstracts of them! How is a man to keep up his acquaintance with all that is doing in the physiological world—in a world advancing with enormous strides every day and every hour—if he has to be distracted with the cares of practice? You know very well it must be impracticable to do so. Our men of ability join our medical schools with an eye to the future. They take the Chairs of Anatomy or of Physiology; and by and by they leave those Chairs for the more profitable pursuits into which they have drifted by professional success, and so they become clothed, and physiology is bare. The result is, that in those schools in which physiology is thus left to the benevolence, so to speak, of those who have no time to look to it, the effect of such teaching comes out obviously, and is made manifest in what I spoke of just now—the unreality, the bookishness of the knowledge of the taught."[7]

At the time that Huxley commented so forcefully, the possibility of a Chair in Physiology had already been considered in Cambridge, and had been rejected, at least for the time being. All was not lost, however: there was support for a more modest alternative, the creation of a Praelectorship of Physiology by Trinity College. It was in relation to this that Huxley was able to exert his considerable influence by suggesting Michael Foster (Fig. 3–3) for the position.[8] Foster, the son of a country doctor, had been taught by Sharpey at University College London and had been the gold medalist for anatomy and physiology in his class. Like Sharpey before him, he had gone to Paris for further studies after graduating in medicine and, at about that time, had conducted some experiments on the heart. Though they were performed on the snail, and must therefore have been technically demanding, the two main conclusions would be shown to have universal application. One was that the force of contraction was proportional to the preceding distention of the heart chamber, and the other was that cardiac muscle fibers were capable of beating by themselves. Even after turning to family practice with his father, Foster continued to do research. It was not long, however, before Foster went back to University College, first as Teacher, and then as Professor, of Practical Physiology. In this position he was able to supplement Sharpey's lectures by directing students in microscopic examination of body tissues, chemical analyses of blood and urine, and basic experiments on nerve and muscle excitation and on the heart and circulation.

Figure 3–3 Sir Michael Foster. (Wellcome Library, London)

Then, in 1870, on Huxley's recommendation, Foster made the move to Cambridge. Huxley knew his man well, having had his assistance in the practical biology classes given in London, and the two would remain on close terms until Huxley's death in 1895. To begin with, the premises for the new physiology laboratory in Cambridge were hardly inspiring—a room in the Philosophical Library with three tables, some chairs, and a few microscopes and bottles of reagents. Yet it was from this single room and its newly appointed Praelector that the great tradition of Cambridge physiology and, indeed, much of British physiology, began. Foster's wide knowledge and lecturing skill and his insistence that students perform the experiments themselves all helped to make the new enterprise successful. Among Foster's early students were two who would, in their time and in very different ways, dominate much of British neuroscience—Henry Head[9] and Charles Sherrington.[10]

To those familiar with present-day Cambridge, with its industrial park, its busloads of tourists, and its shopping mall, it is difficult to conceive of the town as it was in Foster's time. Much smaller then, it was both a market town and a center of learning, two entirely different worlds with each tolerating the other. Cows were still driven past Queens' College along Silver Street, while the Master of St. Catherine's College would be seen riding his small black pony and the workmen their penny-farthing bicycles. And in the warmer months, while undergraduates and Fellows punted along the Cam, the town boys, unabashed, would bathe naked in the same river. It was a world memorably captured by Gwen Raverat, one of the great Darwin's granddaughters, in *Period Piece.*[11] It was a world in which there was no academic distinction between a Professor of Hebrew or Greek and a scientist attempting to unravel the structure of the atom. In such a world a newly appointed physiologist, the first of his kind, would have attracted only the mildest curiosity.

Apart from his witty and genial disposition, Foster had one other gift: an uncanny ability to spot talent in others. Within a few years he attracted Walter Gaskell and John Newport Langley (Fig. 3–4) into the physiological fold, both of whom initiated outstanding research, Gaskell on the excitability of the heart and Langley on salivary secretion and drug actions. Both men investigated and for the first time clearly described the two divisions of the autonomic nervous system, the systems of nerve cells whose responsibilities include the regulation of the heartbeat, blood flow, and the secretions of the stomach and intestines.

It was Langley who, in 1903, succeeded Foster as Professor and head of the Cambridge Physiological Laboratory and made appointments of the utmost significance for the development of nerve and muscle physiology, including the nature of the nervous impulse. By this time the Laboratory was expanding rapidly and was already attracting scientists from abroad, including some from Germany—the tide was beginning to turn. There were other developments too. There was now the *Journal of Physiology*, which, started by Foster, had been taken over and nurtured by Langley. There was also a national Physiological Society, again an initiative of Foster's, though it had been brought to fruition largely by John Burdon Sanderson, Foster's successor at University College London and subsequently

Figure 3–4 John Newport Langley. The second head of the Cambridge Laboratory of Physiology, an authority on the autonomic nervous system, and at one time the owner of the *Journal of Physiology*.

Oxford's founding Professor of Physiology. It was in Sanderson's London home that a small group of physiologists had met in 1876 to discuss the formation of the Society.[12]

Clearly, after years of disinterest and neglect, physiology in Britain was on the move. But who would take up the challenge of the nerve impulse? There was another challenge, too, one that faced all those who were hoping to obtain knowledge as to how the brain worked. As they looked through their microscopes at the complicated tangles of nerve fibers and at the countless numbers of nerve cells, they must have despaired of ever understanding the paths that the impulses might follow.

Burlington House, London, March 8, 1894

The heads turned towards the tall man with the solemn expression and the bushy beard. Michael Foster got slowly to his feet. Now in his 13th year as Secretary of the Royal Society of London, he was known to the members as an intelligent and amusing after-dinner speaker. He had given some thought to what he might say but, as was his custom, had not troubled to write the words down. He knew that he could bring the evening's events to a fitting close.

This year's Croonian Lecture had not been an easy one to organize. Once he and Sherrington had realized the importance of the Spaniard's work, they had been determined to make him this year's lecturer, but they had barely had enough time to make the arrangements. The Society's invitation, received only the previous month, could so easily have been declined. Sherrington, bless him, had done his part. Foster looked with affection at the small man with his pince-nez, sitting next to the Spaniard. Soon, perhaps, it would be Sherrington who would be giving the Croonian Lecture. It was Sherrington who had been looking after the Spaniard during his time in England, letting him stay at his home and helping him prepare his talk. It had been Sherrington who had politely suggested to the Spaniard that, rather than invite the Society members to look down microscopes at his preparations, photomicrographs could be projected on to a screen for the entire audience to see. And Sherrington, in the few days available, had not only made the lantern slides—most of which Cajal had declined to show—but had even found time to procure the Spanish flag, now intertwined with the Union Jack on the presidential dais.

All this went through Foster's mind as he stood behind his chair. Despite the revelations in the pictures, and the printed summaries of the most important conclusions, it had not been an easy lecture for the audience. Although the stocky Spaniard with the olive complexion and the deep brown eyes had been easily heard by everyone in the room, he had been obliged to speak in French. On the other hand, there had been no doubt about the passion and the extraordinary insights that had infused his work. And then there had been those astonishing pictures of nerve cells in the spinal cord and the different regions of the brain.

Foster indulged his habit, and tugged at his beard one more time. Then he began. First, as was customary, he acknowledged the President of the Society, Lord Kelvin, and then the guests and members of the Society, many of whom had traveled down from Oxford and Cambridge for the Croonian Lecture. Then, turning to the Spaniard, sitting beside the President, he began to thank him. This year's Lecturer, Foster said, had converted "the impenetrable forest" of the nervous system into "a well laid out and delightful park." Further, his researches had "established connecting collaterals and motor endplates between the souls of Spain and of England, formerly kept apart by centuries of misunderstanding and indifference."[1]

It was a witty speech, one of Foster's best, and the central point was true. Ramon Santiago y Cajal had indeed shown how the nervous system might be organized.

4

The Anatomist's Eye

At the time of his invitation to the Royal Society of London, Cajal (Fig. 4–1) was 41 and Professor of Normal Histology and Pathological Anatomy in the Central University of Spain. Prior to his appointment in Madrid, he had spent several years as Professor of Anatomy in Valencia and then as Professor of Histology in Barcelona, and it was in those smaller cities that he had commenced his studies of the nervous system. Though he had not, and would not, ever record an impulse in a nerve fiber, he had nevertheless succeeded in the seemingly impossible, in showing how the impulses must flow through the nervous system. With the possible exception of his father, it was a level of achievement that neither he, nor anyone who had known him in his youth, would ever have predicted.

Cajal was born in a poverty-ridden village in the foothills of the Pyrenees and had spent his entire youth in northeastern Spain as his father, a surgeon, moved from one small town to another. It was an unsettled time for the country, as attempts were made to replace the Bourbon monarchy with a republican government. Though his father had plans that the boy would follow him into medicine, the young Cajal, from an early age, had made it clear that he would do exactly what he wanted, which frequently involved flouting one kind of authority or another. At religious schools, unsuccessful attempts to teach him Latin had resulted in beatings and solitary confinement without food. Outside school, his sense of adventure and fearlessness had, on several occasions nearly cost him his life. He had been kicked unconscious by a frantic horse, immobilized on a cliff face after climbing up to inspect an eagle's nest, trapped under the ice after falling through a hole, and struck in the eye by the explosion of a homemade cannon. Though his exploits, added to his skill with the sling, had made him a natural leader among boys of a similar age, Cajal often preferred to be by himself so that he could better indulge his two main interests—the observation and exploration of nature, and art. For his art, he bought his own pencils and sheets of paper but, as he had insufficient money for paints, he would scrape colors from walls or leach them from cigarette packets. In his autobiography, *Recuerdos de Mi Vida* (*Recollections of My Life*),[1] there are two examples of watercolors executed by Cajal as a 9- or 10-year-old—a laborer drinking wine from a *porron* in a tavern, and a small historic church. Both show a precocious talent, but it was one that was strongly discouraged by his father. For the older Cajal, there was only one possible career for his son: medicine.

Figure 4–1 Santiago Ramon y Cajal.

The turning point came when Cajal turned 16 and his father, an excellent anatomist, decided to instruct him in osteology. To obtain their bones, the father and son climbed over the walls of a cemetery at night and scavenged among the exhumed remains in a pit. Inspired by his father's teaching, and drawn by the practical nature of the subject, the younger Cajal, for the first time in his life, excelled as a student. Next came enrollment in the medical school in the provincial capital, Zaragoza, a move supported by his father transferring the family to that city and himself becoming a temporary professor of dissection. The elder Cajal's new position gave him the opportunity to continue his son's education in anatomy, pushing it to a level far beyond that achieved by his fellow students. In other subjects, however, the standard of instruction varied, and there were no opportunities for laboratory work in physiology.

Having graduated from medicine, and faced with conscription into the national army, the 21-year-old Cajal took a further exam that enabled him to serve as a military doctor. Posted to Cuba, then in the process of fighting for its independence from Spain, Cajal, like many of the conscripts, developed malaria and was eventually repatriated in poor health. Worse was to follow, with the advent of pulmonary tuberculosis, but Cajal not only survived but also recovered fully from both disorders. Reinstated in Zaragoza, Cajal embarked on further studies, with the aim of applying for a chair in anatomy in one of the Spanish schools. Not only did he broaden his reading to include the current research publications of the German and French medical scientists, but he also bought a second-hand microscope to begin his own studies. Eventually, through sheer ability rather than the intervention of a patron, Cajal succeeded in obtaining the chair in anatomy in Valencia. At first his research was on bacteria and the response of the tissues to infection, and he was called upon by the authorities in Zaragoza to investigate a devastating nationwide epidemic, possibly of cholera, and the claim by a Spanish physician to have devised a successful vaccine.

In his report, Cajal confirmed that the disease was cholera and gave reasons why the vaccine was unlikely to be effective. In return, the grateful authorities in Zaragoza presented him, unexpectedly, with a new Zeiss microscope, thereby putting him on a par with the medical scientists in Germany and France whose work he had admired and followed. He had already acquired a Reichert microtome for cutting sections of tissue and, on a visit to Madrid, had seen slides of nervous tissue stained by a new method. The slides were so remarkable that he would later write:

> "What a fantastic sight! On a yellow, completely transparent background, there appears sparsely scattered black fibers, smooth and small or thick and pricky, as well as black triangular and star- or rod-shaped bodies! Just like fine India ink drawings on transparent Japanese paper. The scientist gazes upon it in astonishment. . ."[2]

The staining method, pioneered by Camillo Golgi in Pavia, Italy, made it possible to see, for the first time, the fine branches of the nerve cells—the dendrites and the terminations of the axons. Previous staining techniques had allowed only the nerve cell bodies and the myelinated axons to be visualized. The Golgi method had another advantage, equally important. In the brain and spinal cord, the processes of the different nerve cells are tightly packed and interwoven. The only way to distinguish the features of a single neuron had been to dissect it out from macerated tissue under the microscope, using a pair of fine needles. It was an extraordinarily difficult task that required great patience and, even when successful, the connections of the cell had been lost. However, the Golgi staining method, for a reason even now not fully understood, selected only a very small proportion of the cells in the specimen (Fig. 4–2). As a consequence, without having to do any microdissection, it was now possible to distinguish all the complex branching of the axons and dendrites of individual nerve cells.

To make his work easier, particularly in understanding the relationships between one type of neuron and another, Cajal began to study the brains and spinal cords of embryos and chose small animals or birds to investigate nervous systems less developed than the human one. Since the axons in the embryos had yet to acquire their myelin sheaths (that is, their fatty insulation), their full lengths could be shown up with the Golgi stain. It was an idea that Golgi himself had also had but had never pursued to any great extent.

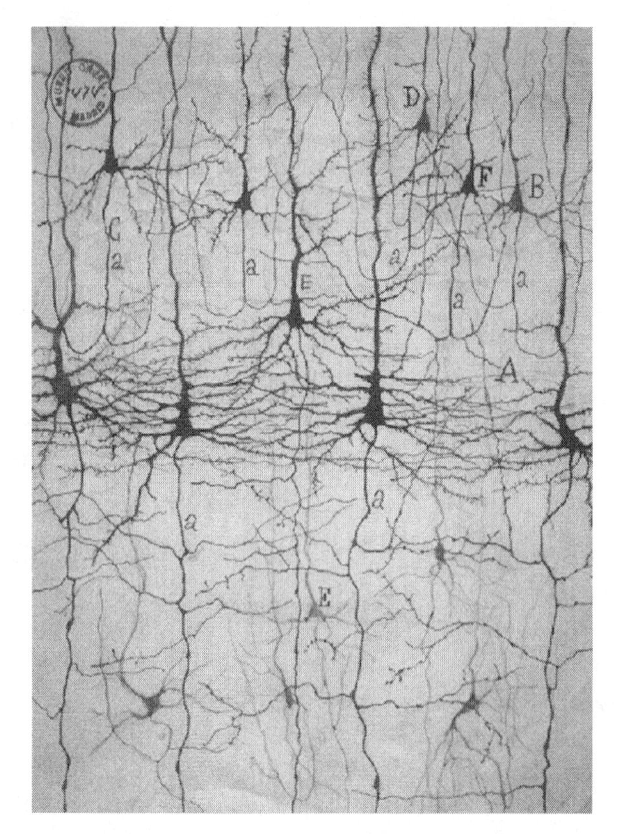

Figure 4–2 Cells of the human cerebral cortex, prepared by Cajal with the Golgi method. The upper half of the photograph contains 10 dark, triangular-shaped objects, four of which have been labeled by Cajal (*B, D, E, F*)—these are the bodies of the pyramidal cells (neurons). Running vertically downwards from the bottom of each pyramidal cell is a fine filament—this is the axon (labeled *a* in several instances). Ascending from the top of each pyramidal cell is a thicker process, the apical dendrite. Both the axons and dendrites are seen to branch. One of Cajal's insights was that "the transmission of the nervous impulse is always from the dendritic branches and the cell body to the axon." Note that it is easy to make out their structures in the photograph since the Golgi method stains only a very small percentage of the cells present in the specimen.

Cajal had discovered his mission in life:

> "Realizing that I had discovered a rich field, I proceeded to take advantage of it, dedicating myself to work, no longer merely with earnestness, but with fury. In proportion as new facts appeared in my preparations, ideas boiled up and jostled each other in my mind. A fever for publication devoured me."[3]

Cajal's personal qualities—the fierce independence, the determination to prevail over his opponents, the ingenuity in overcoming restraints—the very characteristics that had made him the despair of his teachers and the feared but admired leader of his schoolmates, now served him well. Further, his artistic talent, so long suppressed by his father, enabled him to depict the nerve cells and their processes with accuracy and beauty. And then, perhaps most important of all, there was that inquisitiveness, that enquiring mind that had taken him on so many expeditions into the countryside as a boy, and led him to study Nature. Because of the great pace at which he worked, Cajal became increasingly frustrated with delays in publication and started his own Review, making the lithographs for his illustrations himself. The years 1888 and 1889 were extraordinarily productive.

Prior to Cajal, the only clear separation of function that had been discovered among the fibers in the nervous system was in the spinal cord. Independently, Charles Bell in Edinburgh and, more emphatically, François Magendie at the Collège de France in Paris, had shown that the fibers in the dorsal and ventral nerve roots were sensory and motor respectively, a distinction that came to be known as the Bell–Magendie Law. Cajal, in comparison, took the analysis of nerve cell structure and function to a much higher level. For example, in the cerebellum, that organ lying below the posterior cerebral cortex, Cajal described, for the first time, different types of incoming fibers, as well as the way that the axons of

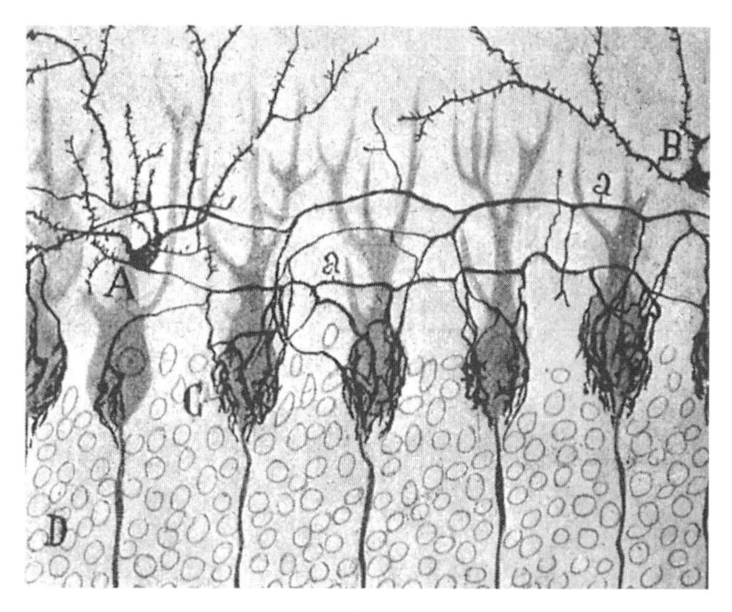

Figure 4–3 Transverse section of a cerebellar lamina. This black ink drawing by Cajal was superimposed on a photograph, and shows a large, Purkinje cell (*C*) surrounded by fine branches from the axons of stellate cells (*A*,*B*). (From Cajal SR. *Recollections of My Life*, translated by EH Craigie & J Cano. MIT Press, 1989)

the small basket cells terminated around the bodies of the large Purkinje cells (Fig. 4–3). In the retina Cajal described new types of cells in the different layers and suggested how nerve impulses would be transmitted to the ganglion cells and then to the optic nerve. In the spinal cord, Cajal was able to recognize different types of neurons in the gray matter, and to show the various ways that axons divided and sent branches across to the other side of the spinal cord or into ascending and descending tracts of fibers. His descriptions were unusual, not only because of the novelty of the findings, but also because of the language. It was as if Cajal had himself entered the microscopic world, imbuing the nerve cells with their own intentions. The climbing fibers of the cerebellum, for example, "broke up into twining parallel networks which ascend along the protoplasmic branches, to the contours of which they apply themselves like ivy or lianas to the trunks of trees."[4] The growth cone, the expanded tip of a growing nerve fiber, "could be compared to a living battering-ram, soft and flexible, which advances, pushing aside mechanically the obstacles which it finds in its way."[5] This last description, a very apt one, was made before the invention of the phase-contrast microscope had made it possible to see the growth cone in living tissue!

Regardless of which part of the nervous system he studied, Cajal was convinced of two underlying rules of organization, both having to do with the way that the nerve impulse must travel. The first rule, termed the principle of dynamic polarization, was that the impulse always propagated from the dendrites to the cell body and then along the axon. The function of the dendrites, Cajal said, was to act as receptors, in the case of sensory nerve cells, or, elsewhere in the nervous system, to collect impulses from the axon endings of other nerve cells. The second rule, which became known as the neuron doctrine, was that nerve cells, although coming into contact with each other through their axons and dendrites, remained separate entities.

However, the problem with all the new work coming from Cajal was that, outside Spain, no one was aware of it. It was annoying, even a source of anguish, for Cajal craved recognition and praise for what he had been able to accomplish: "It is natural that every author should aspire to the approval, and if possible the applause, of his public."[6] Never one to accept unjust fate, Cajal responded to his neglect by sending copies of his publications not only to the distinguished anatomists and histologists in Germany, France, and Italy, but also to the crowned heads of Europe. Indeed, the research papers that had been sent to Queen Victoria had eventually found their way to the Royal Society of London, and thence to Sherrington's attention, resulting in the invitation for Cajal to deliver the Croonian Lecture in 1894.

The Spaniard's other strategy was to set off for Berlin to attend the annual meeting of the German Anatomical Society, taking his microscope and slides with him. During the time allotted to demonstrations, he could then persuade the other scientists to look at the stained specimens of brain and spinal cord and to verify that the architecture of the tissues was just as he, Cajal, had described.

The approach had the same forcefulness that he had displayed so frequently as a youth, but at first it, too, seemed unlikely to succeed: "Some histologists surrounded me, but only a few, for, as happens in such competitions, each member of the congress was looking after his own affairs; after all, it is natural that one should prefer demonstrating his own work to examining that of someone else."[7] The young neurologist van Gehuchten, already a professor at the University of Louvain in Belgium, later recalled seeing Cajal standing alone, "exciting around him only smiles of incredulity."[8] Why indeed should anyone, not having encountered his work, have been interested in someone from Spain, a backward country with no scientific tradition of any kind? What happened next, however, was to prove decisive. As van Gehuchten described it, Cajal took aside Kolliker "who was then the unquestioned master of German histology, and drawing him into a corner of the demonstration hall, showed him under the microscope his admirable preparations, convincing him at the same time of the reality of the facts which he claimed to have discovered."[9]

Cajal had chosen his target well. Albrecht Kolliker was an older man, a professor at the University of Wurzberg, and greatly respected by other anatomists and histologists. After the meeting was over, Kolliker, with true Bavarian hospitality, took Cajal by carriage to dinner at the luxurious Berlin hotel where he was staying and introduced him to some of the other eminent scientists who had also been at the meeting. Further, Kolliker undertook to repeat some of Cajal's studies, using the same modification of the Golgi method, and, by confirming and publishing the results, to make the Spaniard's discoveries widely known. And this Kolliker did. Cajal was deeply grateful:

> "There is no doubt that the truth would have made itself clear eventually. Nevertheless, it was due to the great authority of Kolliker that my ideas were rapidly disseminated and appreciated by the scientific world. A noble exception among great investigators, Kolliker united a great talent for observation, aided by indefatigable industry, with enchanting modesty, and exceptional rectitude and calmness of judgement. It was to this great Bavarian master that I alluded particularly when, . . . deploring the intolerable egotism and vanity of certain men of science, I declared that there were others most learned and at the same time upright, impartial, and honourable."[10]

So Cajal's findings became accepted, and many of the previously skeptical anatomists and histologists began to use his modification of the Golgi method for their own studies. But while there could be little argument over the different types of nerve cell that Cajal had described, or the way that the nerve fibers branched, or the arrangements of the nerve cells in the various regions of the brain and spinal cord, it was a different story with his two principles—the dynamic polarization of the nerve cell, and the neuron doctrine (see above). In relation to the

Figure 4–4 Camillo Golgi. (Wellcome Library, London)

former, his main opponent was van Gehuchten of Louvain, who doubted that sensory nerve endings should be regarded as a form of dendrite. Cajal responded to the challenge by examining sensory fibers in less highly developed species than mammals, such as earthworms and shellfish. So far as the neuron doctrine was concerned, the opposition was very much greater, since the majority of scientists believed that nerve cells were continuous with each other. One of the staunchest believers in this reticular theory was Camillo Golgi (Fig. 4–4).[11]

Golgi, the son of a physician, was 9 years older than Cajal and was born in Brescia, a historic city in northern Italy famous for its violin and cello makers. After graduating in medicine at the University of Pavia, previously the intellectual fortress of Alessandro Volta, Golgi took a post in the city's Hospital of St. Matteo and, inspired by his teacher, Bizzozero, started his research studies. One of his earliest interests was malaria, endemic in the marshlands bordering the tributaries of the River Po. Golgi may well have been the first to recognize the different types of fever, and, using his skill with the microscope, he was able to distinguish the three forms of the malarial parasite, which, after many trials, he eventually succeeded in photographing. His other main interest, and the one that would occupy him for the greater part of his working life, was the nervous system. In his second hospital appointment, at the Hospital for the Chronically Sick in Abbiategrasso, he converted a small kitchen into a laboratory, and it was probably there that he developed the silver staining method that would bear his name and that would become such a powerful instrument for studying the brain.

In 1881, at the age of 38, Golgi returned to Pavia, this time as the Chair in General Pathology, the same position that his mentor, Bizzozero, had occupied. With this appointment Golgi was able to devote more time to his research, as well as to teach and supervise younger physicians with similar interests to his own. Unlike Cajal, who had been obliged to fight for recognition, Golgi's reputation increased as his work progressed. Eventually he would become the Rector of Pavia University and, like Volta many years before, a Senator of the realm.

It was unfortunate that Golgi had been away from Pavia at the time that Cajal had visited that city on his way home from the Berlin meeting of the German Anatomical Society in 1889. Had they met, it is likely that each would have been impressed with the skill, sincerity, and erudition of the other, and that Golgi would have been won over to the Spaniard's conclusions, as many of the German anatomists and histologists had been. Instead, it is probable that Golgi was becoming irked by Cajal's success with the staining method that he, Golgi, had invented. The remarkable discoveries that Cajal continued to make should surely have been Golgi's! The Italian had already complained that Cajal had neglected to give him priority in describing transverse branches in nerve fibers running up and down the spinal cord. It had not been a deliberate omission on Cajal's part, for Golgi's description had been no more than a brief mention in a paper published in an obscure local journal.

In promoting the reticular theory of nerve connections, the concept that all the nerve cells were in protoplasmic continuity with each other, Golgi had carried the majority of anatomists and histologists with him. While some of the latter considered that the connections between nerve cells were through fine dendritic branches, Golgi believed it was the axon twigs that formed an interlacing network. Cajal, however, was able to show that axons were present in regions thought to consist only of dendrites. Further, the axons ended not on other axons, as Golgi maintained, but on the cell bodies or dendrites of other nerve cells.[12] One also had to explain how the Golgi stain would restrict itself to what appeared to be a single nerve cell, rather than staining interconnected networks of fibers. There was, in addition, the old observation that, when nerve fibers were cut, the subsequent degeneration, instead of spreading diffusely in the nervous tissue, ended rather sharply. In later years, Cajal would bolster his arguments by observations on regenerating nerve fibers. To Cajal, the whole idea of a reticular network was nonsense:

> "It has rightly been said that the reticular hypothesis, by dint of pretending to explain everything easily and simply, explains absolutely nothing; and, what is more serious, it hinders and almost makes superfluous future inquiries regarding the intimate organization of the centres."[13]

So forceful were Cajal's arguments, and so convincing was the histological evidence on which they were based, that, once again, his opinion prevailed among the anatomists, and also among physiologists such as Sherrington. It was not long before the honors and awards appeared. One of the first had been the invitation to give the Croonian Lecture to the Royal Society of London. Though the invitation had come from Michael Foster, the Secretary of the Society, the initiative had been Charles Sherrington's. A lecturer in physiology at St. Thomas' Hospital and the Director of the Brown Institute, Sherrington insisted on Cajal being his guest during his stay in London. The lodger caused his hosts some concern. Mrs. Sherrington, puzzled that Cajal, before leaving the house for the day, would lock his bedroom door and pocket the key, discovered that the Spaniard had turned his room into a histology laboratory. The neighbors of the Sherringtons, meanwhile, were intrigued to observe bed linen being hung out to air from one of the bedroom windows each day. And then, much more serious, was the episode of the missing guest. Cajal had been invited to receive an honorary degree at Cambridge, and it was Mrs. Sherrington's responsibility to make sure that the hansom cab taking the Spaniard to the London railway station did so in good time. Sherrington, already in Cambridge, was disconcerted not to find Cajal on the train. The lunch in Cajal's honor, presided over by the university vice-chancellor, had to proceed without its guest of honor. A subsequent enquiry revealed that the police had indeed found someone who answered to Cajal's description, a foreigner who appeared lost and could not speak English. It transpired that Cajal had arrived at the London station in time to catch an earlier train. Finding no one to meet him in Cambridge, he had wandered off and, struck by the beauty of the university buildings, had settled down in the middle of the road to sketch King's College.[14]

The year after the Croonian Lecture Cajal was awarded the Helmholtz Medal by the Royal Academy of Sciences of Berlin. It was a huge medal, made of gold, and, by its presentation to Cajal, it formed a connection between the man who had first measured the speed with which an impulse traveled, and the person who had shown the obligatory direction that the impulse must follow. In the next year, 1906, there was a still greater honor, announced by a single-line telegram from Stockholm—the Nobel Prize in Medicine or Physiology. Sharing the award was the scientist who had developed the staining method that had made Cajal's work possible, Camillo Golgi. It was the one time that the Prize had been given to two people with mutually contradictory opinions. Strangely, in view of his self-confessed desire for adulation, Cajal was not pleased to have been selected. He felt that there were others who were more deserving, that he, as a histologist, hardly qualified for a prize awarded to physiologists, and that, inevitably, his success would become a cause for envy. On the other hand, the award was an honor for his country, and Cajal was a patriot. In the end, as congratulations poured in, he had no alternative but to accept and to travel to Sweden for the ceremony.

He did not have far to look for a source of envy. In his own lecture in Stockholm, Cajal had acknowledged his debt to Golgi. Unfortunately, the "savant of Pavia"—as Cajal referred to him—was unable to reciprocate. Having chosen to talk about the reticular theory, Golgi studiously ignored the large body of work, especially Cajal's, that made the theory untenable. Worse, "he made a display of pride and self-worship so immoderate that they produced a deplorable effect upon the assembly . . . all the Swedish neurologists and histologists looked at the speaker with stupefaction."[15] Golgi, apparently, behaved equally poorly at the official banquet.

Having returned to Madrid—taking the Nobel money with him[16]—Cajal continued to work as hard as ever. He was determined that there should be no respite from the laboratory, detesting "the unpatriotic egoism of those who, having

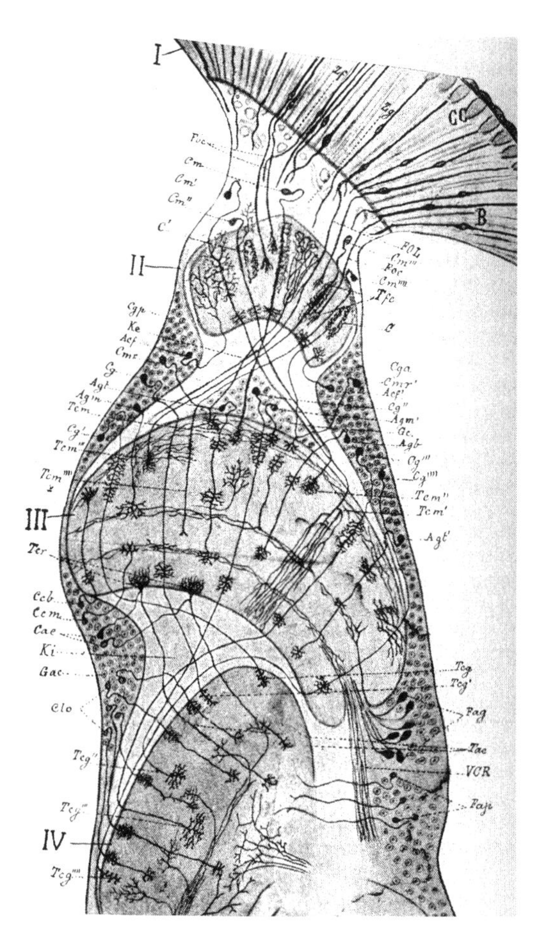

Figure 4–5 Cajal's drawing of the retina of a bee. He had anticipated that the insect retina would be simpler than those of birds and mammals and was surprised to discover the reverse. Indeed, given the enormous complexity of the insect retina, it is remarkable that Cajal was able to make out the patterns of nerve fiber connections. (From Cajal SR. *Recollections of My Life*, translated by EH Craigie & J Cano. MIT Press, 1989)

attained the pinnacle, think only of lying down without further effort."[17] Among other studies, there would be new morphological investigations of the cerebellum and brain stem, and, in a remarkable change of direction, some highly original experiments on the regeneration of sectioned nerve fibers in the brain and spinal cord. Some idea of the extraordinary quality of this later work is given by his drawing of the different types of nerve cells and axons ending in the retina of the honeybee (Fig. 4–5). Cajal's own work was not his only interest, however. Especially important to him was the training of a new generation of Spanish scientists who would be able to continue and expand the studies he had pioneered. Towards the end of his life, he would write:

> "I consider it certain and even desirable that in the course of time my insignificant personality will be forgotten; and with it will, doubtless, perish many of my ideas. Nothing can escape from this inexorable law of life. In spite of all the allegations of self-love, the facts associated in the first place with the name of one man end by being anonymous, lost for ever in the ocean of Universal Science."[18]

At the time he wrote them, he may have suspected that his words would not prove entirely true, and so it was. The name Cajal would continue to be cited whenever the structure of the nervous system was considered, and, so far as the propagation of the nerve impulse was concerned, the principle of dynamic polarization of the nerve cell, and the neuron doctrine, would become two of the fundamental truths on which the new science of neurophysiology would be built. Cajal's opponent, Golgi, would also be remembered. Not only was there a type of nerve cell named after him, as in the case of Cajal, but there was also his staining method. And, making him especially widely cited, there was the Golgi apparatus—the complex intracellular structure that is now known to sort proteins out and direct them to their proper locations within the cell.

Cajal's legacy, however, was the greater.

5

Cambridge, 1904: The Engineer

While studies on the nature of the impulse were being vigorously pursued in the latter half of the 19th century in Germany—by Helmholtz, du Bois-Reymond, Hermann, and Bernstein—very little was happening in Britain. But there was one achievement that, although largely ignored subsequently, was nevertheless quite remarkable at the time. It was made by two youthful brothers-in-law in Oxford in 1888. Slightly the elder of the two, Francis Gotch was a demonstrator in the university's Physiological Laboratory, while Victor Horsley was Professor of Pathology at University College London. Using a relatively sensitive capillary electrometer belonging to Burdon Sanderson as their recording instrument, and with an induction coil for stimulation, the two managed to detect impulses in the sciatic nerves of a variety of animals. True, the deflections were extremely small, but they were definitely there (Fig. 5–1, *left*), and the two investigators were quick to point out their significance:

> "There is thus no doubt that the movement that we obtained and photographed was due to the electromotive change which accompanies the propagation of an excitatory state along the mammalian nerve when this was evoked by the application of a single stimulus."[1]

While the speed of conduction and the shape of the impulse had been worked out by the German physiologists, using ingenious techniques and custom-built apparatus, and while du Bois-Reymond had seen the momentary swing of the galvanometer needle that signaled his negative Schwankung, the Oxford experiments were the first occasions that the nerve impulse had actually been recorded. It is likely that, having made this breakthrough, the two brothers-in-law realized that little more could be done with the equipment at their disposal. Besides, Victor Horsley, the younger of the two and already a professor at 31, was more interested in the role of the motor cortex in producing jerking of the limbs during an epileptic seizure; the experiments on peripheral nerve had been a necessary supplement. And so Gotch went on to become the Waynflete Professor of Physiology at Oxford, while Horsley (later Sir Victor), after original neuroanatomical studies and important work on the thyroid gland, would develop the new discipline of neurosurgery.

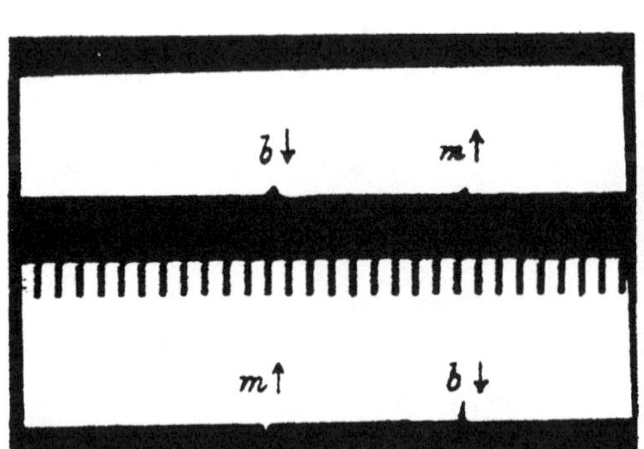

Figure 5–1 *Left*: Gotch and Horsley's recording of action potentials in the cat sciatic nerve (two small triangular deflections on the top trace). The authors explored and rejected other explanations for the deflections, including the possibility that they were artifacts from the "make" and "break" stimuli from the induction coil (*m* and *b* on bottom trace). The time ruler shows intervals of 10 ms. (From Gotch & Horsley, 1888). *Right*: Sir Victor Horsley (1857–1916), probably at a later stage of his distinguished career; he was knighted in 1902. (Wellcome Library, London)

By 1904, however, a potential answer had been found to the question as to who, in Britain, might take up the challenge of the nerve impulse. It was in that year that Langley brought a young man into his department at Cambridge, a man whose presence would prove critical for the development of neuroscience. The appointee, Keith Lucas (Fig. 5–2), was 25.[2] His grandfather had fought at Waterloo, and Keith Lucas's father had been a leading engineer with the company that had laid the first underwater telegraph cables across the Atlantic and Pacific oceans, sometimes devising machinery to make the operation more efficient. Nor was the science all on the father's side of the family, for on his mother's side there had been accomplished mathematicians and astronomers. While excelling in classics at school, Keith Lucas also became very skilled in the workshop at home, and it was science that he would study at Cambridge. Midway through his studies, depressed over the death of a friend, he went to New Zealand to measure the depths of lakes previously regarded as bottomless. On returning to Cambridge, Lucas completed his degree and started his study of impulses in nerve and muscle fibers. It was the sort of problem that appealed to his engineering mind, and it lent itself to systematic experiments, ones in which clearly defined questions could be asked and answered.

One of his first successes was in showing that the contractions of muscle fibers had an "all-or-none" quality—that is, once initiated, they were of constant sizes.[3]

Figure 5–2 Keith Lucas. (From Fisher WM, Keith-Lucas A. *Keith Lucas.* Cambridge: Heffer, 1934)

Lucas showed this by cutting a small muscle in the frog until only a few fibers were left in continuity, and then applying progressively larger electrical stimuli. As the shocks increased in intensity, the muscle response grew in well-defined steps, each step corresponding to the excitation of an additional group of fibers. The proof of the all-or-none "law" was that, between steps, the amplitude of the muscle response stayed the same. Thus, increasing the stimulus did not produce larger responses in fibers that had already been excited. In a subsequent study, Lucas applied progressively larger stimuli to a small number of nerve fibers rather than to the muscle fibers directly. Once again, the response increased in discrete steps, confirming the all-or-none law for muscle fibers (Fig. 5–3). The results of this second experiment have often been interpreted as showing that the all-or-none law applied to nerve fibers, but Lucas was quite clear on that point:

> "We must therefore regard the question whether the response of a nerve-fibre is capable of gradation as being at present undecided."[4]

Though it was a side issue, the last experiment also showed that each of the nerve fibers running from the spinal cord to a muscle supplied a separate colony of muscle fibers within the belly of the muscle. Lucas referred to such a colony and its nerve fiber as a "unit," while Sherrington would later introduce the term "motor unit."[5]

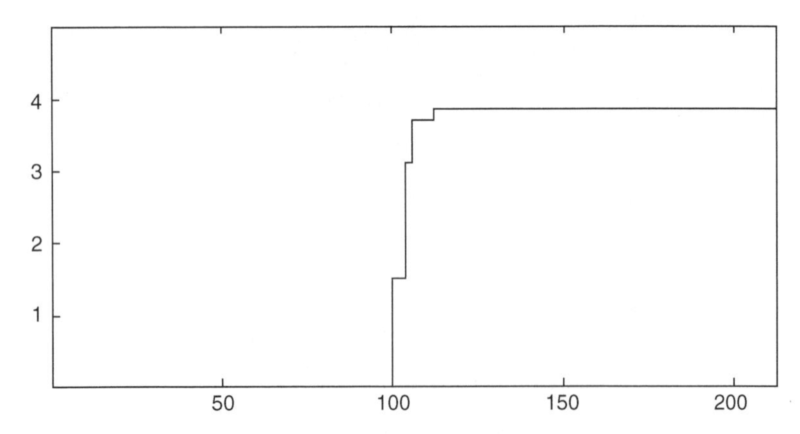

Figure 5–3 "All-or-none" responses of muscle to graded stimulation. As the stimulus to the nerve is increased (horizontal axis), the response of the muscle grows in four steps to a maximum (vertical axis). (From Lucas, 1909)

Within a very few years Lucas was not only well established in his research but, in the field of muscle and peripheral nerve, had already become known internationally through his publications in the *Journal of Physiology*. Lucas's was not the only success, however. It was a golden age for the Cambridge Physiological Laboratory. True, the conditions in which the members had to work were deplorable. Instead of Michael Foster's single room in the Trinity Library, there was now a series of dark and damp cellars in another building, long since demolished, which the physiologists shared with cages of rats and tanks of frogs. A photograph of Sherrington and Roy,[6] lounging outside the Pathological Laboratory during a break from their experiments, gives some idea of the poor state of the Cambridge buildings. Whenever it rained, Keith Lucas's room would flood, and, if he was to experiment, the country's leading neurophysiologist was obliged to walk around on duckboards. There was no workshop or mechanic, nor even a common set of tools for the scientists to build their apparatus. Undeterred, Keith Lucas made his own equipment at home.[7] Yet within the wretched surroundings the small group of physiologists performed outstanding research. In addition to Lucas, and to Langley and Gaskell, already mentioned, they included Fletcher, working on lactic acid production in muscle, Hopkins, the discoverer of vitamins, Hardy, an authority on cell proteins, and Barcroft, an all-rounder specializing in the composition of the fluid bathing the body tissues. There was also Archibald Vivian Hill, who, like Gaskell, had studied mathematics at Cambridge only to fall under the spell of the physiologists; Hill had been encouraged to work on muscle by approaching it as a machine and applying engineering concepts.

The contrast between the working conditions and the quality of the research was not the only remarkable feature of the physiology at Cambridge. There was also the *laissez-faire* attitude of the staff. There were no timetables or expected hours of attendance, nor did Langley oversee his troops. Hill might simply disappear for a week, returning with a sensitive device that he had made to measure the heat produced by a contracting muscle. At other times he might be out of the laboratory on a long-distance training run. In the summer, Hardy would simply hang a "Back Tomorrow" notice on his door and go sailing in the English Channel. Visitors to the department might be invited by Lucas to join him in his large motorboat, *Tiglath*, on the River Cam.[8] Langley himself was elusive, not much given to conversation with his staff and ever eager for weekends, when he could "ride to hounds" (i.e., foxhunt). It was all very different from the situation in Germany, where the research and teaching were organized and tightly controlled by the professor in charge of the department. It was a difference that immediately struck those who visited both countries. Cajal, who had received an honorary degree in Cambridge in 1894, would remark:

> "In that nation [Britain], however, the most eminent scientists and thinkers owe little to the universities; they are privileged natures, which open up ways for themselves in spite of the defective and incomplete organisation of the educational centres. For the investigator, there is not, as in Germany, the direct product of the school, but it is the indirect result of the cultivation of the individuality and of the building up of all the energies of the spirit. . . It would be desirable to know whether, in dealing with a race so admirably endowed as the English, the German method of instructing much and educating little would not yield even better fruit than the Anglo-Saxon method of educating much and instructing abstemiously."[9]

It was, perhaps, the difference between the amateur and the professional. In Britain, it was considered "bad form" to try too hard, or to receive any form of organized training. The amateur cricketer and soccer player were regarded as superior to those who were paid to play the same game. It was an attitude that would be captured brilliantly by Sir John Gielgud in his portrayal of the college Master confronting the over-eager Harold Abrahams in the retrospective film *Chariots of Fire*.

Within the peculiar working environment at Cambridge, Keith Lucas's research flourished. Moreover, the logic and clarity of his thinking made him an influential teacher of students, the majority of whom were only a few years his junior. Among the latter was one who would make his own mark, first as Lucas's collaborator and then as an independent investigator.

Edgar Adrian (Fig. 5–4) was the youngest son of a London barrister who worked for local government. Like his mentor, Keith Lucas, Adrian had excelled in Greek and Latin, but been attracted to science at university. In Part I of the Cambridge Natural Sciences Tripos, he had taken five subjects rather than the usual three, gaining first-class marks in each and reaching an aggregate score 30 percent higher

Figure 5–4 Edgar Douglas Adrian in his Cambridge laboratory, preparing a demonstration for the Physiological Society. (Reproduced with permission of Yale University, Harvey Cushing/John Hay Whitney Medical Library)

than any previous one. It was an academic performance that led his tutor, the muscle biochemist Walter Morley Fletcher, to remark prophetically:

> "I now find it most embarrassing to know in what direction to advise him to steer, seeing that he has first-rate ability in so many different directions. If he has original powers at all comparable with his powers of reception, he ought to do magnificent work in scientific medicine."[10]

There was more to Adrian, however, than a first-class mind. As an undergraduate he had shown daring and skill in climbing the roofs of the various colleges at night. Later, after qualifying, he and a friend achieved a considerable coup by dashing off a number of paintings, exhibiting and selling them as examples of post-Impressionist art, and fooling experts in the field as they did so.

Adrian joined Lucas in the dark and damp cellar. By now it was 1911 and Lucas was well advanced in his study of the impulse, recognizing that a stimulus produced a local response in a nerve or muscle fiber that, if it exceeded a certain threshold, would then travel along the fiber. That there was a local response, he showed by giving two weak stimuli in rapid succession, neither stimulus being strong enough by itself to set up a propagated wave.[11] Under Lucas's supervision, Adrian then carried out some investigations on frog nerve, segments of which were narcotized with alcohol vapor. By comparing results obtained with either one or two regions of narcotization, he was able to show that, if a diminished impulse was able to emerge from a treated region, it would regain its full amplitude.[12]

This type of experiment also showed that, once it had been initiated, the traveling impulse derived its energy from the nerve fiber itself rather from the stimulus. In this respect, the impulse was like a train of gunpowder that had been lit:

> "A disturbance, such as the nervous impulse, which progresses in space must derive the energy of its progression from some source; and we can divide such changes as we know into two classes according to the source from which the energy is derived. One class will consist of those changes which are dependent on the energy supplied to them at their start. An example of this kind is a sound wave or any strain in an elastic medium which depends for its progression on the energy of the blow by which it was initiated. . . A second class of progressive disturbance is one which depends for its progression on the energy supplied locally be the disturbance itself. An example of this type is the firing of a train of gunpowder, where the liberation of energy by the chemical change of firing at one point raises the temperature sufficiently to cause the same change at the next point. . ."[13]

Adrian and Lucas next examined the changes in the excitability of the nerve membrane that followed the passage of an impulse. It was already known that there was a very brief period, lasting a millisecond or so, during which the nerve fiber was incapable of responding to another stimulus. This period of absolute refractoriness was followed by one in which the fiber could be excited again, but only if the stimulus intensity had been increased. The impulse, once initiated, would travel with a reduced velocity during this relatively refractory period. Adrian and Lucas found that there was a third period, in which the nerve fiber had supernormal excitability, responding to a weaker-than-normal stimulus and conducting impulses with a higher velocity.[14]

In all of this work, the experimental design was ingeniously simple and tailored to answer a clearly defined question. Whether or not an impulse traveled, and what proportion of the fibers were conducting, were determined by the force of the ensuing muscle contraction, as recorded on the smoked drum of the kymograph. It would be better, Lucas and Adrian knew, if they could record the nerve impulses directly, rather than relying on the muscle response. At the time, there were two instruments, the capillary electrometer and the string galvanometer, that were capable of detecting the electrical activity of the heart (the electrocardiogram [EKG]), though not the very much smaller signals generated by nerve. The capillary electrometer had been invented first, in 1872, by Lippmann in Germany. It consisted of a narrow (capillary) glass tube, half of which was filled with mercury and the other half with sulfuric acid (Fig. 5–5). When an electric signal was applied across the ends of the tube, through a pair of wires, the mercury meniscus moved in proportion to the voltage. A permanent recording of the

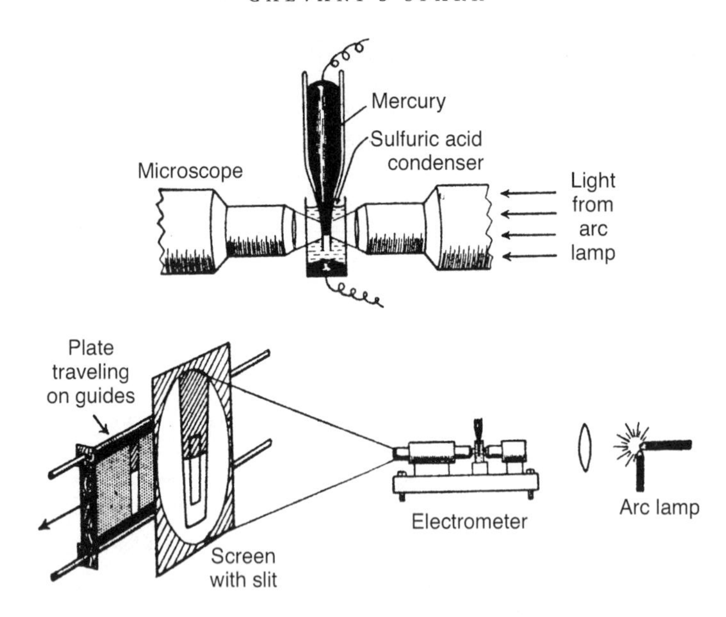

Figure 5–5 Capillary electrometer. Other forms of the electrometer existed, the differences being in the arrangement of the glass tubing containing the sulfuric acid and mercury. (From Adrian, 1928)

size and time-course of the deflection was made by optical magnification and projection onto a falling photographic plate.

The string galvanometer was the creation of Willem Einthoven at the University of Leiden in Holland, the same city where the Leiden jar had been invented,[15] and was described by him in 1901.[16] In the galvanometers used by pioneers such as Matteucci and du Bois-Reymond, an electric current induced a magnetic field as it flowed through a coil of wire, and hence a deflection of an iron needle in the center of the coil. In Einthoven's device, however, a very powerful magnetic field was obtained by passing a large current through two massive coils. A very thin gold-plated filament of quartz ran between the poles of the electromagnets and carried the biological signal in the form of a minute current. Under the influence of the magnetic field, the quartz string moved in proportion to the signal it carried, casting a shadow onto a moving photographic film (Fig. 5–6).

The string galvanometer was vastly more sensitive than the earlier instruments, though the water jacket needed to prevent the coils from overheating made the first string galvanometers extremely bulky, weighing several hundred kilograms; also, the slender quartz threads tended to break or to lose their gold coating. On the other hand, the capillary electrometer, though less sensitive than the string galvanometer, did not overshoot or oscillate and in this respect was the more reliable of the two instruments. For this reason it was chosen by Lucas for

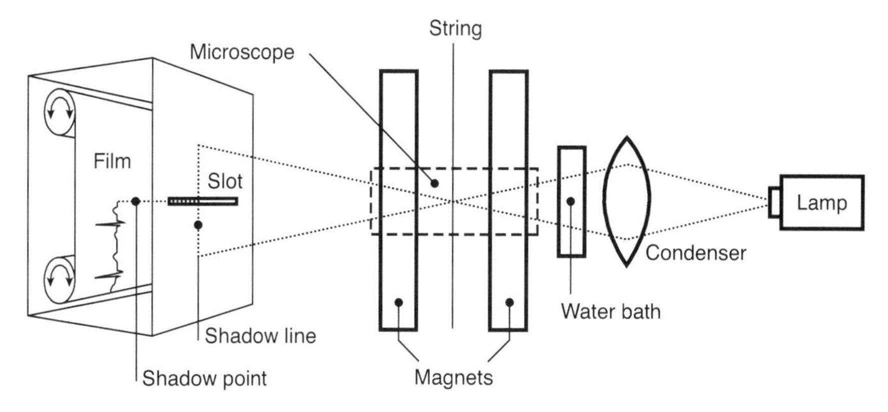

Figure 5–6 Einthoven string galvanometer. Following its invention, the string galvanometer quickly became the instrument of choice for recording the electrocardiogram.

his experiments. The capillary electrometer did, however, have the problem of not being able to respond fast enough to capture the full extent of a sudden voltage change, such as that generated by a nerve impulse. Lucas's solution was to design and build a hand-operated machine that would make the corrections necessary for seeing the true response.[17] In the laboratory, the new apparatus joined the rotary contact-breaker that Lucas had designed and built earlier, and that had enabled very small time intervals to be selected between one nerve stimulation and the next.

By the summer of 1914 Lucas and Adrian had had three very productive years together, and the younger man was full of admiration for his accomplished mentor:

> "Though he [Lucas] had a great deal to do every day he was never too busy to be friendly. He seemed to take the ups and downs of research work philosophically, and the strain of a long unsuccessful experiment made very little obvious difference to his temper. It was the ideal scientific temper, for he was never out to prove a theory or to show his superiority over a rival physiologist; all he was after was a clean experiment that could settle a point one way or the other. He worked with a swift deliberation, without fuss: if the apparatus broke down he regarded it without malice and mended it at once. . . A successful experiment would evoke a pleased chuckle and he would light his pipe, a workshop specimen bound up the stem with copper wire, and after turning off all the switches he would come out of his room into the fresh air and bicycle back to Trinity or Ditton thinking out some new modification for the next day."[18]

The stage was now set for the next phase of their collaborative work, but instead there was a break. Adrian decided to complete his medical training, and in the summer of that year commenced clinical studies at St. Bartholomew's Hospital in London. Hardly had he started, however, than the First World War broke out. Lucas also stopped his experiments and, having passed his medical examination, was about to enlist in the infantry as a private when, because of his scientific skill, he was diverted to the Royal Aircraft Factory at Farnborough, on the south-western outskirts of London.[19] Living in a small wooden hut and rising at 4 in the morning, Lucas grappled with a number of problems that had beset the pilots of the early flying machines. One was to improve a bombsight, and another to eliminate the unreliability of the pilot's compass as the plane was made to turn. Once again, just as it had in the Physiological Laboratory in Cambridge, Lucas's flair for analysis and design, and for constructing equipment himself, served him in good stead, and the problems were solved. To gain first-hand experience of a particular problem, and to see if his solution was effective, Lucas would fly himself, initially as a passenger and then as a trained pilot. For this, he transferred to the Central Flying School at Upavon in Wiltshire.

The physiology of the nerve impulse was not entirely forgotten, however. Late in the summer of 1916, Lucas met with Adrian. It is not known where the meeting took place, though if at a military site, it would most likely have been at the Connaught Military Hospital in Aldershot, where Adrian had been posted and, among other activities, was treating soldiers with different types of war neurosis. The meeting between Lucas and Adrian could equally well have been in a London restaurant. Regardless, it is not difficult to imagine the two men, both of them slender, erect, and smartly dressed in their military uniforms, with Lucas, only ten years the elder, the one with the moustache. It is certainly known that, in relation to their neurophysiological research, the main topic of discussion was the possibility of using the newly invented thermionic valve as a means of amplifying the nerve impulse and making it that much easier to record. Both men must have looked forward to the end of the war and to the resumption of their collaboration.

But would the wish be fulfilled?

Upavon, Wiltshire, October 5, 1916

He pushed the goggles up onto his forehead and began walking across the field. Though it was still too early in the autumn for frost, there was a crispness in the air, and he was glad that he had put the thick sweater on beneath his flying jacket. It was a bright, clear morning, free of the mist that so often descended from the Downs during the night and clung to the valleys, obscuring the villages till late morning.

For once he had slept well. On waking, he had been amused by his recollection of the dream. Experimenting, as always, but this time with strange apparatus that so filled the laboratory that he and Adrian had been obliged to crawl about on their knees. How much better everything would be in the new laboratory. Smiling, he thought of Adrian and his sharp, inquiring mind. They had both enjoyed their last meeting and Adrian had shared his excitement over the idea of the valve amplifier. Once they had the valves, it would be easy to build, and he had already thought of the circuit they would use. Finally, they would be able to see the action current, the nerve impulse. Yes, he must invite Adrian for another discussion, perhaps this time in Upavon. They could have lunch together at the Ship Inn—Adrian would be sure to approve of the thatched roof, not to mention the local ale.

He arrived at his own plane and looked up as one of the other pilots flew low over the field and the hangers at the far end. . .

The mechanic, his work for the moment completed, stood with his back against the wooden hut and watched the lumbering plane lift into the air. God, they were slow, the B.E.2s, not even as fast as his motorcycle. Why, indeed, were they still being produced? The only people who could like them were the Germans who shot them down. "Kaltes Fleisch," their pilots called them. Cold meat, indeed.

As he continued to follow its flight, the mechanic saw the biplane bank and move away from the airfield while, at the same time, a second B.E.2 came into view and started its descent. Careful, careful—surely the pilots could see each other? Or was the lower plane in the blind spot of the other pilot?

Five hundred feet above the ground, the pilot of the lower plane noted the sudden increase in engine noise and quickly checked his gauges. No problem there. Ever logical, he realized that there must be another source of the noise. As he began to turn his head, the propeller of the other plane sliced into his body.

In the official report it was said that Keith Lucas must have died instantly.[1]

A B.E.2 biplane such as that flown by Keith Lucas. With the pilot in the rear cockpit, forward visibility would be restricted by the nose and the wings—a likely cause of the fatal midair collision.

6

The Cathode Ray Oscilloscope

"... he was cast in a special mold; he had a tall loose limbed figure with long arms and a large head. He had finely chiseled features and a smooth skin which gave him a boyish look and he spoke, often rapidly, in a voice with treble overtones ..."

(Description of Herbert Gasser by Lord Adrian)

The war that was supposed to end all wars, the Great War, took its terrible toll. On the western front, as the battles raged through Belgium and northern France, the names of towns, cities, and rivers acquired a new and sinister significance—Mons, Verdun, Ypres, Liège, and the Somme, the Marne, and the Sambre. To the onlookers, and even to the majority of those in the participating countries, it had seemed inconceivable that the peace and promise of the early 20th century would be recklessly, even eagerly, thrown away. What was the value, indeed the whole meaning, of civilization if nations willingly undertook to batter each other to death? Watching events from neutral Spain, Cajal asked himself, of his own work, "What is the use?" Despondent, Cajal continued his own study of the nervous system while training a new generation of histologists, but at the same time doubting whether their results and conclusions would ever be read: "In the face of the formidable struggle of Europe for world supremacy, what could the persistent labour of a group of modest Spanish biologists matter?"[1]

Inevitably, the effects of the war were felt by the universities. In Belgium, one university, Louvain, the home of Cajal's supporter, van Gehuchten, was utterly destroyed in the fighting. In that country and elsewhere, universities saw their younger staff, and the majority of their students, abandon their studies in order to enlist. At Cambridge, the departing physiologists included Keith Lucas, Adrian, and A. V. Hill. While Lucas joined the Flying Corps and Adrian became an Army neurologist, Hill distinguished himself by directing a team of scientists into improving anti-aircraft gunnery. Those left in the universities were not untouched. Some had to deal with the death of sons. In Oxford, Charles Sherrington, appointed to the Waynflete Chair of Physiology in the year before the war broke out, attempted to console Sir William Osler, the Regius Professor of Medicine, on the loss of Osler's only son, Revere, mortally wounded in the fighting on the Ypres Salient. Sherrington himself, too old to be in uniform, abandoned his study of the nervous system and, instead, determined for his government the energy expended by those engaged in factory work, laboring in a munitions factory in Birmingham in order to do so. Sir Victor Horsley, who, with Francis Gotch, was the first to record the nerve action potential, and who went on to make a brilliant career in neurosurgery and medical politics, died from "heat exhaustion" in what is now southern Iraq. In Germany, Julius Bernstein and Ludimar Hermann, like Sherrington too old to fight, died of natural causes. So, too, did a surprisingly large number of the European anatomists and histologists who had, like Cajal, attempted to unravel some of the mystery in the structure of the nervous system. With their deaths, and with the economic ruin and collapse of order after the war, Germany would lose its supremacy in medicine and science.

At first America was undisturbed by the conflagration in Europe, apart from concerns over friends and relatives in the participating countries. Even in 1916, the year that Keith Lucas was killed in the air over Salisbury Plain, young American men could still take stock and plan their careers. Among them was the 28-year-old Herbert Gasser (Fig. 6–1).[2]

Gasser's father had come from the Austrian Tyrol, eventually practicing as a physician in Wisconsin, while Gasser's mother had come from an old Connecticut family. On completing high school, the younger Gasser's intention had been to study engineering, for he had already proved himself skilful with his hands,

Figure 6–1 Herbert Spencer Gasser. This seems to be the only photograph available of Gasser as a young man, and was taken while he was working in St. Louis. (Wellcome Library, London)

both in building furniture and, on a smaller scale, in making photographic equipment for his Kodak box camera. Having obtained bachelor and master degrees at the University of Wisconsin, Gasser, under pressure from his father, started medical school courses—reluctantly. In physiology, however, he discovered that the human body was a physicochemical machine that, though enormously complex, was open to study by rigorous scientific methods. As Gasser would write many years later of his lecturer, "The subject matter he presented differed so widely from what was expected that it amounted to a conversion." There might, Gasser reflected, be a place for someone like himself after all, someone with a strong interest in engineering. It was the same kind of conversion that some of the Cambridge physiologists had experienced.

Gasser's lecturer in physiology was the newly appointed professor, Joseph Erlanger (Fig. 6–2).[3] Like Gasser, Erlanger was the son of European immigrants, his penniless father having arrived from Germany in time to take part in the great California gold rush of 1849. Having worked in the fields as a miner, the elder Erlanger had settled in San Francisco, started a business, and married his partner's sister. Joseph, the sixth of seven children, showed an early interest in plants and animals and, choosing medicine as a career, enrolled in the University of California at Berkeley in 1891, at the age of 17. Since he was the only child in the family to enter college, and since both parents were poorly educated, it was an ambitious move.

Erlanger completed his medical training in the newly organized Johns Hopkins Medical School in Baltimore, coming second in the graduating class and then spending a year as intern under William Osler. As a student, Erlanger did some research, one project on muscle innervation and a second on intestinal absorption. The experience was enough for him to know that his enquiring mind was more likely to be satisfied in the physiology laboratory than at the bedsides of patients, even though, in William Osler, he was privileged to work for the most renowned physician of his time, at least in the English-speaking world. Erlanger gave lectures at Hopkins on metabolism and digestion, but it was in the practical classes that he had excelled, with his manual dexterity and an intuitive feel as to how biological preparations and different types of recording equipment worked—or could be improved. Once, while setting up a demonstration on blood pressure for the medical students, he uncharacteristically broke the glass chamber used to record arterial pulsation in the finger. Undaunted, he immediately devised a more rugged system, later patented and manufactured in New York, which enabled the pulsations to be measured in the brachial artery above the elbow instead.[4]

At this early stage in his career, the circulation was Erlanger's main interest, and in his own laboratory he studied the way excitation was transmitted from auricle to ventricle in the heart. It was a problem that had attracted the attention of other physiologists, among them Walter Gaskell, in the early days of the Cambridge Physiological Laboratory. Erlanger's approach was to devise a clamp with which he could apply graded pressure to the specialized conducting fibers

Figure 6–2 Joseph Erlanger. The photograph is thought to have been taken in 1914, by which time Erlanger, the newly appointed Professor of Physiology at Washington University, would have been 40.

running between the chambers of the heart (the Bundle of His). As a result of his investigations, Erlanger became an expert on heart block, including that which occurs in humans and gives rise to fainting (Stokes–Adams syndrome).[5] He was also interested in finding out where excitation began in the heart, and in the precise relationship between systolic and diastolic blood pressures, and the sounds heard through a stethoscope, as a cuff around the limb was deflated. The high quality of the research, and its obvious clinical relevance, ensured that it was not long before Erlanger's publications, mostly in the *American Journal of Physiology*, attracted the attention of other physiologists, as well as physicians. Erlanger's reputation in the field of the heart and circulation was ultimately recognized by an invitation to give the Harvey Lectures in New York in 1912.

Even before this honor, however, Erlanger's abilities were evident to others, and in 1906 he was invited to head the Department of Physiology and Physiological Chemistry at the new medical school in Madison, Wisconsin. Among the class of students to whom he lectured was Herbert Gasser. There must have been something about the quiet and serious Gasser, perhaps questions raised after a lecture, that had impressed his professor and made him, six years later, invite Gasser to join Erlanger's new department.

New department? Yes, Erlanger had moved again, this time having been offered the founding chair in physiology in the medical school at Washington University in St. Louis. Although he had been highly successful in Madison, Erlanger probably had not hesitated. Madison, though the state capital of Wisconsin, was then, and still is, a small, attractive city with the university as its main occupant. St. Louis, on the other hand, was second only to Chicago in the American Midwest. Standing astride the Mississippi, and at the mouth of the Missouri, it had long been the gateway to the West, as the huge present-day arch over the Mississippi attests. In 1910, the medical school at Washington University, like those at the University of Wisconsin and Johns Hopkins, was in the early throes of a revolution in medical education. Previously medicine had mostly been taught in a haphazard and uneven way by practicing physicians and surgeons who gave few lectures, least of all in anatomy and physiology, and who imparted their clinical wisdom on the wards and in the operating theaters. In fact, medical education in the United States was little different from that previously found in Britain, where the adverse situation had provoked T. H. Huxley's strong criticism in 1870.[6] It was Abraham Flexner, commissioned by the Carnegie Foundation to investigate the state of medical education throughout the United States and Canada, who had recommended, among other measures, that the students receive systematic instruction in the basic medical sciences, including physiology. After visiting St. Louis, he reported that the medical department in the university was "a little better than the worst, but absolutely inadequate in every essential respect."[7]

Since Flexner's visit, the situation in St. Louis had improved. Not only was there determination on the part of the university to do better, but there was also the promise of funds to make it happen, with grants from the Rockefeller and Carnegie Foundations and substantial donations from local businesspeople. There was, in fact, enough money for the construction of new hospitals for adults and children, and of a basic sciences building.

Gasser, too, had moved on since attending Erlanger's physiology lectures in Madison. Like Erlanger, he had completed his medical education at Johns Hopkins Medical School in Baltimore. Like Erlanger, too, he was offered a position in the physiology department but decided to return to Madison, where he took a position as instructor in pharmacology. Then, soon after, Erlanger's invitation came. Remembering Erlanger's brilliant lectures of six years ago, Gasser accepted.

Hardly had Gasser moved to St. Louis, however, than he was called into the military, the United States having entered the First World War in 1917. It was not until the end of the following year that Gasser was able to return to St. Louis and to start his new job in earnest. By then he had some research experience as a graduate student in Madison, exploring the mechanisms by which exercise speeded up the heart rate, while, at Johns Hopkins, he had been directed into a study of blood clotting. During his short period of military service, he worked with Erlanger on the shock that follows injury, a subject of obvious relevance to the war effort. He would have witnessed Erlanger's skill in animal dissection and would have learned much else from the older man. Though this period was brief, barely a year, a series of papers was later published in the *American Journal of Physiology*, several of which described the benefit of treating the shocked patient (or, rather, the shocked animal) with an infusion of gum arabic and glucose.[8]

Erlanger, when he returned to his department in 1918, was 44, and thus 14 years older than Gasser. Apart from the wartime work on shock, he had, as described earlier, published original research on the heart and circulation while on the staff at Johns Hopkins. Indeed, had he never taken up nerve excitability, Erlanger would have remained a major figure in physiology for the rest of his career.

It must have been the younger man, Gasser, who had suggested nerve excitability as a research topic for the new department in St. Louis. While at Johns Hopkins he had known a fellow medical student, Sidney Newcomer, who had had an interest in physics and, after moving to the Department of Mental and Nervous Diseases at the Pennsylvania Hospital, had succeeded in building a three-stage amplifier. Newcomer had visited Gasser in St. Louis soon after the end of the war, and the two had used the new amplifier to record nerve impulses in the phrenic nerve of the dog.[9] Not only was the nerve long and slender, and free of branches, as it ran through the chest to the diaphragm, but there was no need for stimulation, since there were impulses flowing down the nerve with each breath. Nevertheless, the recordings had not been made easily, since the amplified signals were subject to different types of interference and the string in the Einthoven galvanometer was easily broken.

The physiological study by Gasser and Newcomer was not the first to employ valve amplification. In 1920 Alexander Forbes and another Harvard colleague, Catherine Thacher, had published action potentials recorded from frog sciatic nerves with a single-stage amplifier, string galvanometer, and camera. Although the potentials are difficult to distinguish on the prints, they had fast-rising initial deflections with slightly slower returns and had lasted less than 10 ms (Fig. 6–3).

Why was the amplifier so important in this type of experiment? The answer is that nerve impulses, even when recorded with wire electrodes touching the nerve, are extremely small. Depending on the number of nerve fibers conducting, the signals may be no more than a few millionths of a volt, and therefore difficult to detect with either a capillary electrometer or a string galvanometer. With amplification, however, the problem can be overcome. But amplification was not possible until the invention of the thermionic valve by Ambrose Fleming in 1904. In the simplest valve, a diode, the current is carried by electrons flowing from the negative electrode, the cathode, to the positive electrode, the anode. The value of the diode is that it ensures the current in a circuit flows in only one direction. With the addition of a third, intervening, electrode, however, the valve becomes a triode, and can be used as an amplifier. As in the diode, the cathode releases electrons at the base of the valve, which are immediately attracted to the anode at the apex of the glass envelope. The flow of electrons, however, is controlled by the presence of

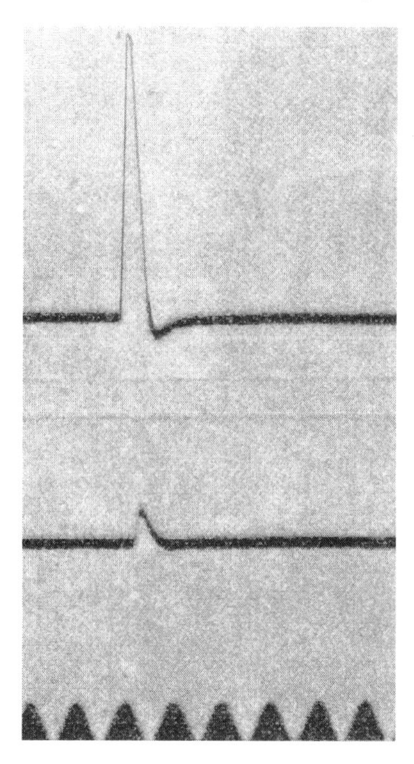

Figure 6–3 One of the nerve action potentials recorded from the frog sciatic nerve by Forbes and Thacher (1920), using a single-stage amplifier and an Einthoven string galvanometer. The action potential, "break" stimulus from the inductorium, and a 100-Hz tuning-fork oscillation are shown on the top, middle, and bottom traces respectively. The small overshoot as the potential returns to the baseline is an artifact of the galvanometer and due to insufficient damping. The action potential has been retouched by the present author.

the third electrode, the "grid," and it was to the grid that the nerve signals were applied in the Newcomer and Gasser experiments, and before those, in the Forbes and Thacher studies. Since the grid is closer to the cathode than the anode is, even small potentials, such as those produced by a nerve, have a major effect on the flow of electrons through the valve, hence the amplifying action (Fig. 6–4). Moreover, once the signal has been amplified by one valve, it can be amplified by another, and then by another, as in the three-stage instrument of Newcomer. Nowadays, thermionic valves are rarely used in electronics, having been supplanted first by isolated transistors and then by integrated circuits. For 40 years, however, after the first successful attempts, amplifiers were built with thermionic valves, and, in the neurophysiology laboratories of the world, it was usually the scientists themselves who did the building. Indeed, the construction of one's own amplifier was to become a rite of passage for an aspiring neurophysiologist.

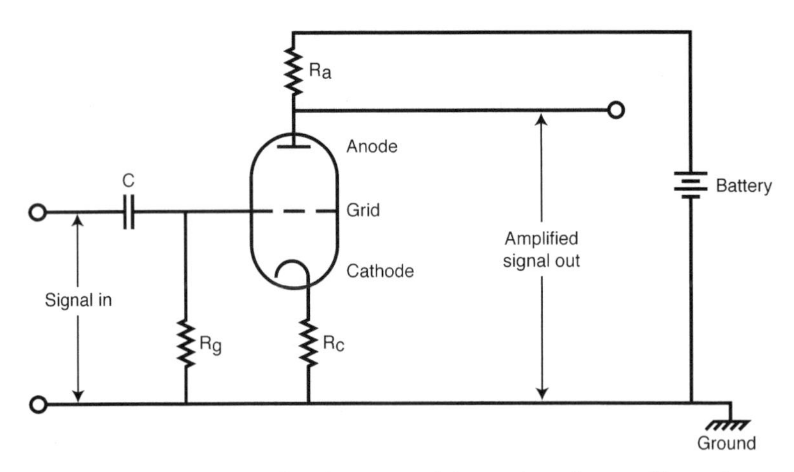

Figure 6–4 A single-stage triode valve AC amplifier such as that used by Forbes and Thacher (1920). The resistors R_a, R_c limit the current flowing through the valve and ensure the anode and cathode are at the correct voltages for operating the valve. The grid resistor R_g is chosen so as to make the grid, to which the biological signal is applied, slightly negative to the cathode. The capacitor C allows only *changes* in potential, rather than steady values, to be amplified.

As Newcomer and Gasser discovered, having sufficient amplification removed only one of the difficulties in recording nerve impulses. The recording device itself, used to display the signals, must be capable of responding very quickly, since the impulse in a single nerve fiber is largely over in a millisecond. As pointed out previously, the two main instruments of the time, the capillary electrometer and the string galvanometer, were too slow, the nerve impulses appearing smaller and broader than they really were. Newcomer and Gasser had used a string galvanometer in their experiments on the phrenic nerve, and this had been adequate to show the complex picture formed by action currents in many nerve fibers, discharging at different frequencies with each inspiration. But how could one make faithful recordings of the impulses, ones in which the amplitudes and time-courses were free of distortion? One solution was to apply a mathematical correction, and this is what Keith Lucas had been able to do for the capillary electrometer, using his specially designed mechanical device.[10]

A much better way, of course, was to use an improved measuring instrument, and, in his influential book of 1912, *Elektrobiologie,* Bernstein had suggested how this might be done. The Braun tube, named after its inventor, had been described two years earlier. Like the triode, used in amplification, it was a valve, but a very special one. To begin with, it was much larger, and shaped like a cone. As with the triode, electrons were released from the cathode and attracted to the anode. In this case, however, the anode was placed close to the cathode, and made in the form of a hollow cylinder, allowing a thin beam of electrons to escape and hit the far end of the tube (Fig. 6–5). Since the end of the glass tube had been coated with fluorescent paint, the electrons, on striking it, caused it to glow momentarily. In addition to the anode and cathode, the cathode ray tube contained two pairs of metal plates, mounted horizontally and vertically. By applying voltages to each pair of plates, the electron beam could be made to move from side to side and also up and down. The amplified signals from the nerve could be put on one pair of plates, while the electron beam was made to sweep across the tube at a known velocity. The tube, and all its attendant circuitry, then became a cathode ray "oscilloscope" or "oscillograph."

Since the oscilloscope used a beam of electrons, rather than a magnetized needle or wire, or a mercury meniscus, to write its picture, there was no inertia and the voltage signals were recorded instantaneously. Moreover, the electron beam could be swept across the tube much faster than a camera film could be wound, so that time measurements could be made more accurately—in fractions of a millisecond, rather than in whole milliseconds. Despite the desirability of using an oscilloscope, there was a catch. Although the Western Electric Company had recently developed an improved cathode ray tube, with a heated cathode to improve its sensitivity, the company was reluctant to sell one to Gasser and Erlanger. Undeterred, the two physiologists decided to make their own, initially with the help of Sidney Newcomer.[11] It was probably Gasser who did most of the design work, having by then acquired a good knowledge of electronics while building his own amplifier for the laboratory. For the tube itself, an Erlenmeyer distillation flask was used, since it had the necessary conical shape. There was room in the flask to install only one pair of deflection plates, the second pair being replaced by two solenoids attached to the outside of the flask. According to Erlanger, "With that outfit we did succeed in seeing the first, presumably perfect amplified action potential, but, after a few sweeps, the filament of the tube fused!"[12] But why should Erlanger have used the adverb "presumably"? And, surely, it would have been remarkable if, at the first attempt, the sizes, shapes, spacings, and voltages of the various tube components had resulted a sufficiently strong, well-focused and adequately deflected electron beam. Even the Western Electric Company, with its considerable resources, had only been able to produce a tube with a beam too weak for photographing single traces. Was Erlanger, mindful of his place in the history of physiology, giving a true description of events? The question is fair, since there was a second, more dramatic, account to the same happenings: "Having got it all made, they turned on the current and there was an explosion and blew it all to thunder—they'd forgotten to put in a central controlling resistor."[13]

This description of events was given by George Bishop in an oral history interview almost 50 years after the event. But there is a possible problem here also, since Bishop may not have been present when the homemade tube exploded. He had been recruited to the department in St. Louis in 1921, presumably arriving in the summer in time for the start of the new academic year. Since the first full account of their work had been written up by Gasser and Erlanger and submitted for publication in late July 1922,[14] it is possible that the first experiment with the homemade cathode ray tube had preceded Bishop's arrival in 1921. If it had not, would Erlanger have invited the new appointee, an expert on the metabolism of bees, to watch such a critical experiment? The alternative is that the story of the exploding tube was told to Bishop by Gasser, or by the departmental mechanic.

As it proved, the failure did not matter. The Western Electric Company decided that it could, after all, make one of its new tubes available to the two physiologists. In addition to having had to make their own amplifier, Gasser had to design a circuit for sweeping the electron beam across the cathode ray tube. Initially, he did it by rotating a variable resistor, but, since the contacts of the motor began to wear out, the discharge from a condenser was employed instead.[15] Even when the apparatus was working properly, a further problem remained, for, although the moving green spot on the cathode ray tube could be seen by the naked eye, it was too faint for the photographic film then available. Either the nerve impulses had to be drawn on tracing paper, or a series of traces had to be superimposed on the photographic film, which, like the tracing paper, had to be held over the end of the tube. There were still other problems, due to circumstances outside the laboratory. The trucks on South Euclid Avenue caused vibrations in the amplifier system, even though Gasser had used rubber shock-mounting, and, despite metallic shielding, the sparks from the trolleybus wires produced spike artifacts in the recordings.[16] If the method for recording was new, that for stimulating was not: Gasser and Erlanger used the device that had been introduced into physiology 70 years earlier—du Bois-Reymond's inductorium.[17]

It is difficult for a modern society, familiar with laptop computers and palm-sized digital cameras, to imagine the size and complexity of Gasser and Erlanger's equipment, as it stood in the basement of the building on South Euclid Avenue. The amplifier alone had 57 switches, while the 300 volts required to operate the cathode ray tube were taken from a stack of dry cells reaching up toward the ceiling, since there was no AC mains supply in the building. The smallest component of the experiment, by far, would have been the excised frog sciatic nerve. A successful day was one in which, by late afternoon, the electronics were working satisfactorily, and it was possible to begin the recordings.[18]

Despite all the difficulties, including that of keeping the nerve in good condition inside the recording chamber, Gasser and Erlanger began to get results. Wisely, they had chosen the bullfrog sciatic nerve as their preparation, since not only were frogs cheap and plentiful, but, compared to mammals, their tissues, including the nerves, were more robust and worked well at room temperatures.

The first paper was submitted to the *American Journal of Physiology* in July 1922 and was published in the same year. Entitled "A study of the action currents of nerve with the cathode ray oscillograph,"[19] it described, with mathematical treatment, the problems encountered in attempting to record nerve impulses, and the limitations of the string galvanometer and capillary electrometer. There followed a description of the cathode ray oscillograph, with a circuit diagram of the entire recording system—nerve bath, amplifier, cathode ray tube, and sweep

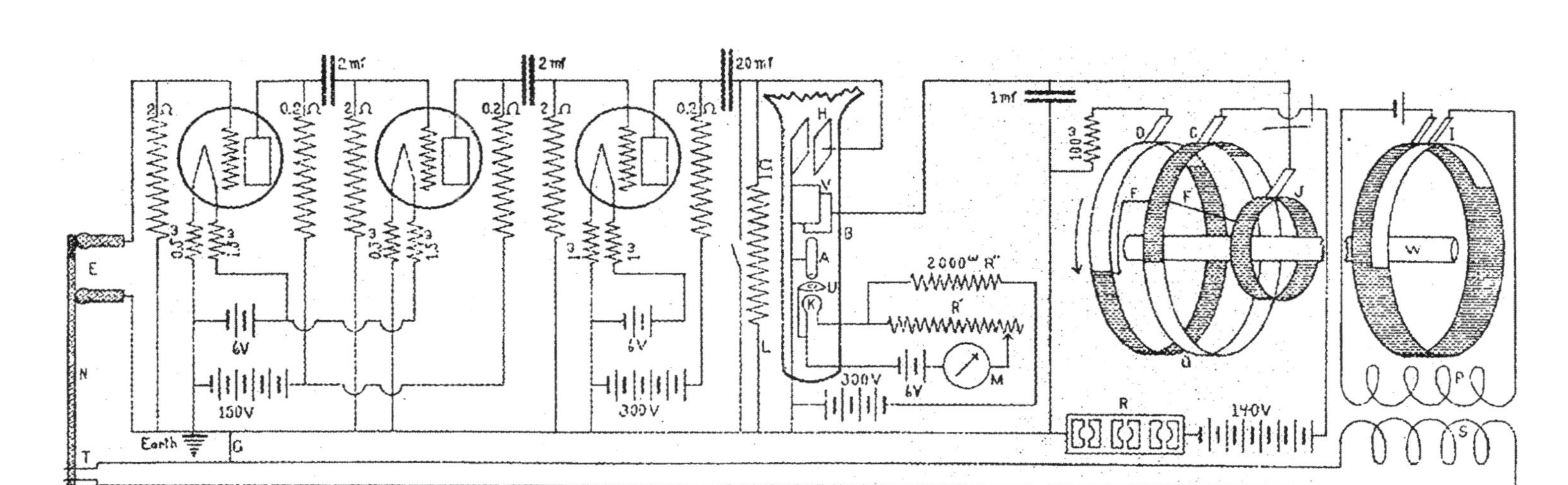

Figure 6–5 Recording equipment used by Gasser and Erlanger (1922) in their first studies, incorporating a three-stage triode amplifier (*left*) and cathode ray tube (*center*). One of the wheels (*right*) charged a condenser, thereby controlling the movement of the electron beam across the cathode ray tube. As the charge on the condenser increased, the velocity of the beam decreased logarithmically.

generator (see Fig. 6–5). Finally, the paper showed the superimposed traces obtained from the bullfrog sciatic nerve with the cathode ray oscillograph, in comparison with the records from a string galvanometer.

What did the nerve impulses look like? The one photograph in the paper, a contact print from the end of the tube, was of poor quality (Fig. 6–6). To begin with, the trace flowed from right to left, which is the opposite direction to the convention in physiological recordings and, of course, in reading a book. There was a bright white spot on the right of the black print, where the electron beam waited to be swept across the tube. The upstroke and downstroke of the trace were faint, and could barely be made out in the white fog. On the bottom of the photograph a series of unequally spaced white dots marked the time elapsing as the electron beam flashed across the tube; each dot was separated from the next by one millisecond. Despite the technical shortcomings of the photograph, the rapidity of the upstroke was evident; and from the baseline the "action current" took just over one millisecond to reach its peak. By raising or lowering the temperature of the nerve, this time could be shortened or prolonged. Gasser and Erlanger also compared the oscillograph recording with the results when a string galvanometer was used instead, and the greater steepness and height of the oscillograph record were immediately apparent.

In one sense, the results were an anticlimax, for other workers, among them Adrian in Cambridge, had already applied mathematical corrections to the records made with the capillary electrometer and string galvanometer, and had obtained

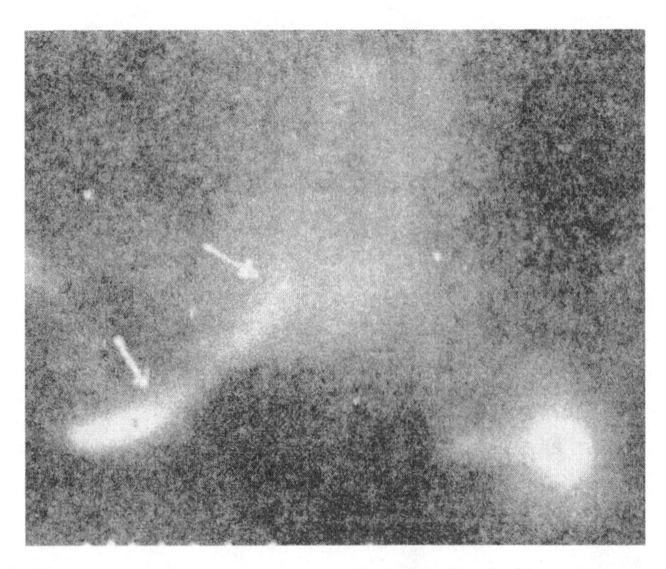

Figure 6–6 Frog sciatic nerve action potential recorded by holding photographic contact paper over the end of the cathode ray tube, and repeating the nerve stimulus many times. The poor quality of the picture is partly due to its reproduction. (From Gasser & Erlanger, 1922)

values for the duration of the rising action current that were very similar to those of Gasser and Erlanger. And before Adrian there had been Bernstein with his differential rheotome. However, it was one thing to estimate what might be happening and quite another to see the event actually taking place, as Gasser and Erlanger were able to, at the end of their cathode ray tube. The rapidity of the action current was not the only feature of interest. On the downstroke, inflections could be distinguished, which were more prominent in some experiments than in others, and there was another inflection at the base of the upstroke. Erlanger was used to seeing tracings with notches in them, for these were characteristic of the blood pressure recordings that he had worked with in his research at Johns Hopkins. In that situation, they signified the opening and closing of heart valves. Surely the irregularities in the action current records were similarly important? As a first step, the notches were identified with arrows on the records and described as "anacrotic" on the upstroke and "catacrotic" on the downstroke.

By the time Gasser and Erlanger's first paper was submitted for publication, Gasser's career had taken an unexpected turn. He had been made the Professor of Pharmacology in St. Louis, this despite a paltry 10 publications, half of them written with Erlanger on shock, and none of them dealing with pharmacological topics. It is not difficult to see Erlanger's hand in the appointment. Erlanger must have quickly grasped the potential of the new work, and he would have appreciated that it could not be accomplished without Gasser. After all, it was Gasser who knew the underlying theory, and Gasser who had designed and helped build the equipment. Tellingly, it was Gasser whose name appeared as first author of the new paper. Further, even though Erlanger may have mastered the full complexity of the new work, he would never have had the time to push on with it by himself. There were lectures to give to the medical students, covering, as they did, most of the body systems, there were the practical classes to organize and supervise, and there was all the administrative work that needed the attention of the head of a busy university department. There may have been other reasons for maneuvering Gasser into pharmacology, for Gasser would have given Erlanger support when important political issues were raised in faculty committees. Also, with Gasser, Erlanger's protégé, as its head, Pharmacology was unlikely to set itself up as a rival to Physiology.

Hardly had the first paper appeared, and Gasser appointed to his new position, than there was a surprising development. Gasser left for Britain. He had already intended to go to the International Physiology Congress in Edinburgh, and the Rockefeller Foundation encouraged him to stay abroad for a further two years to obtain additional experience in pharmacology and to improve his skill in . . . foreign languages! During these two years, Gasser visited several leading British and European laboratories. In London, he carried out work on heat production in muscle with A. V. Hill. Hill had already won a Nobel Prize for his work on muscle contraction and was to remain a major figure in nerve and muscle physiology for the next 50 years. In London Gasser also undertook some research under

Henry Dale, a future Nobel Laureate and one of the leading pharmacologists of his generation. In addition, there were working visits to Straub in Munich, and to Louis Lapicque in Paris. Lapicque had previously studied the relationship between the strength of current used to stimulate a nerve, and the minimum time that it needed to flow. The smallest possible current was termed the "rheobase," and the time taken to stimulate the nerve with a current of twice this strength became the "chronaxie."

Gasser's departure for Britain and Europe left Erlanger in a quandary. For the reasons already given, he could hardly continue the nerve experiments by himself. If he waited for Gasser to come back, the momentum would have been lost, and it was possible that workers in another laboratory would have entered the field and provided competition. It would not have been difficult for them to do so. Not infrequently, in describing a new technique, the authors omit, either by design or by default, a key piece of information. Gasser and Erlanger, however, had done just the opposite; not only had they included a circuit diagram in their paper, but they had given reasons as to why particular components—the resistors, capacitors, and battery voltages—had been selected (see Fig. 6–5).

Inaction did not suit Erlanger. If Gasser could not be there to keep the experiments going, someone else would have to be brought in. It is not known whether Gasser had any part in the selection. What soon became clear, however, was that Erlanger had chosen well, just as he had seven years previously when he had brought Gasser into the physiology department in St. Louis.

Unlike Gasser and Erlanger, George Bishop had not studied medicine, but had graduated first in liberal arts at the University of Michigan, and then with a PhD in zoology at the University of Wisconsin.[20] Before that, however, he had demonstrated his independence and practical abilities by dropping out of high school for two years in order to help his father run a sawmill. Later, on the family farm in Michigan, he had assisted his father again, this time in keeping bees. Ever inventive, Bishop had devised and built many labor-saving devices for the farm, mostly from tin cans, baling wire, and wood. Although he, like Gasser, had attended the University of Wisconsin, it was unlikely that the two had met. Nevertheless, Bishop had some interest in the nervous system, having worked as a technician for the neuroanatomist C. J. Herrick during the First World War. His first research, however, was the outcome of his boyhood on a farm—a study of the sexual organs and the fat body of the honeybee. The achievement of which he was most proud was to have measured the pH of a tiny drop of body fluid from one of his honeybee larvae.[21] For this kind of work, extremely fine manipulations were needed, and much of the measuring apparatus needed for such small samples of fluid had been devised and built by Bishop himself. Having recently been recruited to the physiology department in St. Louis, Bishop was, to his surprise, invited to join Erlanger in the laboratory. At that time, 1923, Bishop had yet to start research on the nervous system, but his dissecting skills, inventiveness, and practical ability, and the two years spent studying engineering as an undergraduate in Ann Arbor, formed an excellent background for a prospective neurophysiologist. At 34, Bishop was only a year younger than Gasser.

So quickly did Bishop settle into his new role in St. Louis that some of his experimental data were included in a publication with Gasser and Erlanger in the following year,[22] and a further five papers with one or both authors appeared over the next three years. Largely because of Bishop's influence, the photographs and the nerve recordings improved. Most of the work had to do with the notches in the action current that had so interested Erlanger. When the recordings were made further and further from the stimulating electrodes, the notches became individual waves, reflecting activity in separate groups of nerve fibers (Fig. 6–7).[23] The authors used the Greek alphabet to designate the different fiber groups, the fastest-conducting being the alpha fibers, and the successively slower ones becoming the beta, gamma, and delta fibers. They were able to show, as had been predicted, that the largest fibers, the A-alpha, were those with the highest conduction velocities and the lowest thresholds to electrical stimulation.

Gasser's preoccupation was in showing that the form of the action current, as viewed on the cathode ray oscilloscope, could be mathematically reconstructed if the numbers and diameters of the nerve fibers were known.[24] Gasser was quite prepared to do the tedious counting and measuring of the nerve fibers himself. In addition, the three investigators were naturally interested in the possible functions of the different groups of motor and sensory nerve fibers. There was also interest in the momentary depression of excitability that followed the passage of an impulse[25]—the absolutely and relatively refractory periods previously studied by Lucas and Adrian in Cambridge.

If, in 1927, one had visited the laboratory in St. Louis, Erlanger would have been quickly identified as the person in charge. Even during lunches, which were brown-bag affairs, the topics for discussion, invariably to do with research, were chosen by Erlanger. Photographs at the time often show him in a white laboratory coat, his dark hair brushed straight back, with a keen intelligence in the eyes and in the half-smile (Fig. 6–8). A tough, wiry man, Erlanger kept fit by walking to work, and by hiking with his family in the Rocky Mountains during the summer vacations. While respected and admired by those immediately around him, to the medical students Erlanger was an aloof figure. As the physiology department in St. Louis was built up, there was less need for him to lecture, but he continued to take a strong interest in the practical classes. Former students would remember Erlanger as being especially ruthless in rejecting kymograph recordings that were less than perfect.[26] The recordings were made by a light lever writing on paper that had been covered with soot in a flame cupboard, and which then had to be dipped in a trough of varnish. Indeed, Erlanger's obsession with smoked drums earned him, behind his back, the sobriquet of "Smokey Joe."

Gasser was altogether different. Photographs taken of him over the years show hardly any change—the same unsmiling face, the spectacles, the hair smoothed down and carefully parted at the side (see Fig. 6–1). Tall, elegant, fastidious in all

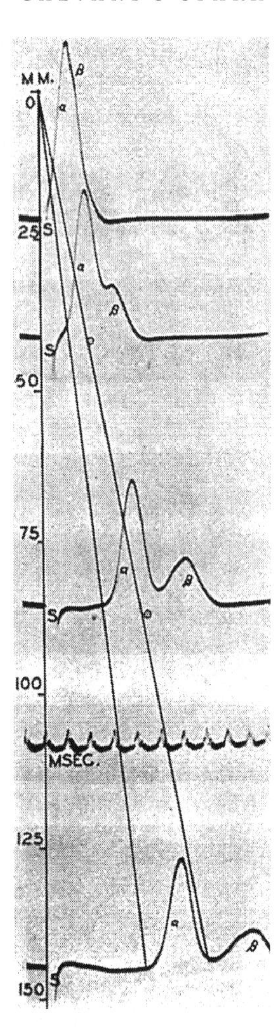

Figure 6–7 Recordings made at different points along the sciatic nerve. As the distance traveled by the nerve impulses increases (*lower traces*), the greater becomes the separation between the alpha and beta components of the potential, due to slower conduction in the beta fibers. This is from a later recording. (From Erlanger J, Gasser HS. *Electrical Signs of Nervous Activity,* 1936, p. 13. Reprinted with permission of the University of Pennsylvania Press.)

matters, and a person of integrity, Gasser was kind, loyal, and amusing to the few who knew him well. To others he remained an enigma. Alan Hodgkin, visiting Gasser some years later, remarked, "Gasser is a queer person. He has a treble voice and looks about twenty except for the eyes which are rather old and sad."[27] Whatever its cause, presumably an endocrine one, Gasser's appearance made for a lonely life. Had he married, he may well have chosen someone like Helen Graham, his junior colleague in the pharmacology department, and one of the few women with whom he was to make an enduring friendship. However, Helen Graham,

Figure 6–8 A later photograph of Erlanger, showing him in the nerve recording laboratory; most of the equipment would have been made in the department. The oscilloscope displays an action potential (in the photograph, immediately in front of Erlanger's head). (Becker Medical Library, Washington University School of Medicine)

attractive, gracious, and intelligent, was already married, and her scientific career was balanced by a successful life as the wife of a university professor and the mother of two children.[28] As a pharmacologist and physiologist, Helen Graham proved herself capable of carrying out original research, initially with Gasser and later on her own. It was Helen Graham who was to study the effects of different drugs on the slow potentials, the "after potentials," which followed the main spike in the action current recordings, and it was she who, with Gasser, was to make some of the first recordings from the spinal cord.[29]

Bishop, in turn, differed from both Gasser and Erlanger (Fig. 6–9). A big, burly man, often unshaven, he lived with his wife, the biochemist Esther Ronzoni, in a log cabin, the two sometimes sleeping on a mattress in front of the large fireplace. One of the rugs in the cabin might have been one he had made in Madison, stitching together the tanned skins of 25 cats from a practical physiology class. Throughout his life, he continued to see himself and others as frontiersmen. Commenting on the early days in St. Louis, he was to say of Gasser and Erlanger, many years later:

> "I can only compare their progress to the trek of the pioneers in oxcarts across the plains and mountains of the West."[30]

Figure 6–9 George Holman Bishop. (Becker Medical Library, Washington School of Medicine)

As he became more independent of Gasser and Erlanger, Bishop was able to indulge his natural curiosity, helped by uncanny insights and a fertile imagination. One former medical student, who had spent 3 months making measurements from the laboratory's oscilloscope records, remembers Bishop as a kind and considerate person, with a pipe always in his mouth, scattering ash over everything in the laboratory.[31]

Erlanger, Gasser, and Bishop—the professor, the scholar, and the frontiersman.

Within a few years of Bishop's arrival, there was enough equipment available for the three investigators to go their separate ways, and this they agreed to do. In 1929, however, there were two important events, of which the first was the XIII International Congress of Physiology. This Congress, held every few years, was to be held in Boston, and it was the intention of Erlanger and Gasser to demonstrate nerve action potential recordings on an oscilloscope to the delegates. To transport their bulky equipment Erlanger and Gasser used Erlanger's car, bought second-hand from Gasser in 1924. If the demonstration was a success, as it evidently was, the other major event in 1929 was a disaster, bringing to a sudden end the peaceful coexistence that had existed among the three investigators. While recognizing the need for independence, Erlanger had not been able to relinquish his control entirely. In particular, he still expected all manuscripts to be submitted for his scrutiny before being sent on to the editor of a scientific journal.

It was first with surprise, and then with mounting anger, that Erlanger must have read the paper that had been sent for publication, without his knowledge or approval, and had now, irony of ironies, been forwarded to him for refereeing. Even worse, the paper described a major breakthrough. Until then, all the nerve recordings in St. Louis had been made from the larger nerve fibers, around each of which there was a fatty covering, the myelin sheath. Mixed in with these nerve fibers, however, was a similar or greater number of much smaller fibers, in which the myelin sheaths were absent. Now, for the first time, the action currents in these small unmyelinated fibers had been detected, as a very small deflection well after the response of the myelinated fibers. The original recordings had been made in the cervical sympathetic chain and vagus nerve of the turtle, but now they were shown to be present in other nerves as well (Fig. 6–10). The authors were Peter Heinbecker, a young Canadian surgeon acquiring research experience in St. Louis, and . . . *George Bishop*![32] To make the recordings, Heinbecker had, on his own initiative, used stimuli far stronger than those routinely employed for the larger, myelinated, nerve fibers. Had they thought to try large stimuli, Gasser and Erlanger could themselves have obtained similar results in their peripheral nerve preparations, and done so years earlier.

Erlanger's violent temper was well known to his family, but at work it had usually been controlled. Now, however, it was unleashed in its full fury. Bishop was sent for, an accusatory letter written, and then came expulsion from the physiology department. Bishop, unperturbed and quite unable to understand the reason for Erlanger's wrath, was probably relieved to move to the sixth floor of the pharmacology building, where he became Professor of Applied Physiology in Ophthalmology, and was to start pioneering studies of the visual system. Within two years, Gasser would leave for New York, accepting positions first at Cornell University and then as director at the Rockefeller Institute for Medical Research. In 1935 there was one last collaboration, a series of lectures to be given by Gasser and Erlanger to the Johnson Foundation in Philadelphia. Meeting beforehand, in the Rocky Mountains, the two reluctantly accepted that, in their work, "there are so many points which we interpret differently."[33]

Notwithstanding the differences, the lectures were extremely well received, and the published version, *Electrical Signs of Nervous Activity*,[34] became a classic. It was Erlanger who dealt with the various elevations in the compound action potential, and with the conduction velocities, thresholds, and diameters of the different types of fiber. When necessary, he drew on the computed action potentials that Gasser had so painstakingly constructed from measurements of fiber number and size. Included, too, were recordings of impulses in single nerve fibers, which Erlanger had been able to make, with Edgar Blair, in the extremely fine nerves of the frog toe.

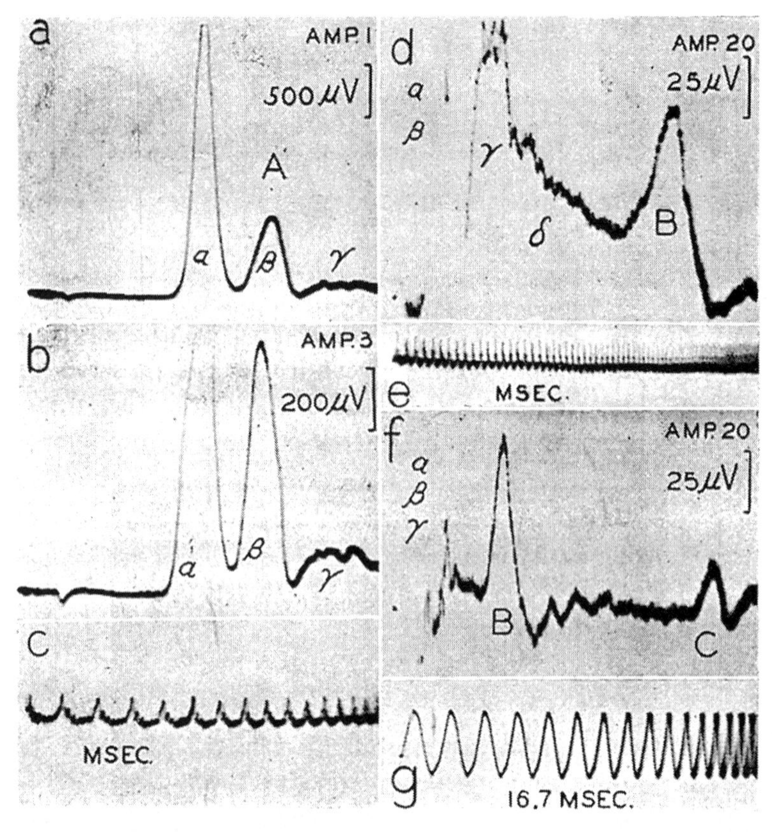

Figure 6–10 The various components of the "compound" action potential recorded from a frog peroneal nerve. The alpha, beta, and gamma components of the A deflection are shown in (*a*) and, at higher amplification, in (*b*). In (*d*), still higher amplification has been used, together with a slower sweep on the oscilloscope—the small B elevation is now apparent. In (*f*) the sweep has been further slowed and a still later, smaller, deflection is visible—this is the C elevation. Though this figure has been taken from Erlanger and Gasser (1936) for its clarity, the discovery of the C elevation (wave) was made by Heinbecker and Bishop (1929), who showed that it was generated by small unmyelinated nerve fibers (the C fibers). The A and B elevations, in contrast, are the potentials in myelinated fibers. The conduction velocity of the C fibers is less than 1 $m.s^{-1}$ (meter per second), while that of the fastest A fibers, depending on temperature, is in the range of 30 to 40 $m.s^{-1}$. In mammalian nerves, at 37°C, the corresponding velocities are higher still. (From Erlanger J, Gasser HS. *Electrical Signs of Nervous Activity*, 1936, p. 6. Reprinted with permission of the University of Pennsylvania Press.)

Gasser, having had much of his work presented by Erlanger, was obliged to cast further afield for his lecture material, and in this he succeeded brilliantly. In a theoretical section, Gasser referred to the work of others, including Ludimar Hermann, in showing how the nerve membrane could be represented as a series of resistors and capacitors.[35] Further, the potential difference across the resting membrane could be a diffusion potential caused by unequal concentrations of ions on the two sides of the membrane, and by the semipermeable nature of the membrane, just as Bernstein had postulated. Gasser then considered how the flow of current across the membrane must in some way activate the unknown mechanism responsible for the action current. He went on to point out that, whereas a small negative potential followed the "spike" in myelinated fibers, it was absent in the unmyelinated fibers; instead there was a positive after-potential that became dramatically larger if several shocks were delivered to the nerve in quick succession, as he and Helen Graham had found. The after-potentials were associated with metabolic activity in the fibers and could be affected by anoxia, changing the hydrogen ion concentration and by applying certain drugs. Next, Gasser described the changes in excitability that followed the spike, namely the absolutely and relatively refractory periods, and the supernormal period. Lastly, and most ambitiously, Gasser attempted to explain the slow waves that could be evoked in the spinal cord by stimulating the incoming sensory nerve fibers. He showed that the negative and positive potentials, recorded from the surface of the cord, corresponded respectively to excitation and inhibition of the motor nerve cells (motoneurons), but he wrongly supposed that the prolongation of excitation was due entirely to activity in intervening nerve cells (interneurons) rather than to slow events in the motoneurons themselves. Even if Gasser's conclusions were partly wrong, as they sometimes were, his arguments were well made, and always stimulating.

But what were the points on which Gasser and Erlanger had agreed to differ? It may have been that, even in the earliest experiments, Gasser had doubts about some of the inflexions and waves seen in the action current recordings. Was it possible that . . . were they . . . *artifact*? Indeed, if one does the experiment of stimulating and recording from a frog sciatic nerve, as Gasser and Erlanger had done, the question becomes obvious. As the stimuli are repeated, notches appear on the downstroke of the action current; these then become waves, which move to the right on the cathode ray oscilloscope screen, as conduction in a group of nerve fibers slows and then ceases altogether. Since Erlanger had set great store on the different waves—alpha, beta, gamma, delta—Gasser could hardly press his doubts. Later, almost 30 years later, in New York, he did, and by a superior recording technique, demonstrated that, at least for the larger myelinated fibers in a mammalian cutaneous nerve, there were only two elevations, corresponding to two groups of nerve fibers.[36] All the other waves were, as he suspected, fictitious. There was also a difference of opinion concerning the relationship between the diameter of a myelinated nerve fiber and the velocity with which it conducted an impulse. While Erlanger argued that the velocity must vary as the square of the diameter, Gasser, correctly, favored a linear relationship.

Bishop, meanwhile, apparently secure in his new laboratory, continued his imaginative research, the scope of which was extraordinary. In 1932 alone, there were papers published on the cervical sympathetic nerves to the eye, the response of the cerebral cortex to stimulation of the optic nerve, the form of the action potentials in peripheral nerves, the functions of the different types of nerve fiber, experimental poliomyelitis in the monkey, and the action potentials of single muscle fibers. But Bishop was to pay dearly for his oversight with the manuscript on the unmyelinated fibers.

Erlanger neither forgot nor forgave.

7

The Code

By 1932 St. Louis had become recognized as the leading neurophysiology center in the United States, especially for those interested in peripheral nerve and in the properties of the nerve impulse. When the Nobel Committee met in Stockholm, however, and decided to award the prize in physiology or medicine for research on the nervous system, it was not to St. Louis that they looked, but to Britain. The prize was given, not for revealing the properties of the nerve impulse itself, but rather for showing the ways in which impulses were used by the nervous system to convey information from one part to another.

If a foreign reporter had wished to interview the two recipients, he or she would probably have started from London and taken a train from Paddington Station for the one-hour journey to Oxford. A different train, this time headed north rather than west, would have been needed to travel from King's Cross to Cambridge. The elder of the two prize winners, the Oxford man, might have been found working in his laboratory in the physiology department, or perhaps sitting behind the enormous desk in his office, discussing science or philosophy with his doctoral students—Sherrington. The younger man might also have been in his laboratory. If not, there was a good chance that he would have been observed riding his bicycle quickly through the streets of Cambridge, weaving his way expertly past the other cyclists and those pedestrians spilling onto the road from the sidewalk. The onlooker would have noted the forward thrust of the nose and chin, and the determined expression. A small, lean man—Adrian (Fig. 7–1).[1]

After leaving the Physiological Laboratory in Cambridge in 1914, the year that the war had broken out, Adrian had undertaken medical studies at St. Bartholomew's Hospital in London, somehow compressing his clinical work into one year instead of the normal three. By the time he had had his last meeting with Keith Lucas in 1916 he was already practicing as a neurologist in the army, having taken additional training at the National Hospital for Nervous Diseases, Queen Square. It is likely that Adrian's reason for studying medicine was that a medical degree would give him an edge over the other young Cambridge physiologists. With his research background on peripheral nerve, and an interest in psychoanalysis as a student, it was perhaps inevitable that neurology was his chosen specialty. During his wartime years, spent mostly at the Connaught Military Hospital in Aldershot, Adrian showed how examining muscle excitability

Figure 7–1 Edgar Douglas Adrian, later Baron Adrian of Cambridge. This photograph was probably taken at about the time he won the Nobel Prize. (Courtesy of the Adrian family)

by passing electric currents through the skin, the "strength-duration" tests, could distinguish between muscles that had become completely denervated, as following a nerve injury, and those in which some nerve fibers had survived or else had started to regenerate.[2] There was also a publication on the successful treatment of hysterical symptoms—usually paralysis, deafness, or mutism—induced by wartime experiences.[3] These clinical years, though few, gave him insights into the disorders of function that resulted from injury or disease of the nervous system, and they had revealed the enormous gaps in knowledge of the ways in which the nervous system worked. Later, he would always recommend to a prospective physiologist that he or she study medicine first.

At the conclusion of the war, Adrian returned to Cambridge as a Fellow of Trinity College. There was no longer the guidance of his mentor, the "Master" as Adrian would refer to him, but the time spent with Keith Lucas before the war had given him the inspiration, some of the techniques, and the proven success of the deductive approach. Adrian took over the laboratory, now in the basement of a new building, and set to work; there would be no time for post-Impressionist painting, only for experiments.

First there was the unfinished business of the action current. What was its true form and how did the latter relate to the changes in excitability of the nerve fiber? To answer these questions, Adrian employed the frog sciatic nerve and the capillary electrometer once more, but this time correcting for the inertia in the system by means of the instrument that Keith Lucas had designed, built, and described in 1912. It was the first real test for Lucas's device and, with it, Adrian was able, in the space of a few minutes, to make corrections that would have taken very much longer, had they been performed mathematically. With the corrections, the compound action potential of the frog sciatic nerve was seen to have a rising phase, at 14°C, of 0.5 milliseconds, and a slower, falling phase, the whole event being over in approximately 2 milliseconds (Fig. 7–2).[4] Cooling the nerve prolonged these times, warming it shortened them. Adrian next showed that the absolutely refractory period, the time that the nerve was totally unresponsive to a second stimulus, lasted until the action current was almost over, and was followed by the phase of diminished excitability (the relatively refractory period).

Intriguingly, for later research on the nature of the nerve impulse, Adrian speculated that the action current was "brought to an end because the membrane cannot remain permeable for more than a very brief space of time, since the action current itself causes a progressive decrease in the permeability of the active region." A similar suggestion had been made in the previous year, by an American physiologist, Lillie, who had developed a model of the nerve impulse by means of an iron wire immersed in strong nitric acid.[5] Adrian was also interested in the "supernormal period" following the relatively refractory period, when the

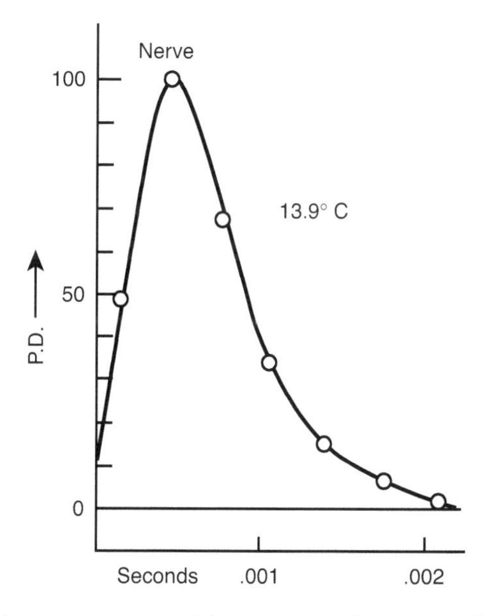

Figure 7–2 Adrian's reconstruction of the nerve impulse. The capillary electrometer recording was corrected with the hand-operated device designed and built by Keith Lucas. The sharp rise and slower decay of the potential are evident, confirming the findings of Bernstein half a century earlier. (From Adrian, 1921b)

nerve fibers became more excitable, and showed that it was influenced by the acidity of the bathing solution, an observation that has yet to be explained.[6]

In describing the true form of the action potential, Adrian, with the capillary electrometer, had narrowly preceded Gasser and Erlanger, with their cathode ray oscillograph. A little later Adrian was to meet Gasser during the latter's visit to the United Kingdom. The meeting would have provided an interesting study in contrasts. On the one hand, the still unsophisticated Gasser from the American Midwest, explaining, in his high-pitched voice, how he and Erlanger had been able to pioneer a new technique for recording the action potential. On the other hand, the slight figure of Adrian, the cultured son of a distinguished family, a Fellow of Trinity College—the same college that an ancestor had entered more than three centuries earlier—rapier-quick in thought and word. Yet each would have respected the seriousness and the ability of the other. It was the unpretentious Gasser who, without hesitation, gave Adrian the circuit for his 3-stage amplifier. It is not clear exactly when this happened, during Gasser's visit or after the latter's return to St. Louis.

Adrian was well aware of the importance of having a good amplifier. He and Lucas had discussed the subject at their last meeting, shortly before Lucas's death in 1916. After the war Adrian had built his own single-stage amplifier, following discussions with the visiting Harvard physiologist Alexander Forbes.[7] It was Forbes who had brought over the triode valves, leaving Adrian to make the resistors and capacitors, and the other parts of the device. From that time onwards, a diagram of the amplifier circuit was to be seen on the wall of Adrian's laboratory in the basement of the physiology building on Downing Street.

Though the single-stage amplifier had worked, Adrian was dissatisfied, especially on seeing the results of Newcomer and Gasser, who had published their recordings from the dog phrenic nerve in 1921. Having now discussed amplifier design with Gasser, and having received a copy of the latter's circuit, Adrian proceeded to get his own 3-stage amplifier made by a local instrument firm.

Though he now had a powerful amplifier, Adrian still had an important decision to make. If he was to investigate the normal flow of impulses in peripheral nerves, as opposed to the much larger volleys evoked by electrical stimulation, which would be better for recording—the newly developed cathode ray oscillograph of Gasser and Erlanger, or the capillary electrometer? In the end, it was the tried and trusted capillary electrometer that was chosen, largely because the image on a cathode ray tube was still too faint for single traces to be photographed. In addition to looking at the nerve impulses, as they were projected from the electrometer, Adrian could also listen to them, by taking an output from the 3-stage amplifier and feeding it into a loudspeaker or into a pair of earphones (Fig. 7–3). If one looks at Adrian's capillary electrometer records, as in Figure 7–4, it is at first difficult to interpret them—which is the nerve impulse, the light deflection or the dark one? And why is the dark region at the top in some illustrations (as in *A, B, C* in Fig. 7–4) and at the bottom in others (*D* in Fig. 7–4)? The answer is that

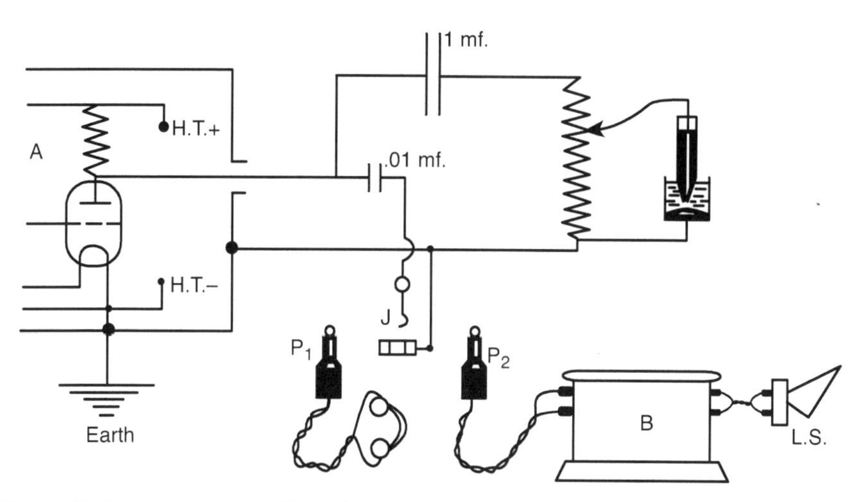

Figure 7–3 Equipment used by Adrian and Bronk in their study of single motor nerve fibers. The last stage of the 3-valve amplifier (*A*) is connected through a 1-mf capacitor to the capillary electrometer (*top right*) and through another capacitor, to the jack socket (*J*). The experimenter has the choice of listening to the discharges through earphones (via jack plug P_1) or, after further amplification (*B*), from a loudspeaker (*L.S*) via jack plug P_2 (From Adrian ED, Bronk, DW. The discharge of impulses in motor nerve fibers. I. Impulses in single fibres of the phrenic nerve. *Journal of Physiology*; **66**: 88–101, 1928. Courtesy of John Wiley & Sons.)

Adrian was showing the nerve impulse as a downward deflection, and these were dark if the photograph had been taken on a falling plate (*A, B, C*), and light if a cinematograph film had been used instead (*D*). Perhaps the most impressive feature of Adrian's method of recording was its sensitivity. With the ability to detect signals as small as 10 μV (10 millionths of a volt), it was easy—with the right preparation—to see the discharges in single nerve fibers, as in Figure 7–4. Indeed, sometimes the amplified signals were too large for the photographic plate or camera film (*A* in Fig. 7–4)!

Most successful scientists, if asked, can point to revelatory moments that determined the courses of their careers. For Gasser and Erlanger, it would have been the first time that they saw the nerve action currents on their cathode ray tube. This was by design, however, whereas for Adrian the moment came, as it so often does, by chance. It also came at a time when he felt he had exhausted all the possibilities for further research on the nature of the nerve impulse, the legacy inherited from Keith Lucas. On this particular day, he had dissected out the sciatic nerve and calf muscles of a frog and had applied a pair of recording electrodes to the nerve and connected them to the newly constructed amplifier. On turning his attention to the capillary electrometer, he was puzzled to find that the baseline was unusually noisy, as if there was a poor contact somewhere in the recording circuit. Unable to find one, he eventually realized that the "noise" was, in fact, the

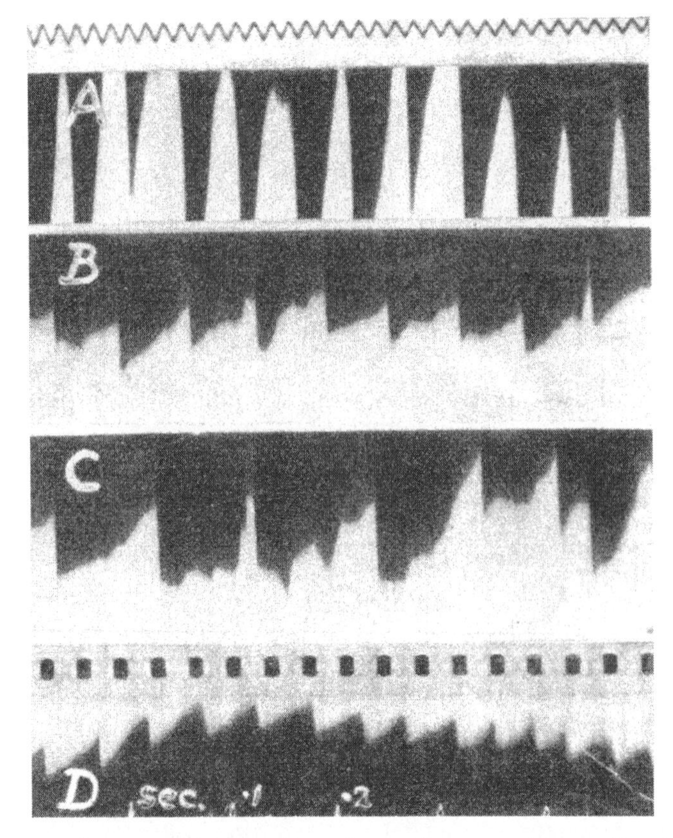

Figure 7–4 Discharges in the fiber of a single stretch receptor in a small neck muscle of the frog, subjected to various loads (1.0 g. 0.5 g, and 0.25 g in *A*, *B*, and *C* respectively). It can be seen that the smaller the load, the less is the discharge frequency. In *D* the impulse discharge quickens as the load is gradually increased. See text for explanation of the rather confusing capillary electrometer photographs. (From Adrian ED, Zotterman Y. The impulses produced by sensory nerve endings. Part 2. The response of a single end-organ. *Journal of Physiology*; **61**: 151–171, 1926. Courtesy of John Wiley & Sons.)

discharges coming from the sensory nerve endings in the muscle that were being stretched by the weight of the muscle belly. Thus, when the muscle was hanging freely, the discharges were present, but when the muscle was supported on a glass plate, they disappeared. In Adrian's own words:

> "The explanation suddenly dawned on me . . . a muscle hanging under its own weight ought, if you come to think of it, to be sending sensory impulses up the nerves coming from the muscle spindles. . . That particular day's work, I think, had all the elements that one could wish for. The new apparatus seemed to be misbehaving very badly indeed, and

> I suddenly found it was behaving so well that it was opening up an entire new range of data . . . it didn't involve any particular hard work, or any particular intelligence on my part. It was just one of those things which sometimes happens in a laboratory if you stick apparatus together and see what results you get."[8]

Quick to seize an opportunity, and working with a visiting scientist from Sweden, the young Yngve Zotterman, Adrian devised an experiment on one of the small muscles in the neck of the frog. The muscle fibers were cut through until only one stretch ending was left attached to the nerve, as in the experiment illustrated in Figure 7–4. It was then evident that the sensory ending in the muscle, the "spindle," discharged impulses in proportion to the extent that the muscle was stretched, firing rapidly at first and then in a steady rhythm (Fig. 7–5).[9] Adrian and Zotterman then examined the sensory nerve endings in the skin, this time using the cat paw, and found that, though the adaptation of the impulse firing rate was more marked than that in muscle sensory nerve fibers, the same rule applied, the slowly adapting pressure receptors firing quickly and in proportion to the strength of the stimulus, and then more slowly.[10] In a third study, Adrian, this time working alone, recorded from nerve fibers likely to be involved in feeling pain.[11] In all three investigations, the published photographs show that the analysis of the capillary electrometer records cannot have been easy, though the application of Keith Lucas's correcting device markedly improved the appearance of the impulses (see Fig. 7–5). Taken together, however, these were the first studies to show how information was sent from sense organs, situated in the periphery, up to the brain for analysis. The general conclusion was that, regardless of the stimulus, all the impulses in a nerve fiber were of the same amplitude and had the same brief duration; they differed only in the frequency with which they were transmitted, and that, in turn, depended on the strength and quickness of the stimulus. Adrian's analogy was with a machine-gun: as long as the trigger was being pulled, the gun would fire a train of bullets, each one exactly the same as the others.

Having demonstrated that information about the skin and the muscles was sent to the brain and spinal cord in the form of a code, in which identical impulses were transmitted along the nerve fibers at different frequencies, he then turned his attention to movements. Was there a similar code for the fibers coming from the motor nerve cells (motoneurons) in the spinal cord, instructing the muscles to contract? First, after carrying out very fine dissection, Adrian and his American colleague, Detlev Bronk, were able to record from single fibers of the rabbit phrenic nerve during normal respiration.[12] Once again, just as with the sensory nerve fiber studies, the impulses were seen to have a constant amplitude, and to vary in their discharge frequencies. Adrian and Bronk then decided to study their own muscle contractions.[13] Rather than attempt to record from the nerve fibers themselves, which would have been very difficult, though not impossible,[14] Adrian and

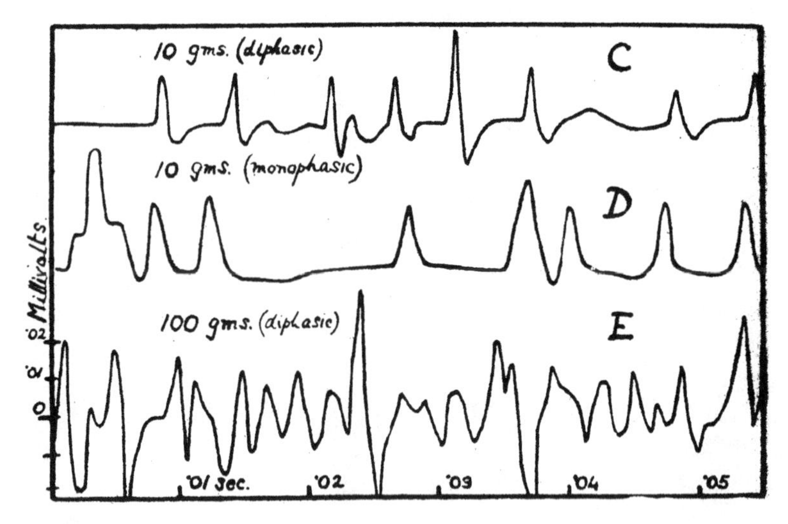

Figure 7–5 Similar experiment to that of Figure 7–4, but with the frog gastrocnemius muscle. This time the capillary electrometer recordings have been corrected by the Keith Lucas device and the nerve impulses are more easily distinguished. The impulse discharge is more rapid with the larger load (100 g) than with the smaller one (10 g). (From Adrian, 1928)

Bronk chose to record from the muscle fibers instead, and to this end they devised a needle electrode they could insert into the muscle belly. Since a colony of muscle fibers was innervated by a single motoneuron (motor nerve cell) in the spinal cord,[15] it was possible, by recording from the muscle fibers, to discover what the motoneurons were doing. Once again, they found evidence of a frequency code. During relaxation, the muscle, and therefore the motor nerve cells, were silent. As the muscle was made to contract weakly, however, a steady discharge of impulses began, which increased in frequency with further effort (Fig. 7–6). As with the sensory fibers in skin and muscle, there were differences in threshold between the fibers, in terms of the forces at which they started firing. Afterwards, Adrian was said to have doubted whether the ability to record muscle impulses with a needle would have any medical application. In this he was wrong, for needle electrodes, built to the same simple design, would enable neurologists to distinguish between those patients with weakness due to inflammation or degeneration of their muscle fibers, and those in whom the motor nerve cells and their fibers were affected. This type of examination became known as electromyography (EMG).

The 1920s were a splendid period for Adrian. Indeed, could there be a better, more satisfying, life than this? To do the experiments during the day, with their appealing mixture of dissection, electronics, and photography. To see the answer immediately with the capillary electrometer, or to hear it in the sounds coming from the loudspeaker. To find out how the different parts of the nervous system worked. To dine, both for lunch and sometimes for dinner in the evening, in the

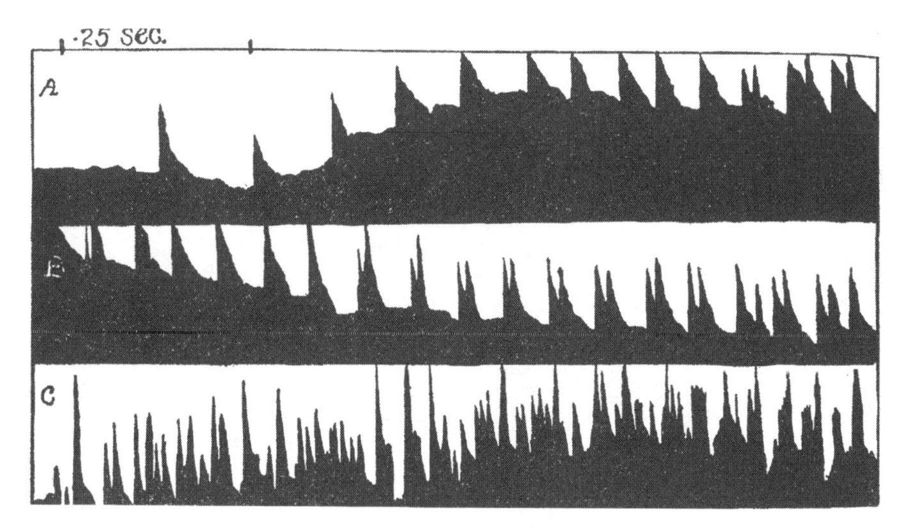

Figure 7–6 Muscle fiber impulses (nowadays referred to as "motor unit potentials") in Adrian's own triceps muscle. As the contraction becomes stronger (from *a*, through *b* to *c*), so the discharges quicken and more fibers become active. (From Adrian ED, Bronk DW. The discharge of impulses in motor nerve fibrers. II. The frequency of discharge in reflex and voluntary contractions. *Journal of Physiology*; **67**: 119–151, 1929. Courtesy of John Wiley & Sons.)

Great Hall of Trinity College, sitting at the High Table with the other Fellows. And there—the old oak tables, the gleaming silverware, and, looking down from the walls, the portraits of Henry VIII, the statesmen, the great scientists and writers, and, of course, the past Masters of the College (Fig. 7–7). For the community of scholars, it was a very different world, a privileged world, in comparison with the one in the country at large. For in the decade following the 1914–18 war, the effects of the lagging British economy were all too evident, with 10 percent of the work force unemployed. There had been a major crisis in May 1926, during the period when Adrian was making his single nerve fiber recordings. Then, the entire country had been brought to a halt by a General Strike, an event precipitated by the determination of the colliery owners to cut back the wages of the coal miners.

Inside the College gates, though, it was a tranquil and beautiful world, and especially so in the spring and summer. The College gardens are then at their best, not with formal arrangements of annual bedding plants, but with luxuriant mixtures of perennials, the clumps of lupins, columbines, Michaelmas daisies, and delphiniums, that have been gradually put together over the years. Contained within each quadrangle is a well-tended lawn, its thick grass kept short and free of weed, the equal of the finest bowling green. At the rear of the Colleges are the "backs," where the land slopes gently down to the river. There on the water are to be seen men in straw hats, propelling their shallow boats by pushing down on the

Figure 7–7 The Great Court at Trinity College, Cambridge. Of the 32 Nobel Prizes awarded to members of the College, 5—including Adrian's—were for studies on the nervous system.

river bed with long wooden poles. And, likely as not, young women reclining in the punts, their faces upturned to the sun, their hands trailing in the water.

In the winter, however, Cambridge can be damp and bitterly cold, with strong winds blowing in from the North Sea across the flat marshland. In an especially cold winter, one in which the dykes had frozen over, Adrian would temporarily abandon the laboratory, and instead, get out his skates, speeding for miles over the long straight stretches of ice, and gliding past the centuries-old villages of the Fens. Adrian was, in fact, an excellent athlete. As an undergraduate, he had been one of those daring enough to climb the college roofs at night, and he had later become an avid mountaineer. He had represented the university in fencing, and he enjoyed sailing. On many of his outdoor ventures, Adrian would be accompanied by his wife, the former Hester Pinsent, a descendant of the philosopher John Hume.

But what was it that made Adrian such a great neuroscientist? Apart, that is, from the enormous intelligence and the probing curiosity? One quality seems to have been an intuition, as if Adrian somehow *knew* how the nervous system functioned. He also seemed to know, instinctively, just as Keith Lucas had known, which animal preparations would be the most favorable for the experiment—not necessarily the usual ones, the cat, dog, and frog, but the more esoteric ones, the hedgehog, the caterpillar, the water beetle, and the conger eel. Another talent was his ability to make his experiments work, whatever difficulties might arise as he went along. In this he resembled Helmholtz and many of the later physiologists. Just as Helmholtz had not hesitated to use bits and pieces of this and that, so in Adrian's cluttered laboratory there would have been a strip of metal or a piece of wire that could be bent into the right shape, a length of string that could tie parts together, and, to hold something in place, a lump of plasticine ("the neurophysiologist's friend," as he would refer to it). Also present in the laboratory, and often indispensable for the experimental apparatus, were pieces of Meccano, a child's construction set. As long as the experimental setup worked, and the results came, it was good enough. Indeed, there was an attitude among the Cambridge scientists, and not just those in physiology, that there was special merit in doing the experiment as simply as possible. Ratcliffe, a physicist at the Cavendish Laboratory, put it best:

> "There was, I think, a feeling that the best science was done in the simplest way. In experimental work, as in mathematics, there was 'style' and a result obtained with simple equipment was more elegant than one obtained with complicated apparatus, just as a mathematical proof derived neatly was better than one involving laborious calculations. Rutherford's first disintegration experiment, and Chadwick's discovery of the neutron had a 'style' that is different from that of experiments made with giant accelerators."[16]

Sometimes, for a year or so, Adrian would have someone working with him, usually a visitor from overseas—Zotterman from Sweden, Bronk from the United States, or Moruzzi from Italy. Such collaborations were not without their tense moments, as Adrian could become irritable under pressure. Zotterman, for example, writing in 1925, described Adrian as:

> ". . . [having] an unheard of capacity for work. All the time he has two lectures a day, besides a mass of private teaching in the College and he has also been temporary editor of the *Journal of Physiology*. He now gives the impression of being somewhat exhausted so that it has been a little difficult to work with him this last week as he can become beside himself if one merely leaves a tap dripping."[17]

After two months, however, which included a winter holiday in Switzerland, Adrian had recovered, and Zotterman was able to write:

> "Adrian who has a very volatile disposition is now very happy again and the work has become very much more pleasant."[18]

By all accounts, Adrian was most content when working by himself, even when the experiments were particularly arduous, sometimes involving the dissection of a large animal. His equipment, too, was mostly of his own making, and, as for other physiologists at the time, the money for the bits and pieces came from the experimenter's own pocket and, in Adrian's case, also from a small grant from the Royal Society.

Although Adrian was a natural "loner," there was one person whose company he always welcomed, and that was the American neurophysiologist Alexander Forbes (Fig. 7–8). They had first met during the period 1910–12 when Forbes had visited Lucas and Adrian for 3 weeks, in between investigating reflexes with Sherrington in Liverpool. On returning to the United States, Forbes had invited Adrian to pay a working visit to Harvard, and Adrian had regretfully declined. The correspondence had resumed after the end of the war, and had resulted in Forbes spending some months with Adrian in Cambridge in 1921, an experience that both men had enjoyed. Forbes had been the first neurophysiologist to use a valve amplifier for recordings of the nerve action potential[19] and had actually preceded Adrian in detecting sensory nerve impulses while stretching a muscle.[20] Already Forbes had shown a gift for originality of thought and for innovative experimental design, and he was a master of electronics. Indeed, in every important respect, he was the equal of his countrymen, Gasser and Erlanger. As Adrian would write later of his friend: "If Alexander Forbes had been less adventurous his many talents might have added an entirely new chapter to neurophysiology."[21]

And therein lay the problem, if problem it was. Alexander Forbes, immensely rich and from an old New England family, had an adventurous spirit that, among other achievements, impelled him to sail his own boat across the Atlantic. At a time when he was visiting Sherrington in the 1930s, Eccles would recall Forbes

Figure 7–8 Alexander Forbes, late in his career. Friends were struck by the resemblance to his paternal grandfather, the philosopher and poet Ralph Waldo Emerson. (Photo by Bachrach)

piloting a rented plane to different parts of Europe at weekends, returning on Monday mornings "full of stories about what he had done, and where he had been, and who he had met, and so was ready for the week's work again."[22] Other than making the first amplified recordings of nerve action potentials, Forbes, under the influence of Sherrington, had studied hindlimb reflexes in the cat, and had demonstrated the importance of seeing the electrical responses of the nerve in addition to those of the muscle. He had also become interested in the slow potential changes that could be detected on the exposed cortex of an animal brain and, in time, would be regarded as a pioneer of electroencephalography (EEG) in humans. With his gifts, Forbes might have done even more, and perhaps it was not merely a matter of adventuresome distractions. It is possible that, on his visits to Cambridge, he was overawed by Adrian's brilliance, for Eccles has drawn attention to Forbes' deferential nature towards his peers.[23] Further, despite his talents in the laboratory, when it came to writing up his work, Forbes lacked the lucidity of Adrian, the skill the Englishman displayed in making a complex story simple and succinct.

A particularly fine example of Adrian's writing style is his first monograph. Appearing in 1928 and entitled *The Basis of Sensation. The Action of the Sense Organs*,[24] it summarized his experimental work and his thinking at the time. Later would follow *The Mechanism of Nervous Action* (1932)[25] and, later still, *The Physical Background of Perception* (1947).[26] Adrian was equally good as a lecturer. He enjoyed it, and would often get up early enough to prepare a demonstration to illustrate his talk. Sometimes he would produce a single nerve fiber preparation, a difficult or impossible task for anyone else but one that Adrian could bring off time after time. During lectures to medical students on salivary secretion, he was known to cannulate his own parotid duct in front of the class and then collect the saliva in a shallow brass cup. Those privileged to hear him would have appreciated the eloquence and the knack of making complex issues appear simple. It was a trait Keith Lucas had also had, and it was one that would become characteristic of the later Cambridge neurophysiologists.

With his razor-sharp mind, his quick movements (he was an expert fencer), and his gift with words, Adrian, in another life, could have been, like his father, a successful barrister. Instead, he reserved his most dramatic moments, not for the courtroom, but for the meetings of the Physiological Society. The number of physiologists had grown since the founding of the Society in 1876, and the time was long past when meetings could be held in private homes. It was at a university meeting in Cambridge in 1933 that Adrian could be seen at his most dramatic, in demonstrating the presence of cochlear microphonics in a dead animal.[27]

In the following year, 1934, Adrian was similarly compelling, this time as the subject for a demonstration with his younger colleague in the department at Cambridge, Bryan Matthews.[28] There had been a report from Germany, some years earlier, that it was possible to detect the electrical activity of the human brain through electrodes applied to the scalp. To many physiologists, however, the

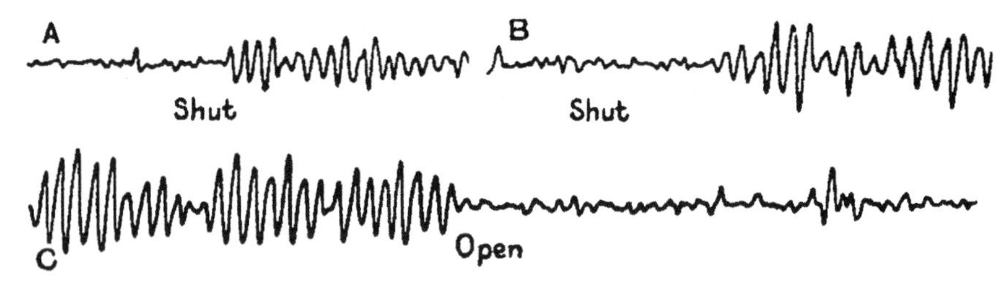

Figure 7–9 EEG recordings made on Matthews' oscillograph. With the eyes closed, the activity ("alpha rhythm") develops promptly in Adrian (*trace A*) and, after a slight delay, in Matthews as well (*trace B*); when Adrian opens his eyes, the activity immediately ceases (*trace C*). (From Adrian & Matthews, 1934)

claim, by Hans Berger, seemed implausible: more likely, the small deflections were an artifact of some kind.[29]

On this occasion Matthews was the demonstrator, and Adrian, who would have had the last word on the planning of the event, was the subject, sitting in a chair with wires running to the saline-soaked pads on his head. With his eyes closed, there were runs of small-amplitude potentials, 10 or so waves a second, at the back of his head. When Adrian opened his eyes, however, the waves—later termed the alpha rhythm—disappeared (Fig. 7–9). Adrian had already made similar recordings from the exposed brains of anesthetized animals, just as Richard Caton had many years earlier, in Liverpool, and George Bishop was then doing in St. Louis. Waggishly, Adrian was heard to remark that his own brain was "almost as good as a rabbit's." The demonstration by Adrian and Matthews was important for, in the space of a few minutes, in front of an attentive and critical audience, Berger's observations were validated, and the serious study of the "brain waves" was launched. It would become the science of electroencephalography, or EEG—still, in the age of functional brain imaging, by far the best way of confirming, or refuting, the diagnosis of epilepsy.

The event sheds an interesting light on the reputation for physiology that Cambridge, by no means through Adrian alone, was gaining. By performing the experiment themselves, the Cambridge neurophysiologists had set a stamp of approval, an *imprimatur*, on work that had already been done in a perfectly competent manner elsewhere. Also noteworthy was the instrument employed in the demonstration: it was an ink-writing device that the ingenious Matthews had designed and built himself. Matthews had also invented an oscillograph, which, unlike the cathode ray tube, was simple and robust, inexpensive to make, and capable of producing superb photographic records. Basically, it consisted of a loudspeaker modified to give a flat frequency response, with a mirror to deflect a light-beam onto photographic paper.[30] Although the apparatus could not respond to very high frequencies like the cathode ray oscillograph, the instrument used by Gasser, Erlanger, and Bishop, it was far superior to the capillary electrometer

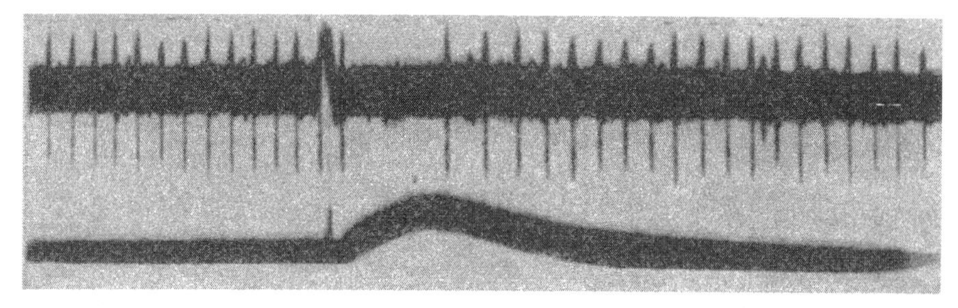

Figure 7–10 *Top trace*: Matthews' recording of the discharge in the nerve fiber of a single stretch receptor (muscle spindle), located in a small toe muscle of the frog. When the muscle is made to contract (*bottom trace*) the discharge momentarily ceases, as there is no longer any stretching of the receptor. Matthews used the oscillograph he had invented and built, and the display of the nerve impulses (vertical lines in the top trace) is far superior to those obtained by Adrian with the capillary electrometer. (From Matthews BHC. The response of a muscle spindle during active contraction of a muscle. *Journal of Physiology*; **72**: 153–174, 1931. Courtesy of John Wiley & Sons.)

Adrian had been so familiar with, and was more than adequate for recording slow potentials, such as the EEG. It also provided beautiful records of trains of impulses, such as those obtained by Matthews himself in a study of single stretch endings in muscle (Fig. 7–10). The Matthews oscillograph was used by others as well, including Hartline in his investigation of the eye of the horseshoe crab, *Limulus*.[31] In addition to the ink-writing device and the oscillograph, Matthews introduced an important modification of amplifier design, the "push-pull" input, which, by reducing the background noise, greatly improved the clarity of the signals.[32]

Supremely confident, a skilful experimenter and a clever inventor, Matthews (Fig. 7–11) was Adrian's natural successor in the school of neurophysiology started by Keith Lucas. As the years passed, Matthews would continue to pick and choose his own research problems, with moments of brilliance, as in the recordings of impulses from single sensory endings in muscles and in the study of slow potentials in the spinal cord. However, in a career that spanned almost four decades, there were only 11 full papers on physiological topics, though, it must be added, there were many interesting notes and communications. In Matthews' later years this paucity of publications may have been related to the adverse effects of his courageous self-experimentation during the 1939–45 war, when he directed physiological research for the RAF.[33] Another reason was that, in common with Adrian's friend and part-time associate Alexander Forbes, Matthews was an adventurer.[34] Like Forbes, he enjoyed ocean sailing and would, whenever possible, arrive at an international physiology meeting in his 56-foot ketch, *Lucrezia*. Typically, when invited to act as head of the department of physiology in Shiraz for a year, he set out for Iran by boat. It was another example of the Cambridge style.

Adrian, in contrast, never wasted a moment. The Physiological Laboratory in Cambridge was the place for work, and his enquiring mind seemed intent in

Figure 7–11 Bryan Harold Cabot Matthews at the helm of his yacht *Lucrezia*. (Photograph by Victoria Hyde, Cambridge)

leaving no part of the nervous system unexplored. So far as sensation was concerned, Adrian gave considerable thought to the ways in which nerve impulses might be initiated in sensory nerve endings. It was one thing to stimulate a nerve electrically, as was so often done in the laboratory, but how did the impulses arise under natural circumstances, in response, say, to a touch on the skin, a stretch of a muscle, or light falling on the eye? He had noticed, at some point or another, that a muscle fiber, after being excised and placed in normal (0.9 percent) saline, became sensitive to mechanical deformation, and would also fire impulses spontaneously. Adrian's speculation was that the deforming stimulus caused a temporary breakdown in the "polarization" of the membrane, and that it would eventually be possible to record such an event taking place in the sense organs.[35]

Much later, well after the Nobel Prize, he would decide to examine the sensory receiving areas of the brain, by mapping out the areas of cerebral cortex responding to stimulation of the various parts of the body surface, as well as those areas involved in hearing and vision.[36] There would also be important studies on the sense of smell,[37] the vestibular apparatus of the inner ear,[38] the eye,[39] and the cerebellum,[40] the highly convoluted part of the brain lying directly below the posterior part of the cerebral cortex. When the Nobel Prize came in 1932, it reached a scientist at the height of his scientific abilities, a pinnacle that, remarkably, would be sustained for another quarter of a century.

Though Adrian was ever-prepared to acknowledge Keith Lucas as his master, as he did in his Nobel lecture, to others it must have seemed that the pupil had, at the very least, become his master's equal. In addition to working out some of the properties of the nerve impulse itself, Adrian had—more than anyone else, before or since—shown what it was that the nerve impulses actually did within the brain and spinal cord.

But what of the other 1932 Nobel Laureate?

8

Excitation and Inhibition

When Sherrington (Fig. 8–1) was awarded the Nobel Prize with Adrian, in 1932, he was 75 years old.[1] He had been born early in the reign of Queen Victoria, before the American Civil War, at a time when Charles Dickens was in mid-career and Charles Darwin had yet to publish *The Origin of Species*. A visitor to the physiology department in Oxford would have found a small elderly man, spectacles perched on his nose, quietly spoken, gentle and courteous in his manner. As with Adrian, there was a reticence, a certain diffidence, toward other people and it was one he had struggled to overcome.

To Sherrington's many admirers, the Nobel Prize had come late. He was, after all, more than 30 years older than Adrian. Yet, if questioned, Sherrington's supporters—and there were many of them, including some who had nominated him unsuccessfully in the past—would have been hard pressed to identify an important discovery made by their hero. Certainly there were no discoveries to do with the nature of the nerve impulse, for there is no evidence of Sherrington ever having recorded a nerve impulse. Even the recording of muscle impulses, a much easier task, did not come until the very end of his extraordinarily long career—and would probably have been at the suggestion of one of his graduate students. Nor had he made any attempt to acquire, or to develop himself, the apparatus necessary to make such recordings. Instead, most of Sherrington's recordings were no more than the contractions of a muscle transmitted by a lever to a smoked drum, with calibration marks added to show the forces developed and the time taken for the response to occur. Only at the end, just before his retirement, did he attempt a more detailed analysis of his findings.

Yet, notwithstanding these shortcomings, Sherrington was regarded as the greatest physiologist of his time. Even Adrian, the brilliant Adrian, whose career sparkled as one success followed another, looked up to Sherrington. Indeed, on learning that he was to share the 1932 Prize with Sherrington, an award for which he had, unknowingly, been nominated by the older man, Adrian had written a congratulatory letter to Sherrington in which he displayed the considerable charm of which he, Adrian, was capable, as well as a rare beauty of phrase:

> "This letter is intended to reach you just as you are starting for Uppsala so as to preclude any reply. I won't repeat what you must be almost tired

Figure 8–1 Sir Charles Scott Sherrington. The photograph was taken when he was head of the physiology department at the University of Liverpool. (Wellcome Library)

> of hearing—how much we prize your work and yourself—but I must let you know what acute pleasure it gives me to be associated with you like this. I would not have dreamt of it, and in cold blood I would not have wished it, for your honour should be undivided, but as it is I cannot help rejoicing at my good fortune."[2]

Why did Sherrington receive this kind of adulation? Why, excluding the obligatory obituary notices in the scientific journals, was Sherrington the subject of three biographies, while Adrian, who had surely discovered much more about the nervous system, had no biographer? There are probably two reasons. The first is that, unlike Adrian, Sherrington seemed to enjoy the company of young graduate students and visiting fellows. Not only did he and Lady Sherrington entertain them at their home—during the Oxford years there was an open invitation to tea every Sunday afternoon for anyone who wanted to come—but he also educated them in matters outside the laboratory. There was poetry and art, medical history and rare books, and a broad range of philosophical issues, all of which would have been discussed without condescension to the younger men. Not surprisingly, they adored Sherrington and, on leaving him, often kept up the interests to which they had been introduced. Two of the biographies were written by his former students.[3]

The second reason for Sherrington's great reputation during his lifetime extended to those outside his immediate circle, those who had read his papers and heard him speak at meetings. Sherrington was a profound thinker with a phenomenal knowledge of the nervous system, a man who, when young, had known most of the greatest medical scientists of his time, and had worked in some of their laboratories. He also had a special way of expressing himself in his writings, not hesitating to coin a term when necessary, and adopting a style both poetic and complex, as in the description of the brain as an "enchanted loom" or a "great revelled knot," and the nerve impulses as "points of light."[4]

Sherrington was more than a scientist and poet; he was also a philosopher, concerned with such issues as the place of humankind in the universe, with evolution, and with the relationship between mind and brain. In the science of the nervous system, he served as a judge, both of his own work and that of other people. It was his opinion that other scientists sought, and like his contemporary, Pius IX, his pronouncements were regarded as infallible. Sherrington did not need a Nobel Prize, nor was the Prize the appropriate recognition for him. A better indication of the respect in which he was held is the knighthood, the Order of Merit, the Presidency of the Royal Society, the 23 honorary degrees, and the many memberships of foreign societies and academies. Whatever the criticisms that can be leveled at his scientific career, it is impossible to ignore Sherrington in any history of neuroscientific research. Almost every contemporary neuroscientist of note came to work with him, and three of his Oxford students went on to win Nobel Prizes themselves.[5] His influence was enormous. Moreover, his life was so varied, so remarkably full and rich, that it begs to be told.

Born in 1857, Charles Scott Sherrington, like Dickens' David Copperfield, spent part of his boyhood in Great Yarmouth. After his father died, however, his mother married a second time and settled in Ipswich, another of the East Anglian seaports. At the Ipswich School, Sherrington received an excellent education, which included Latin and Greek; he also acquired a love of poetry, and, largely through his stepfather, a physician, an exposure to art. After leaving school, Sherrington began medical studies at St. Thomas' Hospital in London. Then, as the family finances improved, he transferred to Cambridge, still an undergraduate, eventually gaining first-class honors in both parts of the Natural Sciences Tripos. While a student, he attended the International Medical Congress in London in 1881 and was able to witness the forceful Anglo–German debate between David Ferrier and Professor Goltz. The latter had presented a decorticate dog as evidence that the control of movements was not localized in the cerebral cortex. Ferrier, by means of a more selective ablation, this time in the monkey, had shown just the opposite, that there was a distinct motor area. It was at this meeting that the great French neurologist Charcot, struck by the resemblance between Ferrier's hemiplegic monkey and his own patients with stroke, had exclaimed, "*C'est une malade.*"[6] Later, still a student, Sherrington had carried out anatomical studies on one of Goltz's dogs, tracing the degeneration of nerve fibers in the spinal cord that followed the cortical extirpation. It was in the Physiological

Figure 8–2 A much younger Sherrington (*right*), photographed with C. S. Roy at the door of the Old Pathology Laboratory, Cambridge, in 1893. Sherrington collaborated with Roy, who was Professor of Pathology at the time, in a classic study of the control of blood flow to the brain. The two of them, with Graham Brown, also collaborated in an investigation of a cholera outbreak in Spain.

Laboratory at Cambridge that Sherrington found his natural home, just as Adrian was to do some 30 years later (Fig. 8–2).

In Sherrington's time, the Physiological Laboratory still consisted of a few small rooms in an old building. But if the accommodation was poor, the quality of the research and teaching was outstanding, largely though the contributions of Walter Gaskell and John Newport Langley, and, of course, the founder of the Laboratory, Michael Foster. Langley had employed histology and physiology to study the secretion of saliva, before passing on to analyze, for the first time, the organization of the sympathetic nervous system. Like Erlanger, a little later, Gaskell had embarked on an investigation of the heartbeat and, among other discoveries, had shown that the cardiac muscle fibers were capable of generating their own rhythm. Both men had acted as Sherrington's mentors, and it was Gaskell who wisely advised him, if he was to continue studying the nervous system, to transfer his attention from the brain to the spinal cord. It was good advice, for the spinal cord was less complex and could be explored through its reflexes. Further, the recordings of muscle contraction could be simply made.

Sherrington continued to have other interests, however, and throughout his long career, papers would appear on such diverse topics as the white blood cells, blood volume, eye movements and vision, various infections, the heart, medical education, fatigue, and the effects of chloroform. Nor did he entirely give up research on the brain, for, apart from the studies on the projection of nerve fibers into the spinal cord, there were several investigations of the effects of electrically stimulating the brain. For these experiments, Sherrington used three species of primate—chimpanzee, gorilla, and orangutan—and was able to define the extent of the motor cortex in each, observing the importance of the representations for movements of the tongue, thumb, and great toe. Another finding, confirmed many times since in the motor cortex, was the inconstancy of the results, such that stimulation of the same point at various times could produce different movements.[7] It was one of the first demonstrations of brain plasticity.

The papers on the structure and degenerative changes of nerve fibers[8] were partly the outcome of Sherrington's initial work on Goltz's decorticate dog, and partly the result of a period of study in Berlin with Rudolf Virchow, the eminent neuropathologist, in 1886. As a member of the Reichstag, Virchow had been able to take Sherrington as a visitor to the German parliament, where he could witness the bizarre appearances, rather like those of a cuckoo clock, of Chancellor Bismarck.[9] While in Berlin, too, he had visited du Bois-Reymond and had heard Helmholtz lecture. By then the old lion was in his seventies, and it may have been that the claws were no longer sharp, for Sherrington remarked on the long and tedious treatment of mathematical problems, the solutions of which were obvious to the audience. This, of the person who had largely invented non-Euclidean geometry! Given his scientific eminence, his position as Rector of the university, and his newly acquired membership of the German aristocracy, it was remarkable that Helmholtz should have bothered to lecture at all. As it was, his public lectures, on the nature of perception, for example, would come to be regarded as models of clear and logical thinking, with the kind of penetrating insights only a practicing scientist could have acquired. Nor would Sherrington prove to be a particularly good lecturer himself, especially in his later years at Oxford and for those who were largely unfamiliar with the subject in hand.[10] Addressing the blackboard rather than his audience, and wandering off the main path of his argument to become lost in byways, the contrast with the clarity of Adrian or Keith Lucas would have been striking. A high proportion of physiology students would not bother to attend Sherrington's lectures, while others, caught unaware, would leave before Sherrington had finished.

Helmholtz's lectures were not the only object of Sherrington's disapproval. There was the attitude of the country as a whole, smug as it was with the devastating victory over its western neighbor in the Franco-Prussian War. It was an uneasy time for those who lived beyond the German borders. Yet whatever shortcomings Sherrington may have detected in the country as a whole, there was much to be learned in Virchow's laboratory, and Sherrington's expertise in histology and pathology, displayed in many subsequent publications, stemmed from his studies in Berlin. There was also time spent with the famous Robert Koch, who, among other achievements, had discovered the tubercle bacillus.

There are two Sherrington stories to do with infection that are quite extraordinary, and that show the young physiologist at his most adventurous. In the first, he had gone to Spain with two colleagues to examine claims for the successful use of a vaccine in a serious outbreak of cholera.[11] It was the same outbreak that had involved Cajal as an investigator for the provincial government of Zaragoza. Finding the vaccine claims to be spurious, and suspected of spreading the disease themselves, the three young Britons were set upon by an angry mob in Toledo, and were in danger of being stoned to death, had it not been for the dramatic appearance in the street of the British Consul, a gigantic Scotsman with a surprising ability to swear loudly in Spanish.

The second story emerged in the form of a letter written to the Treasurer of the Royal Society Club when Sherrington was 90 and in a nursing home. It referred to an incident when Sherrington was, with Armand Ruffer, attempting to raise, in the horse, an antiserum against diphtheria. It is a magnificent account—with a minimum of words the scenes are vividly described as the drama unfolds:

> "Ruffer & I had been injecting the horse—our *first* horse—only a short time. We were badly in the dark as to the dosage to employ, & how quickly to repeat the increasing injections. We had from it a serum partly effective in guinea-pigs. Then, on a Saturday evening, about 7 o'clock, came a bolt from the blue. A wire from my brother-in-law, in Sussex. 'George has diphtheria. Can you come.' George, a boy of 7, was the only child. The house, an old Georgian house, 3 miles out of Lewes, set back in a combe under a chalk down. There was no train that night. I did not at first give thought to horse, & when I did, regretfully supposed it not yet be ripe for use. However I took a cab to find Ruffer. No telephone or taxi in those days—93 or 94. Ruffer was dining out; I pursued him & got a word with him. He said 'By all means you can use the horse, but it is not yet ripe for trial.' Then by lanternlight at 'The Brown' I bled the horse, into 2 litre flask duly sterilised & plugged with sterile wool. I left the blood in ice for it to settle. After sterilising smaller flasks & pipettes & some needle-syringes I drove home, to return at midnight, & decant the serum, etc.
>
> By the Sunday morning train I reached Lewes. Dr Fawcett of Lewes—he had a brother on the staff at Guy's—was waiting in a dog-cart at the station. I joined him carrying my awkward packing of flasks, etc. He said nothing as I packed them in but, when I had climbed up beside him, he looked down & said, 'You can do what you like with the boy. He will not be alive at tea-time.' We drove out to the old house; a bright frosty morning. Tragedy was over the place—the servants scared & silent. The boy was very weak; breathing with difficulty & he did not seem to know me.

> Fawcett & I injected the serum. The syringes were small & we emptied them time & again. The Doctor left. I sat with the boy. Early in the afternoon the boy seemed to me clearly better. At 3 o'clock I sent a messenger to the Doctor to say so. Thenceforward progress was uninterrupted. On Tuesday I returned to London, & sought out Ruffer. His reaction was that we must tell Lister about it. The great surgeon (not Lord Lister then) had visitors, some Continental surgeons, to dinner. 'You must tell my guests about it.' he said. & insisted—so we told them in the drawing-room, at Park Crescent. The boy had a severe paralysis for a time. He grew to be 6 ft & had a commission in the 1st World War."[12]

If Sherrington had done nothing else in his professional life, this first successful use of diphtheria antiserum (antitoxin) would have ensured him a place in medical history. The work was carried out while Sherrington was the Physician-Superintendent of the Brown Institute in London, an establishment that had a mandate to improve the health of animals and humans. Prior to this, Sherrington had been a lecturer in physiology at his alma mater, St. Thomas's Hospital. Later, in 1885, came an invitation to the chair of physiology at University College, Liverpool, and it was there, in Liverpool, that much of his finest work was carried out. One of his predecessors was Richard Caton, who, so far as is known, was the first, in an animal, to study the spontaneous electrical potentials at the surface of the brain, the recorded activity subsequently referred to as the electroencephalogram (EEG). Caton had done this 40 years before Berger succeeded in making similar recordings in humans. During his own time as Professor in Liverpool Sherrington's reputation had grown, and in the 19-year period a succession of students and postdoctoral fellows, including Alexander Forbes and others from overseas, came to work with him.

Not all the visitors had favorable opinions of Sherrington, at least initially. In 1901 the young Harvey Cushing, in whose capable hands the future of American neurosurgery would come to rest, confided to his diary that Sherrington was attempting too much and too quickly, and made insufficient notes of his experiments. Further, Sherrington's daily appearances in the laboratory after 10 a.m. were unseemly late, especially to a surgeon who would already be well into his work in the operating theater or in the hospital wards by that time. At the end of the day, however, Sherrington would linger on, "puttering" around. It was not so much that Sherrington was a great physiologist, thought Cushing, but that the work of his predecessors had been so uniformly poor.[13]

Cushing's view was that of a minority, and even Cushing came to have second thoughts. Meanwhile, Sherrington's reputation continued to grow. He was invited to give the Silliman Lectures at Yale in 1904, and these were subsequently published as *The Integrative Action of the Nervous System*, a book that quickly became a classic.[14] At one point, Sherrington accepted, in writing, an invitation to become the head of physiology at the University of Toronto. After a sleepless night, he changed his mind and then spent the next two days sifting through the transatlantic mail in the main Liverpool Post Office before retrieving and destroying his letter.

In 1914 Sherrington did move, this time to accept the Waynflete Chair of Physiology at Oxford. His immediate predecessor, as at Liverpool, was Francis Gotch, who, with Burch, had used the capillary electrometer to search for impulse activity in the nervous system and who, with his brother-in-law, Victor Horsley, had been the first to record the nerve impulse.[15] During most of Sherrington's time at Oxford, the laboratories remained poorly equipped, with a hazardous AC mains supply. Stimuli were delivered from a du Bois-Reymond induction coil at times determined by the manual release of a heavy pendulum, which tripped switches as it swung down. Muscle contractions were recorded by a lever writing on a smoked drum, the same instrument that Carl Ludwig had invented in the middle of the previous century. Eventually Sherrington was given a valve stimulator, to replace the induction coil, by A. V. Hill, and at the end there was an improved myograph, with a mirror attached to a torsion wire, deflecting a beam of light onto a camera plate. It was the invention of Sherrington himself.[16] Eventually, too, there was a string galvanometer to record the muscle impulses. But all of this came late, almost too late.

If there were no major discoveries, what were Sherrington's achievements in the laboratory? One of them had to do with the muscle spindles, the stretch receptors that gave Adrian his first recordings of naturally occurring nerve impulses. The muscle spindles had been described before Sherrington turned his attention to them, but it was he who showed that they were sensory in nature[17,18] and that, when stretched, they evoked the "tendon" reflexes employed by neurologists in their examinations of patients. More than this, the spindles induced the reflex muscle contractions needed to maintain normal posture, and, in patients with damage to the brain or spinal cord, they initiated the exaggerated reflexes responsible for spasticity.[19] Sherrington also helped the neurologists by mapping out the areas of skin supplied by different segments of the spinal cord.[20]

Sherrington was by no means the first to make a serious study of reflexes, for Leonardo da Vinci had done that long before, in showing the repertoire of movements that could still be induced in decapitated frogs. Around 1730, Stephen Hales, the enterprising vicar of Teddington, near London, had gone a step further in demonstrating that the reflexes were lost if the spinal cord was destroyed. Since then, there had been important contributions by others, among them François Magendie in Paris, who, in the early part of the 19th century, had demonstrated that sensory information entered the spinal cord in the dorsal nerve roots, while the ventral roots conveyed messages out to the muscles.[21] A few years later, Marshall Hall, a successful London physician and physiologist, was able to write, on the basis of his own findings: "The spinal cord is a chain of segments whose functional units are separate reflex arcs, which interact with one another and with the higher centres of the nervous system to secure coordinated movement."[22]

Hall had reached his conclusions after a series of experiments in which he had carefully observed the movements that could still be elicited in a variety of decapitated animals—hedgehogs, turtles, snakes, frogs, eels, and others. After Hall's work, however, little of significance was published on reflexes until Sherrington entered the field towards the end of the 19th century.

A keen observer, Sherrington was able to show, in the cat and dog, some of the important features of reflexes, such as the ability of a stimulus to the skin to cause contraction of one group of muscles and simultaneous relaxation of another, opposing, group.[23] He also showed how the effects of two weak stimuli could add together, postulating that the first stimulus had induced a "central excitatory state" in the spinal cord, on which the effect of the second stimulus was superimposed (Fig. 8–3). Each stimulus, he realized, not only was causing some motoneurons to discharge but was producing subthreshold responses in others. The motoneurons with subthreshold responses formed "subliminal fringes." Where the respective subliminal fringes overlapped, additional motoneurons would discharge. This combining of subthreshold responses he termed "spatial summation" (Fig. 8–4).

By analogy with the excitatory effects, those stimuli that caused muscle relaxation, Sherrington reasoned, had induced a "central inhibitory state."[24] It is a reflection of Sherrington's conservative attitude to methodology that, at the time when he was still limited to recording the reflex contractions of muscles on a smoked drum, Alexander Forbes was able to record the associated impulse activity in animals and, perhaps even more remarkable, the young German, Paul Hoffmann, was studying the electrical responses in human subjects.[25]

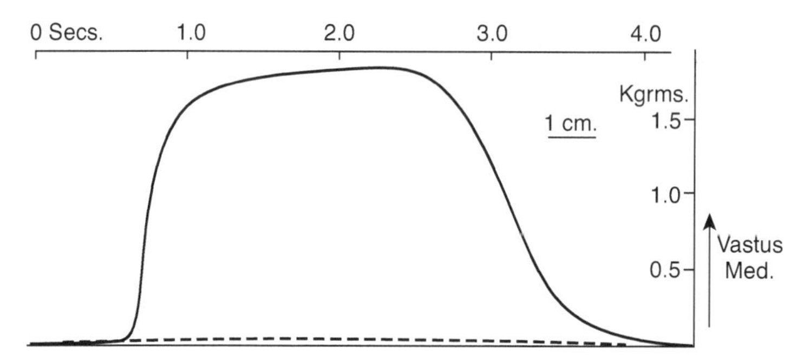

Figure 8–3 Spatial summation, illustrated by two of Sherrington's graduate students, John Eccles and Ragnar Granit. The figure, which has been modified, shows the reflex contractions of the cat medial vastus muscle (a knee extensor muscle) recorded on a kymograph drum. When the two muscle nerves are stimulated separately in the opposite leg, the reflex contractions are barely detectable (*dashes just above the baseline*). When the two nerves are stimulated at the same time, however, a large contraction results (*solid line*) because of overlapping subliminal fringes in the spinal cord. See text and Figure 8–4. (From Eccles JC, Granit R. Crossed extensor reflexes and their interaction. *Journal of Physiology*; **67**: 97–118, 1929. Courtesy of John Wiley & Sons.)

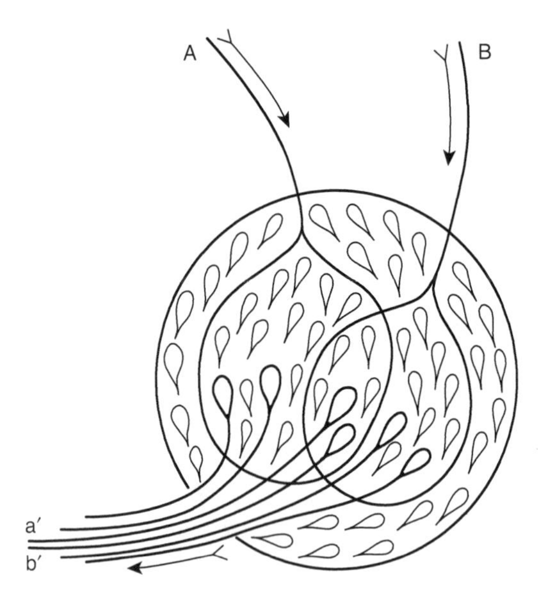

Figure 8–4 Subliminal fringe phenomenon. The large circle contains a population of motoneurons, none of which are excited sufficiently to discharge when each of the two sensory nerves, *A* and *B*, are stimulated separately; instead, the two nerves produce subliminal fringes of excitation (*smaller circles*). When the two nerves are stimulated together, however, some of the motoneurons, *a′* and *b′*, discharge because the overlap of the two subliminal fringes provides enough excitation. This figure was conceived by Sherrington and is taken from Fulton (1938).

Towards the end of his career, Sherrington became interested in the numbers of motor nerve fibers supplying various muscles, and he termed each nerve fiber and the colony of muscle fibers that it supplied a "motor unit" (Fig. 8–5).[26] The concept of the motor unit, though not the term itself, was one that Keith Lucas had developed some 20 years before, by virtue of his experiments with graded stimulation of a muscle nerve.[27] Sherrington set out to estimate the number of motor units in a muscle. With his superb dissecting skills, he was able to cut the sensory nerve fibers in the dorsal nerve roots, before they entered the spinal cord, so that they degenerated over the next few days, leaving the motor nerve fibers to be counted in the nerve to a particular muscle. The histological preparation, entailing the cutting and staining of thin sections of nerve, he did himself, often taking the specimens home and working in the bathroom. The counting of the nerve fibers he also saw as his responsibility.

During this last phase of his career, his closest colleague in the laboratory was a young Australian from Melbourne, John (Jack) Eccles. Prior to his arrival in Oxford, Eccles had never carried out a single experiment, but he soon showed a flair for this type of work. It was probably Eccles who persuaded Sherrington to map out the time-courses of the central excitatory and inhibitory states. To do

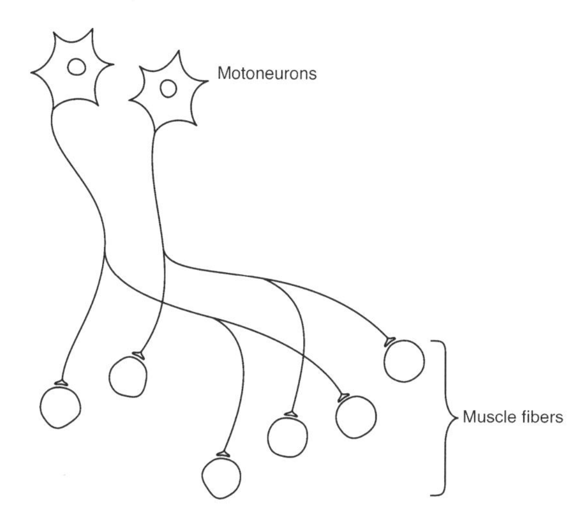

Figure 8–5 Sherrington and Keith Lucas's concept of the "motor unit." Each of the two units shown consists of a motor nerve fiber and a colony of muscle fibers—in reality, a single motor unit usually contains a hundred or more muscle fibers rather than the three shown here. It was shown much later that the muscle fibers of different motor units are intermingled (e.g., Edström & Kugelberg, 1968).

this, it was necessary to deliver a second stimulus at different times after the first, "conditioning," stimulus. To see the effect of an inhibitory stimulus, there had to be a background of excitation set up by the initial stimulus (Fig. 8–6). As with all Sherrington's reflex studies, the effects consisted of a change in muscle contraction. Only at the very end of his long career, and probably at the suggestion of Eccles, were the electrical responses of the muscles recorded as well. The new experiments revealed that the excitatory and inhibitory changes in the spinal cord, following the arrival of an impulse volley, persisted for 10 or more milliseconds.[28,29]

But what sort of changes could they be? For that matter, what happened when the electrical impulses reached the ends of the nerve fibers? When Sherrington had started his career it was still unknown whether all the nerve fibers in the spinal cord and brain were fused together in a syncytium, or whether each nerve cell was a separate entity. The fact that when a nerve fiber was cut, the degeneration of the distant part stopped at a certain point suggested that the nerve cells were separated from each other. The best evidence, however, had come from the nerve cells so beautifully stained by Ramon y Cajal, the same Cajal who had stayed with Sherrington in 1904. Under the microscope each stained cell, with its cell body, its branching dendrites, and its projecting axon, appeared to be a complete, individual entity. It was Sherrington who proposed that the small gap that must therefore exist between a nerve fiber and the next nerve cell in a pathway should be referred to as a "synapse." If a nerve fiber produced an excitatory effect,

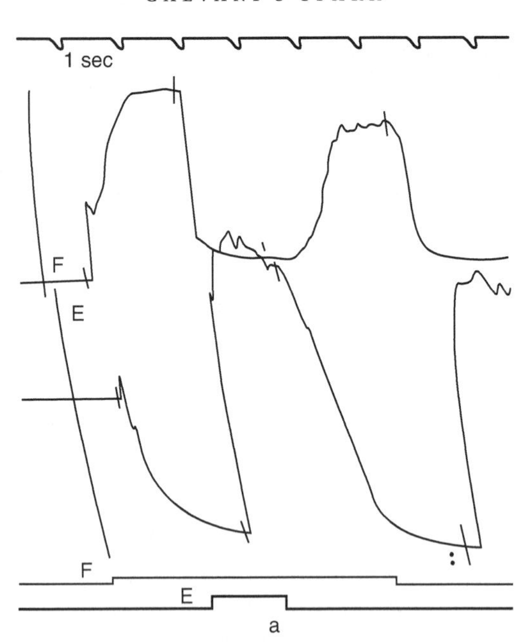

Figure 8–6 Reciprocal inhibition. Recordings are being made on a kymograph drum of the contractions of two cat leg muscles, a knee extensor (*second trace from top*) and a knee flexor (*third trace from top*). Stimulation of the peroneal nerve in the same leg (*fourth trace from top*) produces reflex contraction in the extensor and inhibition in the flexor. When the peroneal nerve in the opposite leg is stimulated (*bottom trace*), the pattern reverses—the extensor is inhibited and the flexor excited (through their motoneurons). (From Sherrington, 1913)

he reasoned, then it must do so by reducing the membrane polarization of the second nerve cell. Conversely, if there was an inhibitory action, it would be produced by a stabilization of the membrane polarization of the other nerve cell.[30]

When Sherrington gave his Nobel address, it was his work on inhibition to which he mostly referred. Eccles had helped to prepare Sherrington's nomination for the prize, which was submitted by Graham Brown, a senior neurophysiologist who had himself made important contributions to reflexology, especially in relation to walking. Graham Brown knew Sherrington well, having collaborated with him both on the motor cortex and on reflexes shortly before the outbreak of the 1914–18 war. Graham Brown also exhibited the same physical recklessness of Sherrington and some of the other early neurophysiologists, having, in his youth, pioneered a new ascent of Mont Blanc.

When Sherrington received his Nobel Prize, in 1932, he was still active in the laboratory and had shown himself capable of dissecting out single nerve fibers no more than 20 microns (thousandths of a millimeter) across. With his new, sensitive myograph, he could have gone on to study the contractions of individual motor units, anticipating the results of others 30 years later. It seemed, though, as if the old gentleman, the very incarnation of physiology, was finally losing his zeal for the laboratory. Eventually, too, there comes a time in the life of a scientist, especially in the biological sciences, when the mind, though still willing, is hampered by eyes that can no longer see detail, and by hands that find delicate manipulations difficult. Experiments that once seemed easy now take longer. Having returned to Oxford, after an enforced absence in a spa for treatment of his rheumatism, Sherrington realized that the work was proceeding perfectly well without him, and so he voluntarily retired. His career, however, was far from over, for in his retirement he wrote what may well be the most imaginative and deep-searching book ever written on the brain, *Man on His Nature*.[31]

Sadly, with Sherrington's departure, the work of the Oxford School came to an end. Eccles, clearly a star in the making, decided to return to Australia. Derek Denny-Brown, a student from New Zealand, who had carried out equally fine work with Sherrington, chose to continue his neurological studies and did so at the National Hospital for Nervous Diseases in London, eventually becoming a world leader in his clinical specialty. The Waynflete Chair of Physiology, which Sherrington had occupied with such distinction for a quarter of a century, passed to the most senior of the younger reflexologists, E. G. T. Liddell.

As Sherrington entered retirement in Ipswich, the East Anglian town of his boyhood, elsewhere a solitary flame, all that was left of a fire that had once burned so strongly, flickered and went out. On October 17, 1934, Santiago Ramon y Cajal, acclaimed as the greatest neuroanatomist in the world, died. His death came at a time when his beloved country, Spain, was teetering on the brink of anarchy, with political assassinations, strikes, and local insurrections. Yet despite the desperate times, tens of thousands of countrymen, citizens of every social class, lined the streets of Madrid as the funeral cortège passed by. Backwards the country may have been in many respects, but it was enlightened enough to give the death of a scientist the respect normally reserved for a monarch.

The last years had not been unkind to Cajal. Though enfeebled, he had still managed to carry on his microscopical studies, often preferring to work in his own home rather than in the new Institute that had been created in his honor. And each day, difficult though it was for him to walk, there would be a meeting and vigorous discussion with his students in one of his favorite cafés.

When by himself, however, there must have been many times when he looked back on his extraordinary life, a life that had taken him from a rebellious youth in an obscure Pyrenean village to international glory. There would still have been doubts, for when he had written his memoirs 30 years earlier, Cajal had wondered whether he and his work would be remembered. The answer was not long in coming. There had already been the new Institute, and after his death, streets and squares would be named after him in cities and towns throughout the country. As to his work, it formed the structural basis on which the neurophysiologists could mount their studies of function. Not the least important question was what happened when the nerve impulses arrived at the endings of the fibers, for one of Cajal's many accomplishments had been to show that the nerve cells were physically separate from each other, as well as from tissues such as skin, blood vessels, and muscle. In the spinal cord it was events in the narrow gaps between the nerve cells—the "synapses"—that must have been responsible for Sherrington's central excitatory and inhibitory states. But what were these events?

The study of the nerve endings would become as important a topic for investigation as the nature of the nerve impulse itself.

Cambridge, August 1933[1]

It was the first time he had been to a meeting of the Physiological Society. The morning's communications had been on a variety of topics, some of them interesting but many of them not. On the whole it was those on the nervous system that had appealed to him. Now, lunch over, it was time for the demonstrations.

He checked the number in his program against that beside the open doorway. There was already a small crowd inside the laboratory, their heads turned towards the dapper figure in the white coat. Rather diffidently, the youth stepped into the room. The man in the white coat appeared not to have noticed his entry and carried on talking to his audience. As he moved farther into the room, the youth saw that there was a cat lying on the table in front of the man. Not breathing, it was obviously dead. Why? What could one possibly demonstrate on a dead animal? And what could be the purpose of the wires running from the head of the cat to the racks of complicated electronic apparatus behind the table?

The slight figure in the white coat glanced at the newcomer, gave a brief smile of recognition, and quickly bent over the table. He then began to whisper in the cat's ear. To the astonishment of those in the room, the man's voice, captured and converted into electric signals by the inner ear of the dead animal, boomed out over the loudspeaker:

"A student appears mystified at what he is seeing and hearing."

Alan Hodgkin, 19 years old and about to start his second year of undergraduate studies at Cambridge, had just seen a brilliant demonstration of the cochlear microphonics. It had been given by the great Adrian, the winner of the previous year's Nobel Prize in Medicine or Physiology.

9

The Messengers[1]

"His temperament was too well adjusted to let him behave unreasonably or unkindly. The four humours, in fact, were so nicely balanced that he was neither too sanguine, too choleric, too melancholy, nor too cold and phlegmatic."

Lord Adrian on Sir Henry Dale[2]

The puzzle as to how nerve fibers produced their effects, once the impulses had reached the endings, was one that had attracted interest well before Sherrington received his Nobel Prize.[1] Du Bois-Reymond may have been the first to appreciate that there seemed to be only two possibilities. The discoverer of the action potential, the *negative Schwankung*, had touched on the matter in his influential 1848 book, *Untersuchungen uber thierische Electricitat* (An Understanding of the Theory of Electricity), and 30 years later he would write:

> "Of known natural processes that might pass on excitation, only two are, in my opinion, worth talking about: either there exists at the boundary of the contractile substance a stimulatory secretion in the form of a thin layer of ammonia, lactic acid, or some other powerful stimulatory substance; or the phenomenon is electrical in nature."[3]

That had been in 1877, and in Berlin. In Paris Claude Bernard had shown that the poison curare could abolish muscle contractions, following nerve stimulation, without interfering with the excitability of the muscle fibers themselves. It was an observation that strongly suggested the normal presence of a chemical intermediary between the nerve endings and the muscle fibers—but did not prove it. In Cambridge there had been Langley and Elliott. The latter, while still a research student, had worked on the effects of adrenal gland extracts on various tissues and on the ability of the ergot alkaloids to block these actions. Since the effects of the extracts, and of adrenaline in particular, resembled those of stimulating the sympathetic nerves (see below), and since ergot blocked the sympathetic nervous system too, it was natural for Elliott to speculate, in 1901, that "Adrenaline might then be the chemical stimulant liberated on each occasion when the impulse arrives at the periphery."[4] A few years later, in 1906, at the time that he was the new head of the Cambridge Physiological Laboratory, Langley reached similar conclusions about chemical transmission, the evidence coming from his own work with nicotine and curare. With the neuromuscular junction in mind, he wrote:

> "The stimuli passing down the nerve can only affect the contractile molecule by the radical which combines with nicotine and curare. And this seems in its turn to require that the nervous impulse should not pass from nerve to muscle by an electrical discharge, but by the secretion of a special substance at the end of the nerve."[5]

There were, however, dissenting voices at Cambridge, in the persons of Adrian and Lucas. So strong was their work on the conduction of the nerve impulse that it was only natural for them to wonder if some of their findings might have relevance to the neuromuscular junction and to synapses in the brain and spinal cord. They had shown, in the first paper they wrote together, that an impulse, although blocked in a region of weakened conductivity (by a narcotic, for example), nevertheless caused a temporary increase in excitability. That this was so, they showed by delivering a second stimulus to the nerve, one that by itself was too weak to initiate an impulse but that was now effective. They speculated that the same sort of summation of subthreshold events might occur at the neuromuscular junction and at synapses in the central nervous system. There was no need for them to postulate any special properties for these regions other than those to be found in peripheral nerve. That had been in 1912.[6]

Despite the work of Elliott and Langley in favor of chemical transmission, it was the electrical theory, supported by the musings of Adrian and Lucas, that came to dominate subsequent thinking on the matter. And why not? If a nerve impulse were to release a chemical that excited a muscle fiber, there were formidable logistics that would have to be overcome. There would have to be exactly the right amount of substance liberated, just enough to cause a single impulse in the muscle fiber. Further, the substance would have to act very quickly, in less than a millisecond, and then be destroyed equally rapidly, so that its effect was not prolonged. To most physiologists, it all seemed highly unlikely. In contrast, there was no doubt about the existence of electrical currents in nerve and muscle: people had been recording them with their capillary electrometers and string galvanometers for years. In Paris Louis Lapicque had gone so far as to suggest that the muscle fiber responded to the nerve because of a match between the excitable properties of the two types of membrane.[7] On the other hand, it was difficult, if not impossible, to explain how the slowing of an animal heart, produced by electrical stimulation of the vagus nerve, could outlast the period of stimulation by several seconds.

The controversy between the two rival ideas, chemical and electrical transmission, took a very different turn after—improbable though it might appear—Otto Loewi had his famous dream.

Figure 9–1 Otto Loewi, shown also with some of his research assistants. The double-drum kymograph, with its considerable length of smoked paper for recording, was commonly used by pharmacologists, whose experiments tended to be long and complex. (Wellcome Library, London, and Österreichische Nationalbibliothek)

Otto Loewi (Fig. 9–1) was Professor of Pharmacology at the university in Graz, the second largest city in Austria. He had been born in Frankfurt-am-Main in 1873, attending high school in that city and then going to Munich, and later Strassburg, to study medicine. After an indifferent start as a student, he had become enthusiastic about the scientific possibilities of medicine, especially after seeing, as an alternative, the dismal results of treating patients with tuberculosis. Loewi's first academic appointments were in pharmacology, first in Marburg and then in Vienna. He was soon able to demonstrate his versatility, his initial papers dealing with the metabolism of carbohydrates and then of proteins, followed by publications on kidney function and the actions of diuretics. Then came more studies of carbohydrate metabolism, this time in dogs that had been made diabetic through removal of the pancreas. Finally, he began an examination of the "autonomic" nervous system, the collections of nerve cells and linking fibers that, automatically and mostly subconsciously, controlled such varied body functions as the heart rate, blood pressure, temperature, sweating, intestinal motility, and the size of the pupil. It was already known that the nerve cells and fibers of this system could be divided, on the basis of their contrasting actions, into two categories—sympathetic and parasympathetic. For example, stimulation of the vagus, the main nerve in the parasympathetic nervous system, caused the heart rate to slow. Conversely, stimulation of the sympathetic nerves to the heart caused it to speed up. The distinction between the two systems of fibers, sympathetic and parasympathetic, had been made by Gaskell and Langley in the years following their appointments to the fledgling Physiological Laboratory in Cambridge.

In his first study of the autonomic nervous system, carried out while he was still in Vienna, Loewi was able to show that the responses of various tissues to stimulation of the sympathetic nerve fibers could be strengthened by small doses of cocaine. However, as already noted, the great problem with all the investigations of the autonomic system was that no one knew for sure how the effects on the different body tissues were brought about. Unlike the situation for the neuromuscular junction, where it was thought that electrical currents provided the linkage, there was the possibility, from Elliott and Langley's work, that the sympathetic and parasympathetic fibers released chemicals that served as the necessary messengers. But that was as far as it went, merely an idea—until Loewi had the first of his dreams.

The dream came in 1921, by which time Loewi had been a professor in Graz for 12 years. By Loewi's standards, they had not been particularly productive years for research, since of necessity there had been much teaching to carry out, befitting the newly appointed department head. However, the situation was about to change dramatically. In his first dream, Loewi was shown an experiment that would enable him to find out whether nerve fibers could release chemicals. Briefly conscious, Loewi wrote down the details and immediately fell asleep again. On waking at his usual time, Loewi reached for the paper, only to discover, to his consternation, that his writing was indecipherable. There was nothing that could be done except, perhaps, to hope that the dream would come again. Astonishingly, it did. This time Loewi dressed hurriedly, dashed to the laboratory, and carried out the experiment of the dream.

A glass cannula was inserted into the ventricle of a frog heart and used to perfuse the heart with a saline solution. The vagus nerve, the route by which the parasympathetic system controlled all the organs in the chest and abdomen, was then stimulated. As expected—for this part of the experiment had been performed in many laboratories before—the heartbeat slowed and became weaker. Now came an innovation of sheer brilliance. It was so simple. Some of the perfusion fluid was removed and pipetted into the cannula of a second heart. The second heart, in turn, began to beat less forcefully. The only possible explanation was that the vagus nerve of the first heart, on being stimulated, had released a chemical that had gone on to affect the second heart.[8] It was the kind of experiment that should never have worked. As Loewi himself put it:

> "On mature consideration, in the cold light of the morning, I would not have done it. After all, it was an unlikely enough assumption that the vagus should secrete an inhibitory substance; it was still more unlikely that a chemical substance that was supposed to be effective at very close range between nerve terminal and muscle be secreted in such large amounts that it would spill over and, after being diluted by the perfusion fluid, still be able to inhibit the heart."[9]

Loewi's paper occupied a mere four pages in *Pflüger's Archiv*, at a time when scientific papers tended to be much longer and more discursive than they are now.

Moreover, the experiment itself was not totally convincing. For one thing, the effects were small, and the muscle fibers in the second heart would have been affected mechanically by the addition of fluid to the perfusate. Also, the electrical shocks would have stimulated the sympathetic (excitatory) nerve fibers as well as the vagal inhibitory fibers. Rather surprisingly, it was five years before the experiment was repeated and improved upon, by Kahn in Prague.[10] This time the two hearts were perfused from the same, Y-shaped, cannula (Fig. 9–2). The arrangement allowed a chemical released from the first heart to diffuse to the second heart without any external manipulation being required. Just as in Loewi's experiment, vagal stimulation of one heart was followed—a minute or so later—by weakening of the contractions in the second heart (Fig. 9–3).

Not knowing the identity of the inhibitory chemical, Loewi simply referred to it as *Vagusstoff* (vagus substance). He and his collaborators in Graz went on to show that the action of the Vagusstoff was abolished by atropine, a poison produced by the deadly nightshade plant. Of all the chemicals that were known to

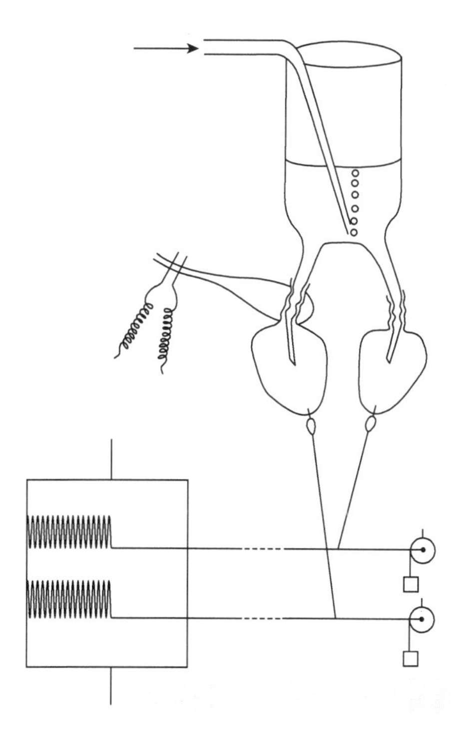

Figure 9–2 Kahn's improved version of the Loewi experiment. The vagus nerve to one of the two frog hearts is stimulated, and if a chemical transmitter is released by the nerve endings, some will diffuse through the perfusing fluid in the Y-tube to the second heart. (From Kahn RH. Über humorale Übertragbarkeit der herznervenwirkung. *Pflügers Archiv fur die gesamte Physiologie des Menschen und der Tiere*; **214**: 482–498, Figure 1, 1926. With kind permission of Springer Science + Business Media.)

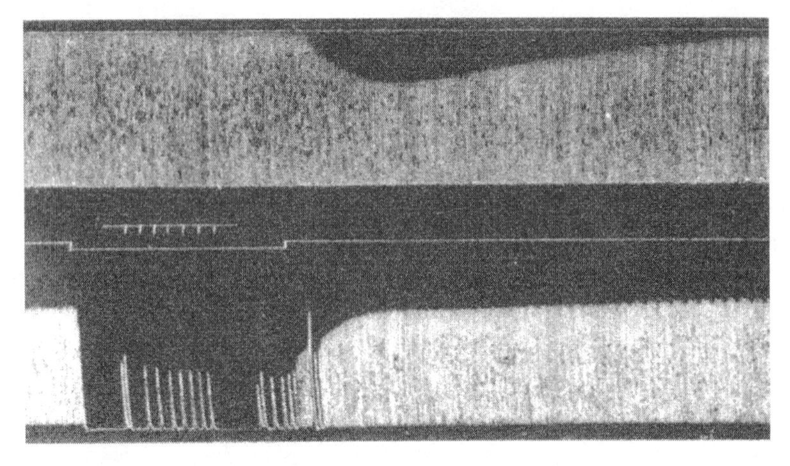

Figure 9–3 Effect of vagal stimulation on a second perfused heart. Experimental setup as in Figure 9–2. The slowing and decreased force of the heartbeat is very marked in the directly stimulated heart (*lower trace*). After an interval of more than 2 minutes, the contractions of the perfused second heart (*upper trace*) begin to weaken (though not to slow) due to the acetylcholine in the perfusate. (From Kahn RH. Über humorale übertragbarkeit der Herznervenwirkung. *Pflügers Archiv fur die gesamte Physiologie des Menschen und der Tiere*; **214**: 482–498, Figure 5, 1926. With kind permission of Springer Science + Business Media.)

affect the body tissues in the same way as the parasympathetic nerve fibers, there was only one that had a similarly brief action and was also inhibited by atropine—acetylcholine. Thus, through a dream, this substance, acetylcholine, was the first chemical transmitter to be identified in the nervous system. It was a major discovery. Loewi next used the same perfused frog heart preparation to investigate any substance that might be liberated by the sympathetic nerve fibers, the fibers that, when stimulated, caused the heart to speed up. In this way he was able to identify adrenaline as the sympathetic messenger.[11]

There were, of course, other physiologists and pharmacologists carrying out related work at the time of Loewi's unexpected triumphs. The most distinguished of these, at least in Britain, was Henry Dale (Fig. 9–4),[12] whom Loewi had met in London in 1902, at a time when the two young men were on the thresholds of their careers. Dale, like Loewi, had trained in medicine and then been diverted into physiology, in this case by his earlier exposure to the gifted group of scientists that Michael Foster had assembled in Cambridge. When he met Loewi for the first time, Dale had just started a two-year research studentship at University College London,[13] where he was working under the direction of Ernest Starling and William Bayliss. The two professors, who were brothers-in-law, had together been the first to isolate and purify a hormone.[14] Dale and Loewi had liked each other immediately and would remain friends for the rest of their long lives.

Figure 9–4 Henry Hallett Dale (later Sir Henry), at his desk in the National Medical Research Institute. Dale was not the only physiologist to smoke a pipe, as is evident from the photos of Langley, Lucas, Erlanger, and Bishop in this book! (Wellcome Library, London)

After leaving University College London, Dale had been approached by Henry Wellcome,[15] the co-founder of the Burroughs Wellcome pharmaceutical business, and appointed to the company as a research pharmacologist. This, despite his protestation that he "had never heard a lecture, or read a textbook, on pharmacology; and that, in fact, had nothing to offer!"[16]

Following his appointment Dale soon showed his ability, his skill in the laboratory being supplemented by a highly critical mind and, especially, by remarkable insights. Indeed, several years before Loewi's dreams and after an exhaustive analysis of the potent substances produced by the ergot fungus,[17] Dale suggested that one of them, acetylcholine, might be the substance released by the parasympathetic nerve endings. He went a step further, by speculating that acetylcholine could also be the transmitter at the synapses in the sympathetic ganglia.[18] In 1914 Dale had moved to the National Institute for Medical Research in Hampstead, London, where he was first the head of the biochemistry and pharmacology program and then the Director of the Institute. The Institute had originally been a tuberculosis hospital, and Dale had converted one of the wards into a large laboratory for his own use, and for that of his associates. There he continued a varied research program, which included definitive studies on the role of histamine in anaphylactic shock. Interestingly, this interest had started by accident, when he was mistakenly supplied with a guinea pig sensitized to horse serum. But so far as research on chemical transmitters was concerned, there was little more that he could do. In 1933 Dale was 58 and his research output had already begun to decline.

In the same year, however, a young man arrived as a refugee from Nazi Germany, and everything changed. Dale had met Wilhelm Feldberg (Fig. 9–5) before, first during Feldberg's visit to London in 1927 and then when Dale had attended a meeting of the German Pharmacological Society at Wiesbaden in 1932.[19] At that meeting some of the work from Feldberg's laboratory had been presented in the form of two communications to the German society. Dale had been impressed. And then, in the following year, had come the bombshell: Feldberg, a Jew, was summoned to the office of the Director of the Physiological Institute of Berlin and summarily dismissed. He would have to leave the Institute at the latest by midnight of the same day, and was forbidden ever to enter it again. Hitler had been Chancellor for only a few months.

That could have been the end for Feldberg, at least as far as his academic career was concerned. A few weeks later, however, he was able to get an interview with a visiting member of the Rockefeller Foundation. The man was sympathetic but not forthcoming—that is, until he realized that the young German in front of him was the very person Sir Henry Dale had asked him to look out for, and to offer a position to![20]

Feldberg landed in England on July 7, 1933, having caught the ferry from Holland to Harwich. A month or so later, Feldberg was in Harwich again, this time to greet his wife and two small children. Now he could really concentrate on his

Figure 9–5 Wilhelm Feldberg. (Royal Society of London)

new work in Hampstead with Sir Henry Dale. It was not surprising that Dale had been eager to take Feldberg. The young German had developed a highly effective method for detecting acetylcholine in tissues. First, a small dose of eserine was given intravenously, eserine being a compound that prevented the normal breakdown of acetylcholine by the hydrolytic enzyme acetylcholinesterase. Then, if any acetylcholine had been present, it could be detected by exposing the venous blood from the tissue to leech muscle in a small bath. Even a very small amount of acetylcholine would cause the leech muscle to contract, the response being recorded with a lever writing on a smoked drum. It was the only way to do it—the mass spectrometer would not be invented for many years.

Before leaving Germany, Feldberg, with Minz, had already shown that the sympathetic nerve fibers to the adrenal medulla released acetylcholine, and it was an obvious question as to whether the same transmitter would be liberated from the sympathetic fibers in the sympathetic ganglia. These small swellings, which run along both sides of the spinal column and are also found in the neck, contain the cell bodies of the sympathetic fibers that supply virtually all the tissues of the body. Using a recently described preparation for perfusing the superior cervical ganglion, Feldberg and Gaddum, working in Dale's laboratory, now showed that, indeed, acetylcholine was released by electrical stimulation and must therefore be the transmitter in the sympathetic ganglia.[21] It was the first time that a chemical transmitter had been demonstrated *inside* the nervous system, as opposed to its presence in an organ such as the heart. Admittedly, it was the autonomic nervous system, rather than the brain or spinal cord, but, even so, it was a very important finding and one that suddenly made the idea of chemical transmission inside the central nervous system an attractive one.

Dale had not taken part in the sympathetic ganglion experiments, but he had nevertheless become rejuvenated and was now collaborating with Feldberg on another project, this to do with the release of acetylcholine by the vagal nerve fibers to the stomach.[22] Before the end of the year, 1934, the pharmacologists in Hampstead were able to give four communications to the Physiological Society, each one based on some aspect of the new work. It was an impressive achievement and it was Feldberg's presence that made it possible.

For Feldberg, money was unimportant and, indeed, for the first six months he received no salary at all. The only thing that mattered was the experiments. Sometimes he would be summoned to breakfast with Dale to make sure the work in the laboratory began early. After a particularly successful experiment, the Feldbergs would celebrate with lobster for dinner. It was an idyllic period.

Although chemical transmission had now been demonstrated, not only in organs such as the heart and stomach but also in the sympathetic ganglia of the autonomic nervous system, there was still one major challenge left, other than the central nervous system. It was the neuromuscular junction. For the reasons given earlier, it was unlikely that a chemically mediated process could operate so quickly and precisely, but the same argument could have been made for transmission in the sympathetic ganglia. Perhaps the kind of approach that had proved so successful elsewhere could be applied to skeletal muscle? Because of its importance, it was a challenge that Dale himself would take up, with the benefit of Feldberg's expertise with eserine and leech muscle. He also involved another newly arrived young German, Marthe Vogt; though not Jewish, she had left her native country because of her intense dislike of the Nazi regime.

Dale acknowledged that others before them, including Otto Loewi, had also tackled the problem of the neuromuscular junction and had obtained results suggestive of acetylcholine release from the motor nerve endings. But the earlier workers had not had the benefit of the exquisitely sensitive biological assay for acetylcholine that Feldberg had developed, one in which leech muscle would contract when the concentration of acetylcholine was as low as one part in a billion. Using a variety of muscles, and obtaining the best results from the cat tongue, the Hampstead workers quickly obtained the answer to the problem of the possible transmitter. It was as Dale had anticipated—acetylcholine *was* released from the motor nerve endings (Fig. 9–6).[23] In reading the published account of the experiments, one is struck by the thoroughness of Dale's approach—before a conclusion was reached, every possible explanation for a result was considered and, when possible, was tested experimentally.

In the very year, 1934, that these experiments had been started, evidence for chemical transmission at the neuromuscular junction came from a totally unexpected source, in the form of a letter in the June 2 issue of the *Lancet*. The author, Mary Broadfoot Walker, was a physician working in a hospital for the poor in the outskirts of London. In her letter she reported that the drug physostigmine,

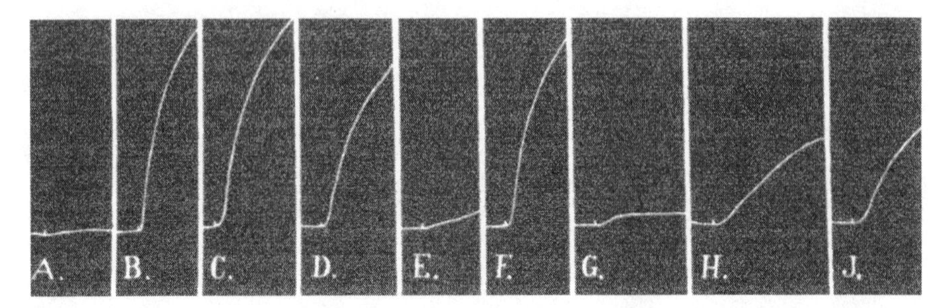

Figure 9–6 The release of acetylcholine from the motor nerve endings in the cat tongue. Each trace shows the result of applying some of the perfused fluid from the tongue to a strip of leech muscle in a bath. *A* shows the absence of effect with fluid collected prior to stimulation, while *C* and *F* show strong contractions with fluid collected after tongue stimulation. Traces *D, E*, and *G* show the effects wearing off after several minutes, while *B* and *J* are control responses when known concentrations of acetylcholine are applied to the bath containing the leech muscle. (From Dale HH, Feldberg W, Vogt M. Release of acetylcholine at voluntary motor nerve endings. *Journal of Physiology*; **86**: 353–380, 1936. Courtesy of John Wiley & Sons.)

when injected into a patient with myasthenia gravis, had produced a quite striking, albeit temporary, improvement in her muscle weakness. It was already known that myasthenia was a disorder of the neuromuscular junction, and that physostigmine potentiated the effects of acetylcholine in a variety of tissues by interfering with its breakdown. The source of Mary Walker's inspiration, one of Sherrington's former graduate students, is not generally recognized.[24]

Dale may or may not have been aware of Mary Walker's discovery. As a pharmacologist, however, he considered that the evidence for acetylcholine as the neuromuscular transmitter was still incomplete. Though acetylcholine was released from the motor nerve endings, it had to be shown to cause muscle contraction. Again, it was an issue that had been explored by others, but the results had been inconsistent, largely because the acetylcholine had been applied inefficiently. Dale's idea was to inject the substance through a cannula into the artery supplying the muscle; in this way acetylcholine would quickly reach all the neuromuscular junctions in the belly of the muscle. It was an idea that, in a less sensitive form, Dale had exploited with Gasser some 10 years earlier, when the American had been sent overseas to acquire additional research experience. Further, instead of simply looking at the muscle to see whether it contracted, measurements would

Figure 9–7 George Lindor Brown, later in his career. After leaving the Institute in Hampstead, Brown became Professor of Physiology, first at University College London and then at Oxford, and was knighted. The studio photograph captures the mischievous sense of humor that was a strong part of Brown's character and made him popular with his staff. (Royal Society of London)

now be made of any tension that developed and of the electrical responses of the fibers. Clearly the experiments would need a neurophysiologist, and so Dale recruited George Lindor Brown (Fig. 9–7).[25]

Brown had previously worked in Oxford with Eccles, at a time when the Australian was emerging from Sherrington's shadow and conducting his own experiments. Although the collaboration had lasted only six months, it was long enough for Brown to appreciate the advantages of making electrophysiological recordings, and in his first month at Hampstead he built a 3-stage amplifier such as Eccles had been using. For the experiments with Dale and Feldberg, a Sherrington optical myograph would be employed for the force recordings and all the results would be displayed on a Matthews oscillograph. The delicate surgery involved identifying and ligating all the fine branches of the popliteal artery in the vicinity of the knee, apart from the branch to the gastrocnemius muscle—which was the muscle to be perfused. The surgery was entrusted to Brown, who, as a medical student, had won the prize for operative surgery,

The new experiments worked perfectly. The injected acetylcholine produced a muscle contraction that was very similar in its size and rate of development to the twitch evoked by a maximal nerve stimulus (Fig. 9–8). The effect of the injection also resembled the response to nerve stimulation in being reduced by tubocurarine and potentiated by eserine. Taken with all the other evidence, there could now be little doubt that impulses in the motor nerve fibers released acetylcholine,

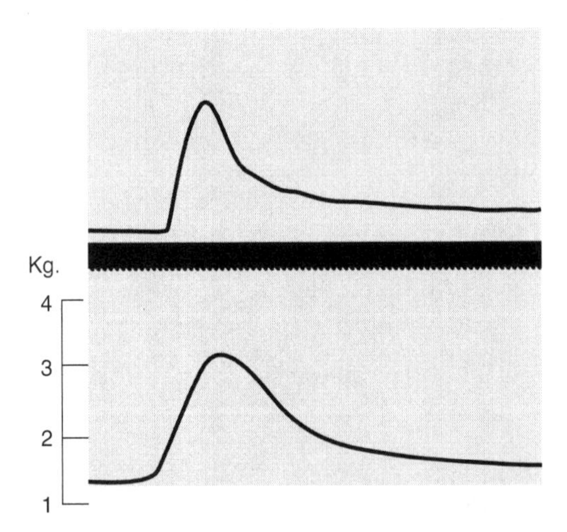

Figure 9–8 Proof that voluntary muscle responds to acetylcholine. The compound was injected into the popliteal artery close to the gastrocnemius muscle and produced a contraction (*lower trace*) that was similar to that obtained by stimulating the nerve to the muscle (*upper trace*). (From Brown GL, Dale HH, Feldberg W. Reactions of the normal mammalian muscle to acetylcholine and to eserine. *Journal of Physiology*; **87**: 393–424, 1936. Courtesy of John Wiley & Sons.)

and that the acetylcholine excited the muscle fibers and ultimately caused them to contract. It was another very important result for an understanding of the nervous system, and the paper describing the results, published in the *Journal of Physiology*, became a classic.[26]

However, not everyone was convinced by the work that had come out of the Hampstead laboratory. The most forceful antagonist to the new doctrine of chemical transmission was Eccles, with whom Brown had previously worked. In a memoir written almost a half-century later, and referring to acetylcholine by its abbreviation, ACh, Eccles would reflect:

> "I can remember being quite outraged when Dale & Feldberg suggested in 1934 that their discovery of ACh release on nerve stimulation of muscle indicated that ACh was the transmitter."[27]

Eccles' sense of outrage would be even stronger in the following year. By then he had completed a study of his own on the superior cervical ganglion and had decided that acetylcholine could be responsible for only the later part of the response to nerve stimulation. This interpretation was consistent, in its nature, with the results of a study he had carried out with Brown prior to the latter's departure for London. They had stimulated the vagus nerve as close as possible to the heart and had found that as much as 100 milliseconds had to elapse before any slowing of the heartbeat was discernible, and that the slowing then persisted for several seconds.[28] Since it had been clearly established by Loewi that the vagus nerve liberated acetylcholine, it seemed logical to Eccles that chemical transmission must be a slow and rather prolonged affair. For the rapid responses that could be seen on the oscilloscope when the superior cervical ganglion was stimulated, there had to be another explanation—electrical transmission.

Eccles presented his new work to a Cambridge meeting of the Physiological Society in May 1935. At the same meeting was yet another refugee scientist from Germany. Bernard Katz would later describe the scene in an autobiographical essay:

> "To my great astonishment I witnessed what seemed almost a stand-up fight between J. C. Eccles and H. H. Dale, with the chairman E. D. Adrian acting as a most uncomfortable and reluctant referee. . . When Eccles had given his talk, he was counterattacked in succession by Brown, Feldberg and Dale . . . what impressed me was Dale's rebuttal of Eccles' criticism. . ."

Katz goes on to say, however: "It did not take me long to discover that this form of banter led to no resentment between the contenders. . ."[29]

But had it really been good-natured banter at the time? While Brown and Feldberg were similar in age to Eccles, Dale was very much older. Further, he was head of a national institute and had been knighted by his king three years earlier.

Figure 9–9 John (Jack) Eccles (*left*) talking to Sir Charles Sherrington in the back garden of Sherrington's house in Ipswich. The photo would have been taken after Sherrington's retirement from Oxford and before Eccles' return to Australia in 1937. From their postures and expressions, it would seem that Eccles was searching for a reply to a question from Sherrington. (Courtesy of Dr. Rose Mason)

And, in what was perhaps the most important scientific achievement of his career, he was being challenged, in front of an audience of fellow physiologists, by Eccles—who, despite having served his apprenticeship with Sherrington (Fig. 9–9), might well have appeared as a brash young man from overseas—a young man who, with a penetrating voice and strong Australian accent, was not prepared to give way! Adrian, who was chair of the session, would have discovered that Dale's four humors were not always perfectly balanced. Contrary to Adrian's previous impression of him, Dale had a temper, as Feldberg had already witnessed,[30] and now it had shown itself. In the heated exchange, Dale accused Eccles of using language more suited to Hyde Park[31] than to a meeting of the Physiological Society. Little wonder that Katz, accustomed to the formal atmosphere of German scientific meetings, would still be able to recall the scene in Cambridge 60 years later.

Despite Dale's onslaught, Eccles was not deterred. He was already a fine scientist, and he was able to use modern techniques—electronic amplification and display—that neither his mentor, Sherrington, nor Dale, his intellectual adversary, had been familiar with. He would carry on devising and performing the experiments that would be needed to show that transmission between nerve cells and between nerve fibers and muscle was electrical. Or so he hoped.

Dale was equally confident, but in the reality of chemical transmission. Yes, Eccles had raised some interesting points that, for the moment, were difficult to counter. Dale was sure, nonetheless, that, in time, the necessary explanations would be found and everything would fit together. His intuition had never let him down. Unlike Eccles, however, he would stop experimenting. It may have been his age, or it may have been that he felt he could do little more in the laboratory on synaptic transmission. There was also the fact that, in Feldberg, he was about to lose his powerful German weapon. Feldberg had accepted a position in Melbourne and would depart with his family in April 1936.

The year was not without its rewards, however. Towards its end, Dale would learn that he had won the Nobel Prize and that he would share it with his old friend, Otto Loewi.

Otto Loewi, the dreamer.

10

The Squid Giant Axon

His new life in New York was much to Gasser's liking. As Professor of Physiology at Cornell Medical College, he was able to revise the entire undergraduate curriculum in physiology, and to combine teaching with his own research. There were other satisfactions. The two years spent in Britain and Europe had given him not only a widened research experience and, in France and Germany, some familiarity with foreign languages (deemed an important acquisition by the Rockefeller Foundation), but also a taste for the arts and for fine dining. New York, with its incomparable array of museums and art galleries, its rich musical life, and its infinite variety of restaurants, allowed him to pursue these pleasures and, in time, to become a connoisseur.[1] In 1935, there was an additional pleasure to contemplate: the forthcoming International Physiology Congress in Leningrad. Gasser had attended the previous congresses in Edinburgh and Boston, and the one in Russia would be especially interesting, not only because of the scientific program but also because it would provide the first glimpse of a communist society. Even the journey promised to be memorable, involving a week in a transatlantic liner, to France, followed by a further three days by boat or train to Leningrad. Gasser, by virtue of his proximity to the Intourist office in New York, had volunteered to assist in making travel arrangements for some of the other American delegates.[2]

And then, suddenly and without warning, as he was preparing to leave for Russia, came an extraordinary invitation: would Gasser accept the Directorship of the Rockefeller Institute for Medical Research? The Institute, founded in 1901 by the hugely wealthy John D. Rockefeller, was the first organization in the United States committed solely to biomedical research. Under the able direction of Simon Flexner, the Institute, centered on a hospital beside the East River in Upper Manhattan, had embarked on a wide variety of projects, many to do with infections, and had already started its formidable collection of Nobel Prizes. Though the Rockefeller Institute was only a few minutes' walk from the Cornell Medical School, the separation between the two positions was great. If Gasser were to accept the invitation, he would be the head not only of the relatively small neurology division, but of the entire Institute. It was a unique opportunity, and one that he could hardly have anticipated.

Yet there was a connection, albeit a tenuous, scarcely visible one, between Gasser and the Institute. The Chair of the Rockefeller Foundation was Alan Gregg, the same Gregg who, with Alexander Forbes at Harvard, had recorded the electrical activity of muscles participating in various reflexes. That had been in 1915, and they had used a string galvanometer. Gregg would certainly have known of Gasser's work with Erlanger, as soon as it had started to appear in print, and there is a good chance that the two had met as fellow neurophysiologists before Gregg had moved into his prestigious administrative position. Gregg's neuroscientific roots had also been evident a few years earlier, when, after solicitation by Wilder Penfield, an American neurosurgeon and former student of Sherrington, he had persuaded the Rockefeller Foundation to provide much of the funding for the construction of the Montreal Neurological Institute.

Gasser accepted the invitation. From the start, it was an enlightened reign, Gasser insisting that the researchers in the Institute be free to choose their own problems and to tackle them in their own ways. To be consistent, he insisted that all the funding came from the Institute, arguing that to apply for outside funding would compromise the independence of the investigator. In addition, he was aware that, as for all grant applications, a proposal inevitably hinted at the likely outcome, as if the answer was already known. If there was also a certain frugality in Gasser's approach, it was appropriate for the time, since America, like most of the developed world, was still firmly in the grip of the Depression, with scant resources available for research.

Even though there was little time for his own investigation of peripheral nerve, Gasser made sure that neurophysiology became one of the developing research interests of the Institute (Fig. 10–1). It is clear, from the report on his first year's work to the Institute's Board of Scientific Directors, that his interest lay in the changes in excitation that might occur at the synapses in the central nervous system. Rather than study the brain or spinal cord directly, however, he saw the solutions, at least initially, to be found in peripheral nerve:

> "The great interest which has been aroused by potentials, spontaneous or induced, appearing in the cerebral cortex is leading to the entrance of many workers into this field. What will come from their efforts is difficult to predict . . . it is, therefore, not our plan for the present to enter this field which promises to be so extensively occupied that our investigations would be only a small part of the aggregate, but to confine our attention to the more fundamental problem of how conduction in nervous tissue is modified by the intercalation of a synapse. . . Potentials resembling those found in peripheral nerve have been recorded from the spinal cord. . . On account of the degree of parallelism between the processes in nerves and neurons, it is essential that the details of the physiology of peripheral nerve be thoroughly understood. Many of the details remain to be elucidated."[3]

Figure 10–1 Gasser's laboratory at the Rockefeller Institute. The racks of equipment, with their hundreds of switches and cables, were typical of electrophysiological laboratories until the advent of transistors and integrated circuits. (Courtesy of the Rockefeller Archive Center)

In using peripheral nerve as his model for synaptic activity, Gasser was approaching the problem in the same way that Lucas and Adrian had done, before Lucas' untimely death.

To further the neurophysiology research, not only did Gasser make some judicious recruitments, but, equally important, he sought out, while in Europe, an electronics engineer, Jan Toennies. It was Toennies who, with Albert Grass, Bryan Matthews, and Otto Schmitt elsewhere, would develop new circuits for the precise stimulation and recording of nerve impulses.

The investigators brought by Gasser to the Institute included Harry Grundfest, with whom he was to collaborate in studies of mammalian C fibers, and, a little later, Birdsey Renshaw, who would become the first to describe a population of nerve cells in the spinal cord that fed inhibition back on to discharging motoneurons. Another recruit was David Lloyd, who, having spent time in Oxford with Sherrington, was to carry out detailed examinations of the central excitatory and inhibitory states described by his former mentor. However, the most interesting, and certainly the most flamboyant, of Gasser's appointees was a 33-year-old Spaniard.

Slight, dark-haired, and bespectacled, Rafael Lorente de Nó (Fig. 10–2), or Don Rafael as he was known to his friends, had been born in Zaragoza in 1902 and had started his medical studies in that city, transferring later to Madrid. He had

Figure 10–2 Lorente de Nó, during his time at the Rockefeller Institute. "Rather too dashing" for Alan Hodgkin. (Courtesy of the Rockefeller Archive Center)

already become interested in the nervous system, having investigated, while in Zaragoza, the response of the spinal cord to compression injury. In Madrid he became Ramon y Cajal's last student, and arguably his most brilliant one, using his master's histological techniques to explore the structure of the cerebral cortex and brain stem. While still in his twenties, Lorente had published papers in German, French, and Russian, in addition to his native Spanish. His very first paper, written and accepted when he was only 15, had been a mathematical treatment of thermodynamics. Out of financial necessity, however, Lorente began to work as a clinician, before deciding to move to the United States and to become a full-time researcher.[4] The move, in 1931, was opportune, for Spanish society, under its Republican government, was rapidly disintegrating. On July 17, 1936, the strikes and political assassinations were succeeded by all-out civil war.

Lorente's first 4 years in his new country were spent in St. Louis at the Central Institute for the Deaf, but it was probably in the university's basic medical sciences building on Euclid Avenue that he was introduced to the cathode ray oscilloscope and its ability to display nerve impulses instantaneously. Lorente immediately realized that it would be possible for him to correlate structure with function, thereby going one step further than Cajal had been able to. With Helen Graham, in the university department of pharmacology, as his associate in some of the early experiments, he soon showed the same instinctive flair in neurophysiology that that he had displayed in his anatomical studies.

It was a natural choice for Lorente to begin his new work in the brain stem, a region of the brain with which he was already familiar through his expertise and training in otolaryngology. By stimulating the floor of the fourth ventricle in the rabbit, and by leading off from the small muscles that move the eye, he was able to distinguish between the early responses due to direct excitation of the motoneurons (motor nerve cells) and the later responses caused by impulses that had reached the same cells through intermediate neurons.[5] Through his anatomical studies, he showed that the intermediate cells could be arranged in chains or in reverberating circuits. Although the neurophysiological experiments enabled an estimate to be made of the time taken for excitation to travel across the small gaps separating the incoming nerve endings from the motoneurons (that is, the synaptic delay), they did not permit an explanation for the changes in excitability that followed the activation of the nerve cells.

Following his transfer to the Rockefeller Institute in 1936, Lorente continued his studies on the brain stem, now with Gasser available to give advice, especially during their often noisy lunchtime discussions. A question obvious to the two men was the extent to which the relatively long changes in neuronal excitability, after the initial discharge, were due to properties that Gasser and Erlanger had demonstrated in peripheral nerve fibers.[6] What was needed, Lorente decided, was a more detailed look at peripheral nerve. It was precisely the approach that Gasser had favored in his report to the Institute's Board of Scientific Directors.

Although Lorente carried out his first experiments on peripheral nerve in the summer of 1937, it is doubtful if he became fully committed to his new enterprise before the following year. Certainly, in the illustrations that were published later, on which the date and time of each experiment were carefully written by pen, nearly all the investigations date from the 5-year period 1940 to 1945. Further, had he been fully engaged on peripheral nerve in 1937, at the time of Alan Hodgkin's one-year stay at the Rockefeller Institute, the latter would have surely mentioned it in one of his letters home. Apart from other encounters, he and Lorente once had dinner together as Gasser's guests, though Hodgkin had found the Spaniard "rather too dashing for my taste."[7]

And so Lorente set to work on his massive project. There would be hundreds of experiments, nearly all conducted by a solitary man in the afternoons and evenings at the Institute, though sometimes in the early hours of the morning as well. The longest experiments would last more than 24 hours. For his studies, Lorente had, like Gasser and Erlanger and many others before them, chosen to work with the frog sciatic nerve, preferring bullfrogs because of their greater length of nerve. Each nerve was mounted in a moist chamber, with three segments immersed in fluid-filled cylindrical chambers. A test solution would be added to one of the chambers and any change in the resting potential of that segment of nerve would be measured with a sensitive galvanometer. In other experiments the effects on the action potential would be examined, using a cathode ray oscilloscope. Though many of the experiments had been done before by others, there would never be such a careful and complete analysis. Lorente was an able mathematician himself, but he enlisted the aid of a mathematical physicist, Leverett Davis Jr., in constructing electrical models of the peripheral nerve fibers.

Though the study was remarkably thorough, there were interpretive problems, as Lorente recognized. Any difference in potential developing between the immersed segments of nerve was a "demarcation" potential and probably only a fraction of the full change in resting potential. There was also the possibility that the connective tissue sheath surrounding the nerve fibers would act as a diffusion barrier. Finally, there was the complication that the observations were being made on thousands of nerve fibers, which were likely to be affected to differing extents, so that the results observed were at best averages. Nevertheless, Lorente reasoned, it should be possible to bring knowledge of peripheral nerve function to a new level and, with it, a better understanding of the synaptic events in the central nervous system.

Only the width of Manhattan Island separates the Rockefeller Institute, now the Rockefeller University, from Columbia University. Had Lorente made the brief journey and discussed his project with one of the assistant professors of physiology at Columbia, he might have hesitated to commit himself so fully to the new enterprise. Kenneth Cole would probably have told him that there was a preparation that might be better for studying the nerve impulse than the sciatic nerve of the bullfrog—something far more exotic. He might even have told Lorente of the man who had drawn his attention to the new preparation, and Lorente would have remembered meeting him in St. Louis in 1936 and hearing him give a lecture.

John Zachary Young (Fig. 10–3), known throughout the scientific world as "JZ," was a large, ungainly, and rather untidy man, with irregular teeth and a shock of hair that fell over his forehead. In later years he was to become Professor of Anatomy at University College London, invariably wearing a red tie to reflect his left-wing sympathies.[8] For a scientist, his ancestry was impeccable: it included Thomas Young, the investigator of elasticity (Young's modulus) and the proponent, with Helmholtz, of the trichromatic theory of color vision. As if that were not enough, Thomas Young had also, with Champillon, deciphered the hieroglyphics of the Rosetta stone. There had also been a noted scientist on JZ's mother's side. JZ himself had studied zoology at Oxford and had been a contemporary of Sherrington's two most able graduate students in physiology, Derek Denny-Brown and John Eccles.

While still a student, Young had made his first visit to the Zoological Station in Naples, returning to start a lifetime's study of the brains and behaviors of the cephalopods, those marine creatures that lack a skeleton and include the squid, octopus, cuttlefish, and nautilus. It has been wittily remarked that the choice of octopus was a fitting one for, like any academic, it reacts to threats by retreating, changing color, raising goose bumps, and squirting ink. The octopus, as Young was able to show, is also capable of intelligent behavior. However, it was not in the

Figure 10–3 John Zachary Young, after his retirement as Professor of Anatomy at University College London. (Wellcome Library, London)

octopus but in the squid that he discovered some remarkably large nerve fibers, so large that they looked like veins.[9] That they were nerve fibers, he showed by a simple test: pinching the structure produced a muscle contraction, while another pinch further down abolished the effect of the higher pinch. It was a demonstration that, had he seen it, Galen, the Roman anatomist and pioneer physiologist, would surely have approved of. Each of the fibers, Young went on to show, was formed by the fusion of several hundred smaller fibers (Fig. 10–4). Whereas the largest fibers in a bullfrog might have a diameter of 20 μm (microns), a giant fiber in a squid could measure as much as 1.0 mm across. The several hundred-fold increase in cross-sectional area made it possible to stimulate and record from a single fiber rather than having to work with bundles of fibers, as in the frog sciatic nerve. With only a single fiber active, the analysis became very much easier.

Young made some simple electrophysiological observations of his own on the giant axons, showing that the speeds with which they conducted impulses were proportional to the diameters of the fibers. He also demonstrated that the rapid conduction was advantageous to the squid. Thus, the axons, either directly or through additional fibers, innervated the muscles that, by ejecting water through the siphon, propelled the squid quickly forward. In the summer of 1936, while the recipient of a Rockefeller Fellowship in the United States, Young met Kenneth Cole (Fig. 10–5) at a meeting at Cold Spring Harbor, on Long Island, and told him of this intriguing preparation. It later transpired that the squid giant axon had

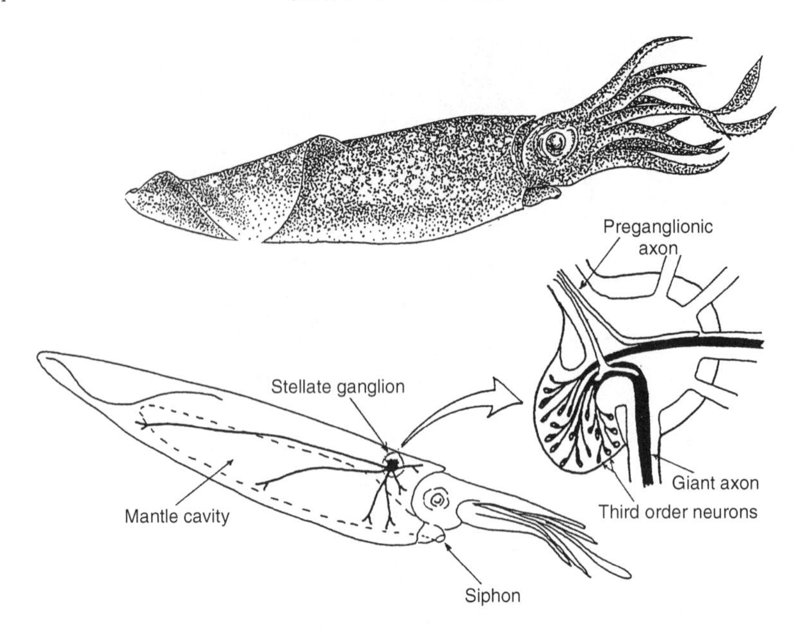

Figure 10–4 Squid, showing the formation of two giant axons in the stellate ganglion, and their distribution to muscle fibers lining the mantle cavity. (From McComas, 1996. Courtesy of Human Kinetics.)

been recognized more than 20 years earlier, by the husband of one of Cole's former landladies, a Dr. Williams.[10]

By 1936, however, Cole had become used to spending his summers away from New York, in the Marine Biological Laboratory at Woods Hole on Cape Cod in Massachusetts. The village of Woods Hole is situated on the southwestern tip of the Cape, and after the founding of the Marine Laboratory in 1888, has continued as a busy fishing port. From the front windows of the red-brick Laboratory one can look across the waters of Vineyard Sound to distant islands, while behind the Laboratory lie Eel Pond and Buzzards Bay. From the time of its establishment, the Laboratory has, in the summer vacations, attracted university staff and students eager to pursue biological problems, especially those involving marine organisms, in an atmosphere likely to be stimulating inside the laboratory and very relaxed outside it. It still does, and in ever-greater numbers.

Prior to the summer of 1936, Cole had been studying the electrical properties of the surface membranes of sea-urchin eggs and of the freshwater alga *Nitella*. He was well equipped for this type of research for, growing up in Oberlin, Ohio, he had shown the same practical ingenuity in his youth that Gasser and Bishop had displayed in theirs. Perhaps the young Cole's greatest achievement had been the building of a licensed wireless station, using a Ford spark coil and, as a detector, a piece of galena from the local geology department.[11] His interest in sea-urchin eggs had come about partly because he had already measured their heat production as a graduate student at Cornell University, and partly because the eggs were

Figure 10–5 Kenneth Cole, early in his career. (Courtesy of University Archives, Columbia University in the City of New York.)

spherical, which made the mathematical analysis of the results simpler. His interest in *Nitella* was for a different reason altogether: the alga had long cylindrical cells known to be capable of transmitting electrical impulses, though these lasted a thousand times longer, approximately one second, than the action potentials in a nerve fiber. The fact that a plant cell should be capable of firing impulses is surprising, but the property is not limited to *Nitella*. *Mimosa* and *Dionaea muscipula* (the Venus flytrap), for example, can do the same thing; in this case, the propagated impulse is a signal for the leaflets to curl up after another part of the plant has been touched.[12] For much of his work, as a physiologist at Columbia University, Cole constructed instruments of his own design, culminating in a very sensitive Wheatstone bridge that could detect changes in impedance over a wide range of alternating current frequencies.[13]

In 1936 Cole was joined by a younger man, Howard Curtis, who had himself experimented on the electrical properties of cell membranes, and the two were to work together for the next 6 years. With the new bridge Cole and Curtis were able to determine the capacitance of the surface membrane of *Nitella* (1 $\mu F/cm^2$) and to show that it did not change during the passage of an action potential. In contrast, the resistance of the surface membrane underwent a dramatic reduction, falling to a few percent of its initial value.[14] Cole and Curtis than proceeded to study the squid giant axon, using external electrodes for stimulating and recording the action potential, and for measuring any changes in membrane impedance. Excitingly, the same kind of result that had been found in *Nitella* could be demonstrated in the giant axon of the squid.[15] The figure of the fall in membrane impedance (or increase

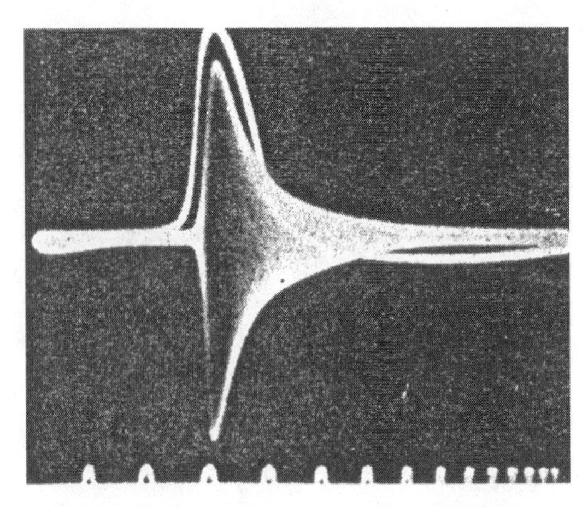

Figure 10–6 Large transient decrease in membrane impedance in squid giant axon, during passage of an action potential (*superimposed*). (From Cole & Curtis. Originally published in *The Journal of General Physiology*; **22**: 649–670, 1939, and reproduced courtesy of the Rockefeller University Press.)

in membrane conductance, which is much the same thing), with the action potential superimposed, was to become one of the classic illustrations in nerve physiology, and is still found in contemporary textbooks (Fig. 10–6).

The importance of the fall in membrane resistance was that it indicated that the nerve membrane had become much more permeable. The ions, which were carrying the current, could now move more freely across the membrane. The finding of an increased permeability was very much in keeping with the theory of excitation put forward by Bernstein in 1902. As described in Chapter 2, he had proposed that the potential of the membrane at rest was due to the membrane being permeable to potassium but not to sodium, and to the much higher concentration of potassium on the outer side of the membrane than on the inner side. The action potential, he thought, was due to the membrane becoming permeable to all ions, so that the potential would fall almost to zero.

But did the potential really fall to zero? It was impossible to find out the true size of the resting potential, and therefore of the action potential, from frog nerve since the best that could be done was to measure the demarcation potential, as Lorente was then doing. But the demarcation potential, the potential difference existing between an intact segment and the crushed end of a nerve, was likely to be only a fraction of the full resting potential. With the discovery of the squid giant axon, however, it might now be possible to try a different approach to the problem of nerve excitation.

We shall never know why, having also been told of the squid giant axon by Young, in 1936, Lorente had declined to use it for his experiments. Perhaps it was disbelief for, at Young's lecture in St. Louis, Lorente had doubted whether small fibers could merge into a larger one, as Young had claimed. Had not Lorente's

Figure 10–7 The harbor at Woods Hole. In the background is the Marine Biological Laboratory, where Cole and Curtis conducted their pioneering experiments on the squid giant axon.

master, the great Cajal, shown that nerve cells were separate entities? How could such a strong doctrine be challenged by a 27-year-old from Oxford, a university that had yet to demonstrate any expertise in neuroanatomy?[16] Perhaps, too, Lorente felt that the properties of frog nerve would be more relevant to the goings-on in the vertebrate brain, the structure with which he was so familiar. He may also have felt that, with Cole already using the giant axon, the opportunity for him, Lorente, to make new observations was diminished. Whatever the reason, Lorente's lack of interest in the squid giant axon would prove a serious error. But then, who, at that time, could have foreseen the rich harvest that the giant axon would bring?

While Lorente went his own way with the bullfrog sciatic nerve, Cole and Curtis commenced their next crucial experiments on the squid axon in the summer of 1939. As they did so, they were unaware that a race was under way, though it was not Lorente who was their competitor. In the meantime, the issue as to what happened at the nerve endings was being examined elsewhere.

11

The Neuromuscular Junction

On the other side of the Atlantic to Curtis and Cole, John (Jack) Eccles was concerned about the probability of a war in Europe. His time in Oxford with Sherrington had been well spent. Together they had, with the older man's optical myograph, documented the time-courses of the excitation and inhibition in the spinal cord following activation of sensory nerve fibers from the muscles.[1] They had also estimated the numbers of motor nerve fibers supplying different muscles in the cat hindlimb.[2] So well had the work gone that by 1934 Eccles had his own, permanent, position in the university as a Fellow of Magdalen College and as a Demonstrator in the Physiological Laboratory. But in the following year, at the advanced age of 78, Sherrington had retired from the Waynflete Professorship, and to Eccles, his departure left a void too large to be filled by others. It was not just the loss of a colleague, someone with whom he had carried out so many experiments. Rather, it was the end of those conversations, when Sherrington would have drawn on his great knowledge of European history, and of art and literature, to continue the education of the young Australian.

Scientifically, however, Eccles was more than ready to stand on his own feet. In addition to his muscle work with Sherrington, he had started his own examination of synaptic function. In the first such study, in 1934, with G. L. Brown, he measured the latency of the slowing of the heart that followed stimulation of the vagus nerve.[3] He followed this by experiments in which cells in the superior cervical ganglion responded to nerve stimulation, in the presence or absence of eserine (the compound preventing the breakdown of acetylcholine). Both types of experiment led him to conclude that the initial phase of synaptic action might be electrical—that is, brought about by currents flowing from the nerve endings through the membrane of the nerve cell or muscle fiber on the other side of the synaptic cleft.[4] Because it was so much faster in its action than any chemical transmitter, the electrical current would act as a "detonator." This view had led him into violent argument with Sir Henry Dale.[5]

By 1937, it was clear to Eccles, as it was to many others, that war with Germany was inevitable. Sherrington had gone and, much as he, Eccles, enjoyed living in England, he would do better to return to the comparative security of Australia.[6] Though it was a decision he was later to regret, there were, at the beginning, several attractive features, of which the most important was the offer of a directorship of a medical institute. Acceptance meant that, for the first time in his career, Eccles was in charge of his own research unit. The second attraction was the city in which the institute was housed. With its tiers of pastel-colored bungalows, its eucalyptus and palm trees, and the sparkling blue waters of its harbor, Sydney was one of the most beautiful cities in the world.

If Eccles had had any hesitation, it might have been to do with the nature of his new kingdom. The Kanematsu Institute was unconnected with research, and certainly had nothing to do with neurophysiology. It was, in fact, the pathology department of a large hospital, and concerned with such routine matters as the culture of infectious organisms, the typing of blood samples, and the diagnosis of tissue specimens. Although Eccles had overall responsibility for all these activities, it was 12 years since he had practiced medicine and he would have been content to leave such matters in the hands of those already in the Institute. He directed his energies instead to the top floor of the building, where he oversaw the construction of a suite of neurophysiology laboratories. As for equipment, he was allowed to bring his old apparatus from Oxford. Two decisions had still to be made, however: which research project should he work on, and whom should he have to help him?

The answer to the first question came easily. If one was to explore the general principles of synaptic action, there were considerable advantages in studying the junctions between the motor axons and muscle fibers. Not only is muscle plentiful in the body of an animal, representing about a third of the body weight, but it is easily accessed; further, because muscles vary so much in their size and structure, it is possible to choose one especially suited to the nature of the project. Finally, the structure of the neuromuscular junction is simpler than that of the synapses in the brain and spinal cord, where, instead of a single motor axon lying on a muscle fiber, the endings of hundreds of axons are intermingled on the surfaces of the cell body and dendrites of a neuron.

The choice of colleagues was more difficult. If there were none that could be lured from Oxford, one would have to look elsewhere. The first prospect to show up had no experience of neurophysiology but was an expert at tennis. Stephen Kuffler was a diminutive 25-year-old who, as a boy, had been forced by the new communist government of Hungary to flee, with his family, across the border into Austria.[7] After qualifying in medicine in Vienna, he had been obliged to flee again, this time to escape the Nazis after their takeover of his adopted country in 1938. Making his way to England, but with no immediate prospect of being able to practice medicine, Kuffler decided to take a boat to Australia. Disembarking in Sydney, he was offered a job, without salary, as a demonstrator in pathology at the Kanematsu Institute. A chance meeting with the new director of the institute led to a game of tennis. It is not difficult to imagine the contest. On one side of the net, Eccles, much the bigger man, hair flying, hitting the ball ferociously. On the other side of the net, the small figure of Kuffler, scampering across the court, returning the ball with a cunning combination of spins, slices, and lobs. Though there is no record of the outcome, Eccles was sufficiently impressed to have appointed Kuffler to his research unit at the Institute. It was an act of faith, for not only had Kuffler never carried out any research, he had disliked physiology as a medical student.

Eccles' second selection was also a refugee, but a more promising one. Bernard Katz had grown up in the German city of Leipzig, where his father was a fur trader.[8] As a Jew, the young Katz had been initially bewildered and then increasingly frustrated by the anti-Semitic attitudes and actions fostered by the Nazis. A brilliant scholar, he had combined his studies in medical school with membership of the Zionist movement. As a preclinical student, he had already been drawn to the nervous system, largely because the frog sciatic nerve-gastrocnemius muscle preparation offered the opportunity for precise measurements of excitability, measurements that could then be used in mathematical equations. During the clinical phase of his training, he had found time to carry out research into the changes in electrical impedance that accompanied muscle stretching, and for this he had won a coveted prize. While completing his medical training, Katz had also considered plans for leaving Germany. But where should he go? One option was to join the new Jewish settlers in Palestine; the other was to go to England. As a student, Katz had been impressed by the Thomas Huxley Memorial Lecture that was given by a British physiologist in 1933, and printed in *Nature*.

The British physiologist was A. V. Hill (Fig. 11–1) who, it will be recalled, had been instructed by Langley, Michael Foster's successor as Professor of Physiology at Cambridge, to investigate muscle as if it were a machine.[9] That was in 1909, when the Physiological Laboratory was still housed in an old, depressing building. Hill had taken the instructions to heart. It had long been known that, when a muscle contracts, it develops force and can do work, by moving a load. This requires the expenditure of energy by the muscle fibers, and additional energy appears as heat—exercise warms the body. Over the next several decades, first in Cambridge, then in Manchester and finally in University College London, Hill made a number of observations on heat production by muscle, using exquisitely sensitive thermopiles constructed by himself and, later, by his technician.[10] Among other things, he found that the heat released by a contracting muscle had early and late components, and he associated these with the production and disappearance of lactic acid. He was also interested in the inverse relationship between the force developed by a muscle and the speed of shortening of the muscle fibers. A heavy load, a box full of books, for example, can only be lifted slowly, while something much lighter, such as a Frisbee or a table-tennis bat, can be flicked quickly by the extensor muscles of the wrist. In the course of his research, Hill proposed a viscoelastic mechanism of muscle contraction, and, in 1923, for his work on muscle, shared the Nobel Prize with the German biochemist Otto Meyerhof.

Throughout all these years, however, there was another side to Hill, a vigorous life spent outside the muscle laboratory.[11] In the 1914–18 war, for example, he had carried out research on anti-aircraft gunnery, and at one time could have been observed pacing the sands, inspecting the practical results of his calculations.

Never one to waste his mental energies, he had immediately started to write a book on the subject at the end of the war, and the two-volume work, issued by the government in 1924–25, became the standard reference. Later, in the 1930s, after Hitler had risen to power and the Nazi persecution of the German Jews had begun, it was Hill, perhaps more than anyone else, who took active steps to bring as many scientists and their families to the safety of the United Kingdom. At the same time Hill did not hesitate to castigate the German government in the British press for its persecution of Jewish scientists and critics of the Hitler regime. Throughout this period, Hill remained an active and highly esteemed muscle physiologist, and from time to time took scientists from overseas into his laboratory for a period of work. Gasser had been an early visitor, as part of his further education ordained by the Rockefeller Foundation.

And so it was to A. V. Hill that Katz turned. Without any warning, Katz arrived, literally, on Hill's doorstep on Gower Street in London. It was February 1935 and Katz had less than £4 in his pocket.[12] He had taken the North Sea ferry from Holland and had nearly been rejected by an immigration officer at Harwich—the same port that had received Feldberg under similar circumstances two years previously. Climbing the stairs to the top floor of the Physiology Department, where Hill ran his small Biophysics Unit, Katz found the tall, austere Englishman. Neither was able to speak the language of the other, at least with any facility, but at the end of the bilingual fumbling Katz had impressed Hill with his need for a job in research, and Hill had agreed to take him on "as an experiment." Though they would later be separated, it was the beginning of a friendship that would last

Figure 11–1 A. V. Hill, in mid-career (Wellcome Library, London)

for the rest of their lives, Katz continuing to look up to the older scientist as "the most naturally upright man I have ever known . . . the person from whom I have learned more than from anybody else, about science and about human conduct."

Even if he had little money, a mere £50 a year, this was a halcyon period for Katz. Among Hill's visitors were eminent physiologists whose names Katz had known from the scientific literature and to whom he would be introduced, among them the head of the Cambridge physiology department, Joseph Barcroft, and William Rushton. There were also the meetings of the Physiological Society, at one of which he had witnessed the violent argument between Eccles and Dale over the nature of synaptic transmission. In the summer there were working visits to the Marine Biological Laboratory at Plymouth, where Katz met the young Alan Hodgkin.

So far as Katz's own work was concerned, he began to work for a Ph.D. under Hill's guidance. Earlier in his career, Hill had attempted to measure the very small rise in temperature that occurs when a nerve fiber becomes active, finally succeeding some years later with the aid of an extremely sensitive thermopile.[13] Now, in the 1930s, Hill had turned his attention, like Gasser, Cole, and Hodgkin, to the events in the nerve fiber membrane that led to the formation of the impulse. He postulated that excitation would occur if a stimulating current caused a critical reduction in the membrane potential of the fiber (the "threshold"). However, this change had to occur quickly, since otherwise the depolarization of the membrane decayed rapidly, and, in addition, the threshold of the membrane would rise if the stimulating current was prolonged ("accommodation"). Hill was not immediately concerned with the possible molecular processes in the membrane responsible for these phenomena, but he devised simple equations for the membrane changes that accounted satisfactorily for many of the experimental observations on nerve fibers.[14]

It was natural, therefore, that Katz, with the prompting of Hill, should have become preoccupied with the electrical excitation of nerve and, in particular, with the local response that must occur before a full-sized impulse was generated. However, while this work was necessary for his Ph.D. thesis, Katz was also thinking about the events that took place when an impulse reached the synapse at the end of a nerve fiber. Having gained his doctorate, following a gentlemanly interview at Adrian's home in Cambridge, Katz started, in 1938, the experiments on the neuromuscular junction that, many years later, would lead him to a Nobel Prize. Hill was perfectly content to let Katz take up a different line of work to his own. In this he was like Gasser, concerned only that the work in his unit should be of high quality. Besides, although he had taken a small number of students and fellows into his laboratory over the years, Hill never really needed their assistance; he liked to do experiments on his own. There was also the fact that Hill himself liked to move between projects. Indeed, at about the same time that Katz was turning his attention to the neuromuscular junction, Hill gave up his own work on nerve excitation, returning to his first love, the energy exchanges underlying heat production in muscle. Remarkably, he was able to do this while continuing to aid refugee scholars and while organizing the air defenses of Britain in anticipation of the war he felt was sure to come.

In 1938 there was another significant event in Katz's life, one that was to mark him out as serious participant in the contest to elucidate the nature of impulse transmission. At the same time it confirmed the correctness of his decision to leave Germany. Through Hill, he was asked to prepare a review article on nerve excitation for the prestigious German scientific publication *Ergebnisse der Physiologie*. Having written and submitted a lengthy manuscript, Katz was astonished to discover that the journal editor had written to Hill, pointing out that publication would not be possible unless Katz had an Aryan co-author. Rather than submit to this indignity, Katz immediately wrote to the editor, demanding the instant return of his manuscript, and then approached Oxford University Press instead. In the following year, 1939, a 145-page monograph appeared. Entitled *Electric Excitation of Nerve,* it did not sell well, leading Katz to reflect that Oxford University Press' "generosity in taking on such a loss-making proposition I have never ceased to admire." The translation from German to English was almost certainly Katz's own, and showed how well, in three short years, he had acquired the new language of his science.

Contained within the book was an extensive section on the excitation of the nerve fiber by local action currents, and the intriguing comment that "when the resting potential [has] been reduced, to a certain extent, a sudden reaction occurs by which the membrane, at that particular point, 'depolarizes itself,' i.e loses the whole or a large part of its remaining e.m.f. . . This in itself is astonishing, and one of the most fundamental properties of nerve. . . It means that the electric stimulus can sum with the response it elicits. This makes it possible for the impulse to be transmitted by its own self-generated local currents, and presumably ensures the all-or-none type of propagation."[15]

How this explosive event took place in the membrane, Katz did not indicate, though he was aware of the significance of Cole and Curtis's work on *Nitella*, showing the marked drop in membrane resistance. After his book was published, Katz did a surprising thing: he left the United Kingdom for Australia.

The decision to move to Australia was not made easily. After gaining his Ph.D., Katz was awarded a Beit Memorial Fellowship, with a generous stipend of £400 a year, enough to live comfortably on and to support his immigrant parents as well. Moreover, he remained content to be working in Hill's biophysics unit on projects of his own choosing. Still, Eccles had written to ask if he would consider joining him in Sydney. Katz accepted, on the grounds that it would not have been right for him, the guest of a foreign country, to refuse a call for help by a person isolated from his former colleagues. While in transit by sea, the European war he had long anticipated finally broke out, leaving him temporarily stranded in Ceylon.[16]

When Katz finally reached Sydney, it was to find not only Eccles but also Stephen Kuffler, the newly acquired research assistant and tennis companion.

Unlike Kuffler's, Katz's debut on the Eccles court was disastrous. Invited to mow the grass, Katz somehow succeeded in cutting through the cord of the electric mower, electrocuting himself and unable, for several painful seconds, to free his involuntary grip from the handles of the mower.[17] In the laboratory, however, it was a different story. Having been trained by Sherrington, it was natural for Eccles to continue using cats for his experiments. Katz, who detested the killing of animals, especially domestic ones, pointed out that the frog was equally suitable for investigation of the neuromuscular junction. The change of preparation was welcomed by Kuffler, who, in the early days, was often in a state of bewilderment in the laboratory, partly because of the new, and often confusing, body of information he was expected to know. He had, however, quickly become expert in setting up the cat muscles, and it was Kuffler who would enliven the proceedings with his irrepressible humor, a trait that would continue throughout his life, even as a senior neuroscientist and establishment figure.

A photograph of the time shows the three men walking to catch the tram to the Kanematsu Institute (Fig. 11–2). One's gaze is drawn first to Eccles, the person in the middle of the group, striding purposefully, swinging one arm and carrying a briefcase in the other, his face a picture of determination. On his left is Katz, also looking ahead, but with a calmness and serenity in his expression. Completing the

Figure 11–2 Eccles (*center*) with Katz (*right*) and Kuffler in Sydney. They were on their way to catch a tram to the Kanematsu Institute. (Photograph courtesy of Dr. Rose Mason)

trio is the smaller figure of Kuffler, hurrying to keep up with the others, his head turned towards them as he makes one of his amusing observations.

What was it that Eccles hoped to show from the new experiments? During his time in Oxford, he had, as already noted, carried out experiments on the heart in collaboration with G. L. Brown.[18] Otto Loewi had previously shown acetylcholine to be the chemical transmitter released by the endings of the vagus nerve, and this was universally accepted. However, Eccles and Brown had found that, when the vagus nerve was stimulated electrically, even very close to the heart, there was a long delay, approximately 100 ms, before any effect was discernable in the heart rhythm. Moreover, the effect of a single shock lasted for several seconds. From these observations Eccles reasoned that there should be similar delays and prolongations of action at all other synapses in which chemical transmitters were released. If the latency was found to be short, of the order of 1 ms, then this must be because the initial excitation (or inhibition) had been brought about by electric currents flowing through the membrane of the nerve cell or muscle fiber on the other side of the synapse—the "detonator" action. Hence synapses could be chemical or electrical, or both. By applying certain drugs to the neuromuscular junction, and by observing their effects on the electrical responses evoked by single or multiple stimuli, it should be possible to distinguish between the three types of transmission.

The neuromuscular junction, because of its accessibility, was an ideal preparation with which to work but, in planning his new studies, Eccles must have been troubled by one important fact. Each muscle fiber is very much larger, and therefore has considerably more membrane, than the fine nerve twig that innervates it. Consequently, an action potential in the nerve twig would be unlikely to draw sufficient current through the muscle fiber membrane to excite it. The electrical transmission hypothesis was doomed from the start.

Nevertheless, the new experiments got under way and, as sometimes happens with ill-conceived plans, results of value were obtained. The main achievement was a detailed analysis of a slow change, the "end-plate potential," recorded when the motor nerve was stimulated and the normal response of the muscle was partially blocked by the drug curarine or by a stimulus that had been delivered a short time (less than 1.6 ms) before.[19] Curare, from which curarine was derived, was a naturally occurring compound used by the aboriginal peoples of the Amazon in their hunting expeditions. Their arrows, coated with the substance, caused their prey to become paralyzed.[20]

The Sydney workers were not the only ones to study the end-plate potential. Gopfert and Schaefer, in Germany, and T. P. Feng, in China, made similar recordings at about the same time. However, by using different concentrations of curarine, or by adjusting the interval between a pair of shocks to the motor nerve, Eccles, Katz, and Kuffler were able to obtain "pure" end-plate potentials of varying size, and to estimate the amplitude necessary to trigger an action potential in the muscle fiber (Fig. 11–3).[21] In later experiments they employed eserine, the same

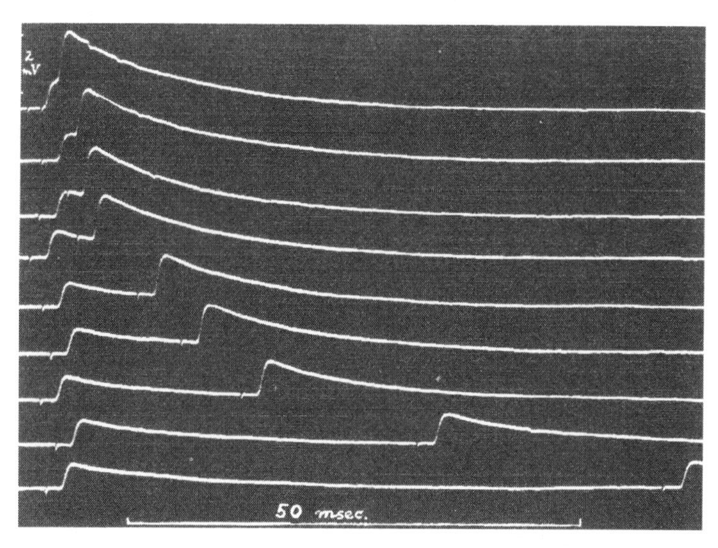

Figure 11–3 End-plate potentials in frog sartorius muscle. Each of the nine traces shows the results of giving two shocks to the nerve. As the intervals between the shocks decrease (*towards the top of the figure*), there is more and more summation of the responses. (From Eccles JC, Katz B, Kuffler SW. Initiation of impulses at neuromuscular junction. *Journal of Neurophysiology*; **4**: 402–417, 1941. Courtesy of the American Physiological Society.)

substance that had been used so successfully by Feldberg in his leech muscle assays of acetylcholine. The effect of eserine on the neuromuscular junction was in sharp contrast to the action of curare. Whereas curare blocked transmission and caused muscle paralysis, eserine, by preventing the breakdown of acetylcholine, exaggerated the effect of nerve stimulation, producing stronger than normal muscle contractions. When the Sydney investigators examined the oscilloscope traces they saw that, instead of a single impulse in the muscle fiber, following a shock to the motor nerve, there were multiple discharges superimposed on an enlarged end-plate potential. The effects of eserine, they concluded, "were reconciliable with the hypothesis that ACh [acetylcholine] is responsible for all the local potential changes set up by nerve impulses."[22] Eccles was not slow to send an air-mail letter to Sir Henry Dale in England, letting him know of his change of mind.

But before Eccles could start on the central nervous system, events outside the laboratory brought the work to a halt, and Eccles found himself obliged to leave Australia.

Plymouth, August 1939[1]

They could easily have been mistaken for holidaymakers.

The two young men, both good-looking and both wearing sports coats and flannels, stepped out of the boarding house and paused briefly before walking on. It was only a short distance to their destination but, apart from the enjoyment of the fresh air, it would probably be their best chance during the day to feel of the sun's warmth. So far it had been an unusually good summer and it seemed that everyone was taking advantage of it. The Royal and the other large hotels fronting the Promenade were full, and so were the smaller buildings, the boarding houses in the streets behind. As in previous years, some visitors had come to Plymouth to use the city as a base for exploring the Devon countryside, but for the majority the attraction had been the sea. For the elderly it was sufficient to sit on one of the benches on the Hoe and to look out over the waters of the Sound. For the more energetic adults, there was sailing from the Marina at the foot of the Barbican, the smaller yachts venturing along the coast and some of the larger ones heading out into the Channel, perhaps making for France or for the Atlantic itself. For the youngest and therefore the most active holidaymakers, there were the amusements on the Pier and the chance to swim and play in the blue waters of the Tinsdale Lido.

The two young men, engaged in earnest conversation, paid little attention to their surroundings or to the holidaymakers. Both knew, as did most who listened to the clipped English voices on the radio or who read the newspapers, that this could be the last summer of enjoyment for some time. It did, indeed, seem that war was inevitable, and the signs were already there—the increased number of warships in the Sound and the army trucks bringing troops in and out of the Citadel.

The young men crested the Hoe and looked at the array of warships in the Sound. Three destroyers at anchor and, further out, a battleship. As they watched, a fourth destroyer made its way past Drake's Island into the narrow entrance to the Tamar and the Devonport dockyards. Alan, slightly the older of the two young men, was used to the sea. As a small boy, he had clambered over the sand dunes and marshes in Northumberland during bird-nesting expeditions with his Aunt Katie, and he had seen the North Sea in all its moods. And then there had been the wide, desolate Norfolk beaches near his boarding school and, later still, the experience of dinghy sailing. Andrew, the younger of the two, was also familiar with the sea. To some extent, it was in his blood, for his grandfather, in his youth, had sailed from Plymouth as a newly qualified assistant surgeon on H.M.S. Rattlesnake, a four-year voyage that had taken him to Australia and New Guinea and had given him the chance to make his name as a scientist. Ninety-three years ago, almost a century. If Andrew's father had not been so adventurous, he had, at least, shown his younger sons something of the sea, making them row the heavy clinker-built boat across Loch Etive in the summer vacations in Scotland.

Crossing the Hoe, the two young men passed close to the spot where, in 1588, so it was said, Francis Drake, the greatest of Plymouth's many famous seafarers, had observed the approach of the Spanish Armada and had calmly continued his game of bowls, confident that he and his fellow captains would beat the enemy when the time came for battle. As they did.

A few more steps and the young men reached their destination, the imposing gray stone building of the Marine Biological Laboratory, situated on the seaward side of the Citadel. They mounted the steps, passed through the entrance, and made their way along the corridors. Having arrived some weeks before, and having already carried out some experiments on his own, Alan now looked forward to having Andrew's collaboration. He opened the door to their room and invited Andrew to step inside.

Alan and Andrew.

Hodgkin and Huxley.

Alan Hodgkin (nearest camera) and Andrew Huxley. This recently discovered photograph is the only known one of Hodgkin and Huxley together. The striking youthfulness of Hodgkin, in particular, suggests that the photograph is a pre-war one, quite likely taken during their stay in Plymouth in 1939. (Courtesy of Professor Bill Harris of the Department of Physiology, Development and Neuroscience, University of Cambridge)

12

The Giant Axon Impaled

Traveling independently, two young men left Cambridge in the summer of 1939 and set off for the Marine Biological Laboratory at Plymouth, the same institution in which Keith Lucas had carried out some of his experiments in the first few years of the 1900s. Driving his own car, and towing a trailer filled with electrophysiological equipment, Alan Hodgkin was the first to arrive. Only 25, Hodgkin (Fig. 12–1) was in the third year of a Research Fellowship at Trinity College and had already had several papers on peripheral nerve published in the prestigious *Journal of Physiology*. Further, he had spent a year in research at the Rockefeller Institute in New York in 1938–39, writing weekly letters home to his mother with comments on the people he had met—Gasser and Lorente de Nó, for example.[1]

Boyishly handsome, Hodgkin could, like J. Z. Young, point to a distinguished relative, in this case the Northumbrian physician and great-great-uncle, Thomas Hodgkin, who had, in 1832, described the malignant disease of the lymph nodes that bears his name. Alan Hodgkin's father had studied at Cambridge, where his best friend had been Keith Lucas, the man who would become the foremost neurophysiologist of his day, at least in the United Kingdom. The two men had, after graduating with first-class honors in the Natural Sciences Tripos, spent six months together on a government survey of lakes in New Zealand. Whereas Keith Lucas had returned to Cambridge and taken up the study of peripheral nerve, Alan Hodgkin's father had become a banker. With the onset of the First World War, however, he had been unable to serve because of his strong Quaker belief in pacifism. Instead, he took part in a fact-finding expedition to Armenia to assess the need for aid following the attacks of the Turks. On a second expedition, this time by way of Baghdad rather than through Russia, he had contracted dysentery and died.

Despite the loss of his father, Alan Hodgkin and his two younger brothers had happy and interesting childhoods, as recounted by Hodgkin in his autobiography *Chance and Design*, published when he was 78. In reading this book one cannot help but be dazzled by the sequence of extraordinary happenings, meetings, and adventures, particularly after Hodgkin started his undergraduate studies at Trinity College, Cambridge. It was a charmed existence, one made possible by the legacy of an uncle. It was a life very different to those of most young men at that time, for whom there was the bitter prospect of unemployment on leaving school,

Figure 12–1 Alan Hodgkin, soon after his election as Fellow of Trinity College, Cambridge, in 1936. (From Hodgkin, Sir Alan. *Chance & Design. Reminiscences of Science in Peace and War*, 1992. Courtesy of Lady Marion Hodgkin and Cambridge University Press.)

a consequence of the Depression. Nor was the economy the only problem, for those attuned to German affairs knew that the sounds of marching feet and breaking glass were the portents of much worse to come. For Hodgkin, however, there were holidays spent on the Scottish island of Islay, or mountaineering in Norway, skiing in Austria, or studying art in Italy. A keen naturalist, he took part in a student expedition to the Atlas Mountains in Morocco, during which he developed a dangerous cellulitis in one of his legs. As to his studies, Hodgkin became interested in physiology, and after undertaking some experiments of his own design as a student, decided to continue his career in that field as a research scholar of Trinity College.

At that time, in 1936, the Physiological Laboratory at Cambridge was especially strong, since it included Adrian, who had won the Nobel Prize four years earlier, and his brilliant young associate, Bryan Matthews, who, among other achievements, had designed and built an oscillograph (without a cathode ray tube). There was also Grey Walter, who was pioneering research on the spontaneous brain activity that could be recorded with electrodes on the scalp (the EEG), and who was an inventor himself. As head of the Laboratory, Joseph Barcroft pursued an interest in the ability of the body to maintain a nearly constant environment for its constituent cells. Once, following an academic challenge from Oxford, his interest had prompted him to lead a physiological expedition to the Andes. The person who was to influence Hodgkin the most, however, at least in the study of nerve, was William Rushton (Fig. 12–2).[2]

Figure 12–2 William Rushton. (Royal Society of London)

While still at school, Hodgkin had heard Rushton, a former pupil, lecture on nerve, and at Cambridge the two were keenly aware of the scientific legacy of Keith Lucas. It was a legacy that, in relation to the nature of the nerve impulse, Adrian had eventually abandoned in favor of his wide-ranging studies on the coding of information by the thresholds and impulse frequencies of sensory and motor nerve cells. Adrian's departure from the field coincided with an attack on his investigation of impulse conduction in a narcotized region of nerve. That study, undertaken during Adrian's time with Keith Lucas, had been interpreted by Adrian in favor of a continuous decrement in the size of the impulse in the treated region of nerve. It was an attack against which Adrian had chosen not to defend himself and, in any case, the outcome of the dispute, one that would favor Adrian, would have to wait for studies performed on single nerve fibers very much later.[3]

With Adrian no longer in the field, it was Rushton who had maintained an interest in the nature of the nerve impulse. Like some of the earlier electrophysiologists—Helmholtz in particular, but also Bernstein, Lucas, and Gasser—Rushton had an affinity for mathematics and especially for Euclidean geometry. Nor was mathematics his only intellectual passion. There was a great love, as well as an aptitude, for music. Indeed, in true Cambridge style, he composed a madrigal in order to attract the attention of his future wife. Because of his strong interest and ability in mathematics, it was natural that Rushton should apply his talent to the theory of nerve fiber excitation, treating the fiber as an electric cable. In this model, a very appropriate one, the axoplasm in the core of the nerve fiber was envisaged as a conductor, with the tissue fluid surrounding the fiber forming the return path for a current flowing down the fiber. Since the core of the fiber was narrow, however, it had an appreciable resistance. The nerve fiber membrane also had a resistance as well as a capacitance, while additional capacitance and a very large resistance were provided by the fatty myelin sheaths around the larger fibers. There was also a battery across the membrane, in the form of the resting potential. The cable theory approach was one that had been used many years before, notably by Bernstein's friend Ludimar Hermann[4] and by Gasser.

In his own application of cable theory, Rushton was able to derive the "space constant," or the "analytical length" as he referred to it, identified as the length of fiber over which an applied electrical signal declined by a critical amount. Similarly, Rushton was able to derive the "time constant" of a fiber, this being a measure of the time taken for a current to charge up the capacitance of the membrane (and myelin sheath, if present). He also showed that, for an impulse to be set up, a certain minimal length of fiber had to be involved. Finally, he drew attention to the importance of the connective tissue sheaths surrounding bundles of nerve fibers in determining the distributions of current flow when nerves were excited.[5] In each case Rushton combined his theoretical insights with experimental validation, using equipment much of which he built himself. His annual research budget for many years was a mere £60.

In a curious reversal of fortune, at the very time that Rushton was carrying out these fundamentally important studies on the nerve impulse, and demolishing any opponents along the way, he was having the greatest difficulty in passing his clinical examinations in medicine. This contradiction in ability had puzzled Katz, a recent immigrant to England, as well as Rushton's contemporaries in Cambridge.[6]

A brisk, rather peppery man and an excellent lecturer, Rushton exerted a strong influence on the more junior members of the Cambridge Physiological Laboratory and, while he was still working on peripheral nerve, on Hodgkin in particular. In the later part of his career, Rushton would leave nerve excitation altogether, making striking advances in color vision and eventually coming to dominate that field.

Hodgkin's undergraduate research was noteworthy for the simplicity both of the ideas and of the apparatus used to test them. Largely through Rushton's influence, Hodgkin was interested in the changes in the nerve membrane that gave rise to the action potential. These changes, he supposed, were due to small electrical currents flowing through the membrane in advance of the traveling impulse. The idea was far from new and had been implicit in Hermann's work in the previous century, as well as in Rushton's. As current flowed through the nerve membrane, either during an electrical stimulus or, under natural circumstances, as an impulse approached, some change must take place in the resting membrane. If large enough, this change would itself give rise to an impulse in that segment of nerve. Keith Lucas had demonstrated the existence of such a "local response" indirectly by applying two weak stimuli to a nerve, one after the other.[7] Though this combination of stimuli evoked a discharge, neither stimulus was adequate by itself. Clearly the first stimulus had briefly changed the nerve fiber in some way. This transient local response would be expected to be of the same nature as the action potential itself.

Theoretically, the change in the membrane responsible for the local response and the action potential could be in the resistance or the capacitance, or even in the battery. Indeed, Rushton, in his theoretical analysis, thought that it was the battery that discharged without any other changes taking place in the system. Otto Schmitt, in contrast, was able to account for the known features of the action potential by assuming that the capacitance varied, increasing as the exciting current started to flow across the membrane. The uncertainty was resolved only when Cole and Curtis, applying their sensitive bridge to the squid giant axon, had been able to demonstrate a huge fall in the membrane resistance; the capacitance, in contrast, was unchanged. This exciting finding, in 1938, came in the year that Hodgkin was spending at the Rockefeller Institute, and four years after his first experiments as a student at Cambridge.

In his attempt to demonstrate a local response Hodgkin had employed a cold block, achieved with a silver rod partially immersed in ice, to prevent the action potential itself from invading the same region, and interfering with the observations. As a student, Hodgkin did not have access to an oscilloscope and it was therefore not possible for him to see the nerve action potentials directly. Instead he had to use the same antiquated classroom apparatus, the drum kymograph, with its smoked paper, that students had been using since the previous century. This enabled him to see the strength of the calf muscle contractions when the sciatic nerve of the frog was stimulated in the cooled region. With this system, it was reasonable to assume that the more vigorous the contraction, the greater the number of nerve fibers that must have been excited. To his delight, the response was larger if the stimulus was given when an action potential, elicited by a slightly earlier shock, had arrived at the block. The result demonstrated that the threshold had indeed been lowered, and the most likely mechanism available was a local current. Later, with the aid of a homemade amplifier and oscilloscope, Hodgkin was able to refine the experiments and to show the local electrical responses of the nerve fibers beyond the block (Fig. 12–3). Further, the time course of the local response mirrored that of the increase in excitability in the segment of nerve beyond the block—as would be expected if the local response was responsible for the latter (Fig. 12–4). It was this work that Hodgkin wrote up for two papers in the *Journal of Physiology* that were published back to back in 1937.[8,9] Not only was the laboratory work first-class, but the reasoning behind the experiments and the maturity of the writing were impressive—especially for a 23-year-old. The simplicity and clarity of expression that had been so characteristic of Lucas and Adrian had been passed on.

Apart from his pleasure in the results, there was satisfaction for Hodgkin in having accomplished the initial work with rather primitive equipment. It was, after all, the Cambridge style—elegance in simplicity. It was after these experiments

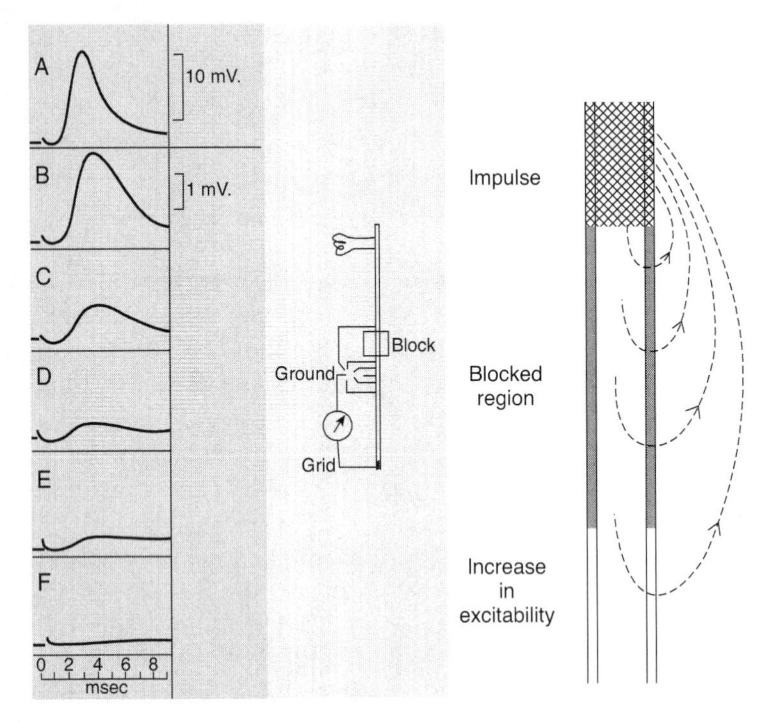

Figure 12–3 *Left*: Local responses in a frog sciatic nerve at 1.4 mm, 2.5 mm, 4.1 mm, 5.5 mm, and 8.3 mm beyond a cold block (*B, C, D, E, F* respectively). *A* is the action potential arriving at the block. The oscilloscope traces have been redrawn. *Right*: Eddy currents extending through and beyond the blocked region of nerve, thereby producing a local response. (From Hodgkin AL. Evidence for electrical transmission in nerve. Part I. *Journal of Physiology*; **90**: 183–210, 1937. Courtesy of John Wiley & Sons.)

that Adrian, drawing on his vast knowledge of animal species, suggested that Hodgkin switch from frog to crab nerve, pointing out possible advantages in the larger sizes of some of the nerve fibers. And, indeed, after some practice, Hodgkin found that he could dissect out individual fibers for study. Using a block and with a cathode ray oscilloscope to observe the results, he found that, in this simpler preparation too, he could record the subthreshold responses directly.[10]

However, when Hodgkin discussed his results with Gasser in the same year, at the Rockefeller Institute, he found the eminent neuroscientist unreceptive. Gasser, who knew the cable theory of nerve well, had looked in vain for a subthreshold response on the oscilloscope. Without the use of a cold or pressure block to prevent the subthreshold response from being swamped by the full action potential, it was not to be seen. Erlanger, in St. Louis, was equally skeptical about the presence of local currents, though without having any other suggestions as to how the action potential might propagate. He did, however, put forward an interesting challenge to Hodgkin: if the latter was correct about local currents, then it should be possible to alter the speed at which the action potential was conducted

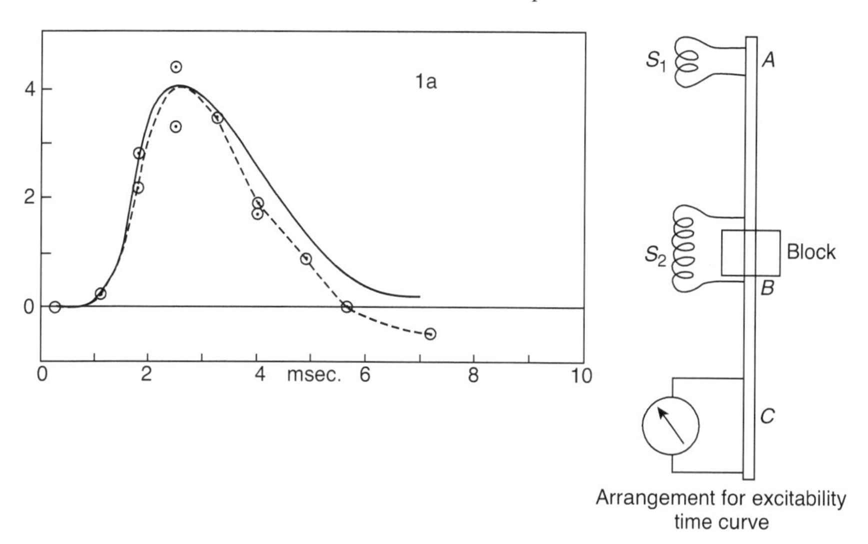

Figure 12–4 *Left*: Similar time-courses of local response (*continuous curve*) and increased excitability (*interrupted curve*) beyond a cold block. *Right*: Experimental arrangement employed. The action potential is fired from one pair of stimulating electrodes (S_1) and the excitability of the nerve beyond the block is tested with another pair (S_2), while the local responses are recorded with the electrodes at *C*. (From Hodgkin AL. Evidence for electrical transmission in nerve. Part II. *Journal of Physiology*; **90**: 211–232, 1937. Courtesy of John Wiley & Sons.)

along the nerve, by varying the electrical resistance of the fluid around the nerve. Hodgkin accepted the challenge by immersing a single crab fiber in oil and thereby raising the electrical resistance surrounding the nerve fiber. As predicted by local circuit theory, the conduction velocity fell. A similar slowing could be demonstrated in a squid axon, after it had been lifted from a chamber of sea water into air.[11]

This last experiment was carried out at Woods Hole, in Cole's laboratory. Hodgkin had met Cole in New York in the previous year and Cole had apparently invited him to visit the laboratory in the following summer. Or, had he? Such an invitation would have been in keeping with Cole's generous nature, but Cole, much later, described Hodgkin's visit as a surprise.[12] The distinction is not trivial, for it influences the perspective of much later events. Whatever the truth of the matter, Hodgkin's arrival at the laboratory was, at least for him, dramatic. There, on the oscilloscope screen, as he entered the room, was the fall in membrane resistance accompanying the action potential in a squid giant fiber. It was the picture that would become famous. According to Cole, Hodgkin literally jumped up and down with excitement.

Having arrived, Hodgkin was made welcome by Cole and Curtis. He learned how to dissect out and clean squid giant nerve fibers. He was lent apparatus to test the local circuit theory again, this time by lowering the resistance on the outside of the nerve fiber. He could now try the experiment on a squid fiber and, instead of the usual practice of bathing it in sea water, place it on a series of platinum strips instead, so as to obtain the low resistance needed. This time, the conduction velocity of the impulse increased, as the theory had predicted.[13] After that, he and Cole were able to work together, measuring the internal resistance of a giant fiber by a method suggested earlier by Rushton in Cambridge.[14] With the acquisition of these results, and with Cole's earlier ones on membrane resistance and capacitance, the analogy between a nerve fiber and a cable became much stronger. Real values for resistance and capacitance could now replace the electrical symbols.

The summer of 1938 proved to be a marvelously productive one for Hodgkin, and the way of life at Woods Hole an appealing one. He was able to do experiments that would otherwise have been impossible for him at that time. He enjoyed working with Cole and was grateful for all the help the older man had so freely given. But then, in July, the year of research in the United States was over. Before he left, there was yet more American generosity in the form of a grant from the Rockefeller Foundation and advice from Toennies at the Institute as to what equipment he would need. No less important, there was also the memory of Dr. Peyton Rous' eldest daughter, Marion, who would eventually become Lady Hodgkin. All that was for the future, however. Now, one year after his return from the United States, with apparatus that had either been purchased or else built in the university laboratories at Cambridge, he arrived in Plymouth to continue his experiments.

Having had one stroke of good fortune, in being able to work in Cole's laboratory the previous year, Hodgkin now had another. The Cambridge student he had invited to help him with his investigations of nerve would, like Helmholtz, who had been the first to measure the impulse conduction velocity many years before, prove to be a genius. Indeed, in his later life, many would acknowledge him as the most intelligent person they had ever met.

Like Adrian and Hodgkin, his seniors at Cambridge, and like J. Z. Young, then at Oxford, Andrew Huxley belonged to a distinguished family—a very distinguished family.[15] Ninety-one years earlier, his grandfather, Thomas, had stood on the deck of H.M.S. *Rattlesnake* as it sailed past Plymouth and up the English Channel, after a four-year voyage of exploration of the waters separating Northern Australia from New Guinea, and those around New Guinea itself. A qualified doctor and a keen naturalist, the first of the eminent Huxleys had taken his microscope with him and had studied some of the marine organisms encountered by the ship, writing up and illustrating his findings and sending the manuscripts back to England for publication. Long after that voyage and his resignation from the Navy had come his spirited defense of the evolutionary theory of his friend Charles Darwin, at a time when Huxley was well advanced in his career as a scientist, writer, and educational pioneer. Among his many achievements was his assistance in founding the Physiological Society, some of whose early meetings he had chaired. He had also recommended that Trinity College appoint Michael Foster as a Praelector in Physiology, the first of the many Cambridge physiologists.

Andrew Huxley's own father was a noted schoolmaster, writer, and editor whose children by his first marriage included Julian Huxley, the biologist, Secretary of the Zoological Society of London, and founding Director of UNESCO, and Aldous Huxley, the visionary author of *Brave New World* and other powerful novels. Andrew himself was the younger of two sons by his father's second marriage. Unlike his stepbrothers, Andrew Huxley had developed a keen interest in mechanical things and intended, like his famous grandfather at a similar age, to become an engineer. A photograph at the time (Fig. 12–5) shows a young man with the looks of a poet but with the confident attitude of a scientist. The dark good looks, and especially the extraordinary piercing eyes, were the legacy of his grandfather, as was the ability to use the microscope that his mother had given him.[16]

As a boy, Andrew Huxley attended Westminster School in London, the same school that Adrian had gone to some 30 years earlier. Situated next to the Abbey and founded in the 12th century, this private school could number a host of distinguished alumni, including statesmen—five of whom became prime ministers—as well as poets, playwrights, and scientists. Henry Purcell, the greatest of the

Figure 12–5 A youthful Andrew Huxley. The structure on top of the microscope is an adaptor for a Leica camera; it contains a mirror that, operated by a cable, deflects the light path to the observer's eye instead. A side tube, containing a piece of glass with scratches on it, is used for focusing the camera. These additions were devised and built by Andrew Huxley himself. (Sir Andrew Huxley, personal communication)

English 17th-century composers, had been a scholar, as had Sir Christopher Wren, the architect of St. Paul's Cathedral. A prize-winning student, well taught in Latin and Greek in addition to the sciences, Andrew Huxley eventually left Westminster School for undergraduate studies at Cambridge. Although physics, mathematics, and chemistry were his main topics, Huxley took physiology as well, as a necessary additional science subject. In addition to Hodgkin, Huxley's senior by four years and a fellow Trinity man, there were, of course, Adrian, Matthews, and Rushton with their strong interests in the nervous system, and all of them influenced Huxley's subsequent decision to specialize in physiology.

As a student, Huxley had already given a paper on the conduction of nervous impulses to the Natural Sciences Club at Cambridge and, still a student, had assisted Hodgkin in some experiments on single nerve fibers in the crab. As Lorente de Nó was doing, in his study of the frog sciatic nerve at the Rockefeller Institute, they measured the demarcation potentials of the fibers, which were still the best approximations to the resting membrane potentials. On stimulating the same fibers, the action potentials were found to be roughly twice as large as the demarcation potentials, a finding that was of considerable significance, as they may have appreciated at the time. The problem was that both potentials, resting and action, had been measured indirectly and would not have reflected the full potential differences across the nerve fiber membrane.

As the summer vacation approached in 1939, it was natural that Hodgkin should have invited Huxley to join him at Plymouth and help him with his new experiments. This time, the experiments would be on the squid giant axon, the same preparation that Hodgkin had discovered in Cole's laboratory the previous year. Because it was so much larger than the largest fibers in a mammalian nerve (Fig. 12–6), the squid axon should lend itself to new kinds of experiment. Huxley was not to arrive until a month after Hodgkin. By then Hodgkin had already encountered a major difficulty in that the squid were in short supply. The only one that he had been able to work with was the one caught when he had gone out in the trawler himself.[17] Such enterprise was necessary, for, once taken from the trawler's net and placed in a tank of sea water, the squid would batter themselves against the walls and quickly deteriorate. Hodgkin had, in fact, carried out the initial dissections of the nerves on the boat. After Huxley's arrival, however, the supply of squid began to improve, and, equally important, they would arrive at the dock in better condition.

As a side project, Hodgkin suggested that Huxley measure the viscosity of the axoplasm, the matter inside the fibers. The method proposed was to somehow introduce a very small drop of mercury into the cut end of a fiber, and to measure the speed with which it fell, the fiber hanging vertically. It was a project that might have occupied an average student for most of the summer. Instead Huxley set the experiment up, using his skill with the microscope to watch the anticipated events inside the fiber. Within a day or two, he had a surprising answer. The mercury droplet did not fall at all but remained on the cut surface of the fiber. The inside of

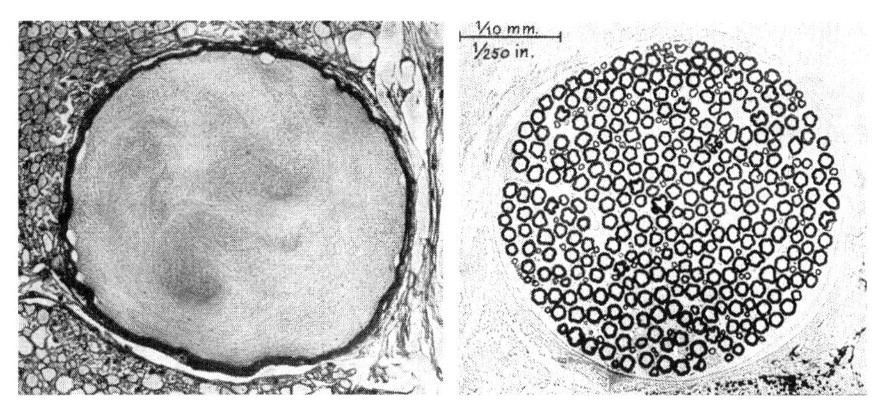

Figure 12–6 Sections through a squid giant axon (*left*) and the many myelinated fibers in a rabbit nerve (*right*), photographed at the same magnification. (From Young, 1951)

the fiber was not fluid, as had been supposed, but appeared to be solid.[18] Indeed, many years later, it would be shown that the axoplasm could be squeezed out of a squid fiber and retain its shape, much like the contents of a tube of toothpaste.[19]

Though it was a negative experiment, the kind that scientists like to avoid, the viscosity study had an important consequence. Huxley, with his enviable dexterity and his gift for devising new techniques, thought that the squid axon was wide enough for a length of very fine glass tubing to be pushed down the inside of the fiber. The capillary tubing, having been previously filled with a conducting salt solution, could then be used as an electrode to measure the true potential across the membrane of the nerve fiber. The possibility of recording with an internal electrode was something Hodgkin had considered the previous year, during his time at Woods Hole. He and Curtis had even tried to insert a wire into the cut end of a squid axon, but without success.[20]

Although Hodgkin was an expert dissector, as he had already demonstrated with the single nerve fibers of the crab, it was probably Huxley who did the initial experiment at Plymouth, cutting halfway through the axon with a pair of sharp scissors, inserting a glass cannula through the cut, and then tying the axon around it with silk thread. The next step was to guide the capillary electrode, 100 μm in diameter, through the cannula and into the axon. The new experiment worked immediately, though on subsequent occasions it would fail if the internal electrode scraped against the membrane. Huxley overcame this difficulty by setting up two mirrors at right angles to each other, which, with the use of a horizontally mounted microscope, enabled the glass capillary to be guided down the center of the fiber (Fig. 12–7, *left*).

It was already known—indeed, had been known for a long time—that, in the resting state, the inside of the fiber was electrically negative with respect to the outside. With the internal electrode it was now possible to measure that potential directly for the first time. The resting potential of the squid fiber was found to be

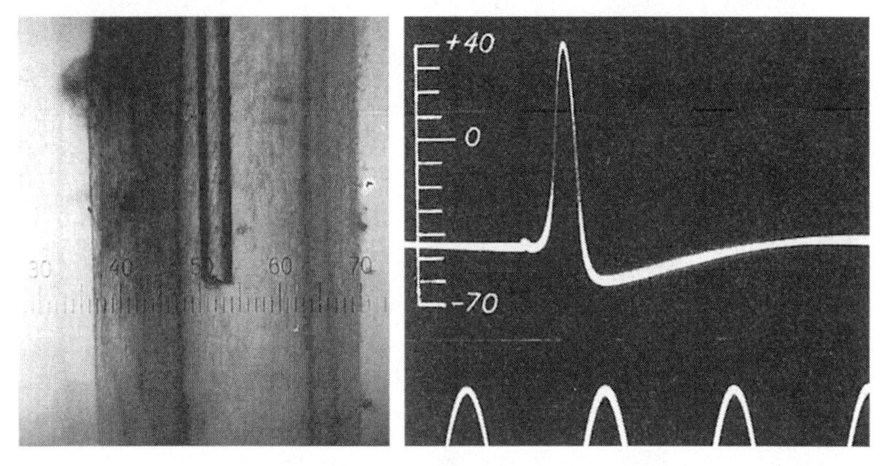

Figure 12–7 *Left*: Glass capillary electrode, 0.13 mm in diameter, inside a squid giant axon (clear space in figure) flanked by small nerve fibers. *Right*: Action potential and resting potential of squid giant axon, recorded with a capillary electrode inside the fiber and a reference electrode in the surrounding sea water. The interior of the axon has a resting potential of approximately –50 mV and an action potential of +40 mV, with respect to the outside of the fiber. Time marker, 500 Hz. This was the first demonstration that the electrical polarity of a nerve fiber membrane briefly reversed during a nerve impulse. (Reproduced by permission from Macmillan Publishers Ltd: *Nature*. Hodgkin AL, Huxley AF. Action potentials recorded from inside a nerve fiber. *Nature*; **144**: 710–711, 1939.)

approximately 50 millivolts (thousandths of a volt). Then came one of the most important observations in nerve physiology: when the squid fiber was stimulated and an impulse ran along the fiber, the potential on the inside of the fiber changed from a negative value to a positive one (Fig. 12–7, *right*). As a result, the action potential was larger than the resting potential, approximately twice as large. It was a result that had been hinted at when Hodgkin and Huxley had carried out their experiments on crab fibers earlier in the year, but now there was no doubt.

For those scientists working on nerve, the consequences of the new finding were enormous, since they contradicted the long-held theory of Bernstein that during the nerve impulse, the membrane potential would fall almost to zero, since the membrane would become permeable to all the ions in its vicinity. However, in formulating his theory, Bernstein had disregarded one of his own observations, that the action potential, as reconstructed with his differential rheotome, could sometimes be larger than the demarcation potential. With the new result of Hodgkin and Huxley, it was clear that something other than a nonspecific increase in membrane permeability was occurring in the nerve fiber membrane.

The intracellular recording of the nerve action potential was a spectacular achievement, one that had been obtained in only three weeks, the time that

Huxley had been in Plymouth (Fig. 12–8). The two young men were, in Hodgkin's words, "tremendously excited." They immediately wrote up their findings in the form of a short note to *Nature*.[21] It included a figure, for Huxley had been able to photograph the glass capillary in one of the fibers, as well as the action potential itself. At the end of the letter, there was a well-deserved acknowledgement to Mr. J. Z. Young "whose discovery of the giant axon in *Loligo* had made this work possible"—and who had also organized the supply of squid in Plymouth. Curiously, there was no mention of Cole, in whose laboratory Hodgkin had been introduced to the new preparation.

The intracellular recordings by Hodgkin and Huxley took research on the nerve impulse to a new level, and it is time to look back and see what had been accomplished. It had certainly been a long journey since that day, in the laboratory in Bologna, when Galvani had observed the unexpected twitch of the frog leg, and it had been almost a century since Helmholtz had measured the conduction velocity of the nerve impulse. After that had come the determination of the time-course of the impulse, a feat achieved by Bernstein with his differential rheotome and confirmed by Adrian with his valve amplifier and capillary electrometer. Gasser and Erlanger, and the enterprising Bishop, had refined the recordings with the cathode ray oscilloscope, and had been able to recognize the contributions of different groups of nerve fibers to the compound action potential. Without the benefit of an amplifier or oscilloscope, Keith Lucas had already deduced that the impulse traveled in an all-or-none manner, maintaining its full size as it passed along the fiber, and that its passage was followed by a brief refractoriness. And it had been Lucas's pupil, Adrian, who had shown that, when sensory information was conveyed to the brain, it was in the form of trains of identical impulses. Yet another Cambridge man, Hodgkin, by recording the local response, had confirmed the local circuit theory of impulse propagation—that current flowed through the membrane in advance of the impulse, depolarizing the membrane as it did so. As to what happened in the membrane after that, Cole and Curtis had shown that the electrical impedance fell, and the likeliest explanation was an increase in ionic permeability. And now Hodgkin and Huxley had demonstrated that, as the permeability increased, so the potential across the membrane reversed, the inner face becoming briefly positive with respect to the outer.

All this had now been established. The next problem was to find out the nature of the permeability change in the nerve fiber membrane. But at the end of August the research in Plymouth stopped. It had to, since events in Europe were about to take the ominous turn that had long been feared. On September 1, as the first light appeared in the sky, Hitler's tanks and infantry invaded Poland. Two days later, the British government issued its ultimatum for the German forces to withdraw. Receiving no response by the stipulated time, the British government, on the same day, declared war on the aggressor nation. By that time, however, the

Figure 12–8 Plymouth Sound, with Drake's Island, seen from the Marine Biological Laboratory.

two Cambridge men had already left Plymouth and driven back to their university. The equipment had been left in the Marine Biological Laboratory, for it was Hodgkin's hope that the war would soon be over and that the experiments could then be resumed.

The resumption was to be delayed for 8 years.

13

The War Years

At first nothing much happened. In Plymouth, its population swollen with children and expectant mothers evacuated from London and the other large industrial cities, there was a sense that the worst horrors of war might be avoided. Though sandbags continued to be piled around the most important buildings, and around the newly created air-raid control centers and first-aid posts, the air-raid sirens remained quiet. On a warm night, there might still be open-air dancing on the Hoe, with soldiers from the Citadel and sailors from the Devonport dockyards vying for the prettiest of the Plymouth girls.

While there was dancing and a sense of cautious optimism in Plymouth and elsewhere in the country, on the other side of the Channel events were taking a depressingly, seemingly inevitable, course. Shelled by the German Panzer tanks, strafed and bombed by the Luftwaffe, the soldiers of the British Expeditionary Force fell back across northern France. Miraculously, in the last days of May 1940, the majority would escape capture through the evacuation from Dunkirk. In Plymouth, as in the other ports along the English south coast, the defeated troops arrived not only in naval vessels but in the fishing boats and yachts that had also set off across the Channel. With the arrival of the soldiers, many of them seriously wounded, the horror of the new war was at last apparent, and there was the threat of much worse to follow.

In comparison with what was to come, the first bomber raids on Plymouth, those in the summer of 1940, were small in scale, a few bombs here and there, little damage to buildings, and casualties in the tens or fewer. In early September, however, a full year after the start of the conflict, the massive attacks on the British cities began, some involving more than 200 bombers. London was the first to experience the Blitz, then Birmingham and the other cities in the Midlands, and then, on the night of March 20, 1941, it was Plymouth's turn. The attack came a mere two hours after the visit of the King George VI and Queen Elizabeth to the city, and, by the time it had finished, the thousands of incendiary bombs had destroyed the heart of the city and killed hundreds. The narrow, winding streets that had led down to the harbor, the same centuries-old streets that had echoed to the steps of Drake and Hawkins and countless other sailors, no longer existed.[1]

The Marine Biological Laboratory, exposed as it was on the seaward side of the Citadel, did not escape either. Among the damaged rooms was that in which the giant axons had been impaled.

As for the neurophysiologists themselves, the war affected them all, but in different ways. One of the most remarkable contributions, but one fully in keeping with his adventurous nature, was that of Alexander Forbes. At the age of 60, the veteran took it upon himself, at his own expense, to fly aerial surveys of the Greenland and Labrador coasts, in the event that the United States would require far northerly bases.[2] It was a task for which he was well suited, having made earlier surveys of Labrador in 1931, at the request of Sir Wilfred Grenfell, the missionary doctor. Had Forbes not left his laboratory, he would probably have taken part in the important work on thalamic activation of the cortex by Morison and Dempsey,[3] experiments that had been his, Forbes,' conception.

On the other side of the world, the outbreak of the 1939–45 war in Europe had little effect at first on the three neurophysiologists at the Kanematsu Institute in Sydney, even though Australia, like the other countries of the Empire, had immediately joined the British side and had sent troops to fight in the Middle East. With Japan's subsequent entry into the war, however, the situation, from an Australian perspective, had suddenly become more serious and, after two years of successful collaboration, the little team of neurophysiologists in Sydney broke up.

Katz was the first to go. He had received his British naturalization papers in Sydney and had decided to repay the generosity of his host countries, Britain and Australia, by enrolling in the Royal Australian Air Force. He had already become knowledgeable about valves and oscilloscopes, and it was perhaps inevitable that he should become a radio officer, spending part of the war on a small island off the coast of New Guinea.[4] Since the latter was mostly in Japanese hands, there was considerable risk in sending wireless messages to the Australian mainland, announcing any flights of enemy bombers. In the excitement and danger of war, it is doubtful if Katz gave much thought to the nerve impulse or to transmission across the neuromuscular junction, even though he had appreciated the importance of the squid work of Hodgkin and Huxley, and, before anyone else, had realized what had probably occurred in the nerve membrane at the time of the action potential.[5]

Kuffler, too, enlisted, but his experience in the Australian army was to last less than an hour. Told that he would spend the war building roads in the outback, he declined to proceed further, later serving as a consulting neurologist to the American forces in their field hospital near Sydney.[6] This position gave him time to continue his research in the Kanematsu Institute, research that had now acquired a momentum of its own. Having been much the junior partner in the team, Kuffler had surprised Eccles and Katz by practicing his dissection skills until he was able to stimulate a single motor axon and record the end-plate activity of a single muscle fiber. It was a considerable technical feat, and not only was the new preparation more elegant, but its simplicity enabled the end-plate recordings to be made with greater resolution. This ability, to devise exquisitely refined preparations to explore his many new ideas, would become Kuffler's trademark. That, and his sense of fun.

After the loss of his two colleagues, Eccles carried on in the laboratory by himself. And a strange thing happened, for having convinced himself, on the basis of the experiments with Katz and Kuffler, that transmission at the neuromuscular junction was purely chemical, he began to have doubts again about other junctions. The doubts arose because of his new investigation of the sympathetic ganglion, a structure that he had studied before, while at Oxford. Just as with the neuromuscular junction, he was able to record a slow potential when he stimulated the nerve, in this case the nerve to the ganglion, in the presence of curare (Fig. 13–1). However, the early part of this potential seemed unaffected by eserine, and therefore, Eccles thought, was unlikely to have been generated by the release of acetylcholine from the nerve endings. Instead, he reasoned, "A preganglionic impulse exerts a brief depolarizing action on the ganglion cell by the direct effect of its action current."[7] There was also the consoling thought that, even if Dale had been victorious in the battle over the neuromuscular junction, there were still all the synapses in the central nervous system to be sorted out. Surely some of those, like the sympathetic ganglion, would prove to be partly or wholly "electrical" rather than merely "chemical."

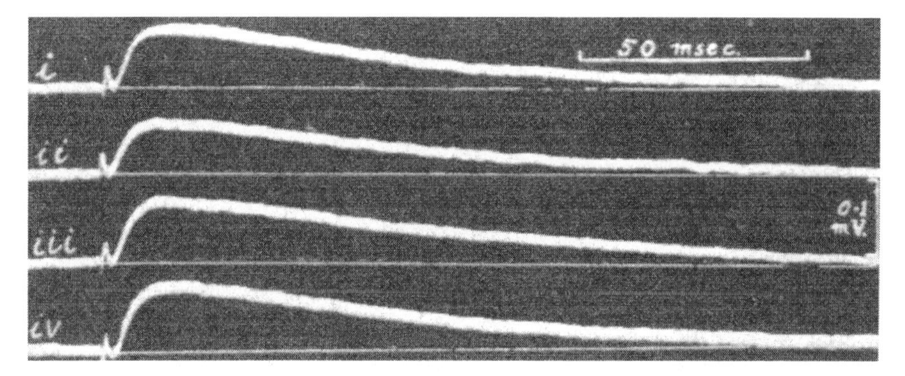

Figure 13–1 Recordings from a stellate (sympathetic) ganglion, in the presence of a small amount of curare (given to prevent the ganglion from discharging). In each of the four traces there is a slow wave, the synaptic potential, following stimulation of the nerve to the ganglion. The four traces are very similar, despite the fact that the lower three were obtained after the application of eserine—a drug that normally potentiates the action of acetylcholine. Despite this absence of effect, Eccles (correctly) attributed the slow potential to the action of acetylcholine, while arguing (incorrectly) that its onset (the "peak") was due to current flowing from the presynaptic nerve endings. (From Eccles JC. The nature of synaptic transmission in a sympathetic ganglion. *Journal of Physiology*; **103**: 27–54, 1944. Courtesy of John Wiley & Sons.)

With his research well established, Eccles would have carried on in Sydney, with the prospect of enlarging his laboratory space at the Kanematsu Institute when the war was over. It was a promise that had been made to him as part of his offer of employment but, as with other promises to Eccles, it would be a promise broken. An extra floor would be built above his present laboratories, yes, but it would not be for Eccles: the medical residents in the hospital would have new accommodation instead. Eccles, sensing that the new hospital administration had turned against him, did the only sensible thing: he left. There was, however, the problem of where to go. He could not move to one of the Australian universities, for there were no positions available. In the end he accepted the Chair of Physiology at the University of Otago in New Zealand, and with an extraordinarily heavy teaching load, gave up that which he loved most, his research.[7,8]

In England, the work on the squid had stopped immediately before the outbreak of war. Following the quick successes in Plymouth, Huxley returned to Cambridge, intending to make his career in physiology.[9] Adrian, as was his custom with promising students, had earlier advised him to obtain a medical degree first. With the outbreak of war, Huxley decided to complete his clinical training as quickly as possible, beginning with an introductory clinical course in Cambridge and then switching to University College Hospital in London for access to patients. Within a few months, however, the German Luftwaffe began its heavy bombing raids on the capital and the teaching stopped. University College, on the other side of Gower Street, was evacuated to the relative safety of the English countryside, and the same was happening to the other London schools and colleges.

Fortunately for science, Huxley never returned to his clinical studies. Instead he was diverted into anti-aircraft gunnery by the influential physiologist and government advisor A. V. Hill. The mathematicians and physicists had already been recruited into various war projects, but Hill knew that there were gifted physiologists such as Huxley also able to carry out the research so urgently needed. Most of Huxley's new work involved adapting the existing gun control systems, which had been designed to work by eye, to systems that would be controlled by radar. After 18 months, Huxley was transferred to the Admiralty and worked to improve naval gunnery. In the process, he learned more about statistics and something about servo systems. There were also gunnery trials at sea, with voyages on the battleship H.M.S. *King George V* in the Atlantic and on another warship in the Mediterranean. Ironically, though he would later invent an interference microscope, a micromanipulator, and a microtome for cutting ultra-thin sections of biological tissues, he was not put into instrument design and so did not learn electronics. Despite these restrictions, he was able to design a model gun sight incorporating an image on a cathode ray tube.[10] As with everything he made, it would have worked well. Typically, Huxley made the model himself in his small workshop at his London home. It is both relevant and of interest that Hill, responsible for moving Huxley into gunnery, had himself helped to develop anti-aircraft gunnery in the 1914–18 war.

Hodgkin, the son of pacifist parents, had an exciting war as a scientist, displaying considerable bravery on more than one occasion, in keeping with his adventurous nature. His description of his own doings, described matter-of-factly and with a touching modesty, is given in the middle section of his autobiography, *Chance and Design*.[11] For the first months of the war, he assisted Bryan Matthews, his senior colleague at Cambridge, in the Royal Air Force's newly created Physiological Laboratory. Situated in Farnborough, the "Laboratory" consisted initially of nothing more than a corrugated iron shed attached to a larger building. Yet it was in these unlikely premises that invaluable research was carried out on the effects of pilots undergoing sudden decompression—as in baling out of their planes at high altitude. Both Matthews and Hodgkin were subjects for the pressure chamber experiments and both were exposed to the "bends" or "chokes," caused by bubbles of nitrogen forming in the tissues and entering the bloodstream. It was dangerous, indeed potentially fatal, work. Not only could there be difficulty in breathing, associated with a choking sensation, and severe joint pains, but there was also the risk of such neurological complications as paralysis and impaired vision. A susceptible subject, Matthews was left with permanently damaged sight.[12]

By the beginning of the following year, 1940, Hodgkin was switched into radar research, of the kind needed by a fighter pilot for the interception of enemy aircraft. Once again the move was engineered by Hill. In addition to mathematics, the new work required an extensive knowledge of the relatively new science of electronics, which Hodgkin rapidly acquired for himself. As he was moved around the country, usually working and sleeping in simple huts, the project brought him into contact with other talented young men, some of whom would distinguish themselves after the war in physics and radio-astronomy. It also presented him with challenges of a different kind. The best, indeed the only responsible, way to assess the effectiveness of the developing radar systems was to go up in a plane and observe the results oneself, and this Hodgkin did. It involved him sitting in the rear of a Beaufighter, the plane used for this developmental work, with his back to the equipment. This arrangement, though obligatory, was dangerous since, in addition to the observer having to cope with the effects of altitude in an unpressurized plane, the equipment could malfunction without his knowledge. In particular, there was a risk that the very high voltages used would "arc" in the thin atmosphere inside the plane and start a fire, as actually happened. Moreover, the addition of the radar equipment to its nose gave the plane an appearance unfamiliar to pilots of other fighters, both friendly and enemy. One of Hodgkin's colleagues had, in fact, been killed when his plane had been shot down by a British Spitfire. Unlike Keith Lucas in the 1914–18 war, Hodgkin survived his aerial hazards.

The seriousness of the wartime challenges, and the intensity of the responses with which they were met, gave Hodgkin little inclination, and certainly no opportunity, to take the squid work any further. In the end, he stopped thinking about it altogether. It was just as well, for the Marine Biological Laboratory, in which he and Huxley had spent those exciting days in the summer of 1939, and in which he had left his prized electrophysiological equipment, was badly damaged in one of the bombing raids on Plymouth. Having already had the good fortune to have seen the squid work in Cole's laboratory in 1938, and to have enlisted Huxley as his research associate in Plymouth the following year, Hodgkin's luck held a third time. He had given permission for the equipment to be borrowed by one of his former university associates and, just in time, it had been removed from harm's way and taken back to the safety of Cambridge.

Much later, in the autumn of 1944, at a time when the Allies were advancing across Europe and Germany's defeat was inevitable, Hodgkin was able to visit Cambridge and to discuss his future with Adrian and with his former colleague, William Rushton. Hodgkin found his old department much diminished, with Adrian shouldering most of the teaching, including the obligatory course of physiology lectures for the nurses in Addenbrooke's Hospital. There had been a family misfortune, too, for Adrian's wife had fallen while the two had been climbing Great Gable in the Lake District. With her femur fractured, she had lain, exposed and in agony, for a full 5 to 6 hours before help had arrived;[13] later, the fractured bone had necessitated an above-knee amputation. Yet, despite having to undertake household work on top of his heavy teaching and administrative loads, Adrian had somehow found time to experiment during the war years.

This time it had been an examination of the sensory receiving areas of the brain, made by mapping out the areas of the cerebral cortex responding to stimulation of the various parts of the body surface. It was already known, from anatomical and clinical studies, that the sensory nerve pathways crossed to the opposite side in the brain. Thus, sensations for the left side of the body were generated by the right hemisphere of the brain, and those for the right side of the body by the left hemisphere. Other neurophysiologists had started to explore the situation further, most notably Woolsey and his group at Johns Hopkins, but Adrian was able to show something that, though unexpected at the time, made perfectly good sense. It was this: that the map of the body varied greatly from one animal species to another, and, in each case, reflected the behaviors of the animal. For example, in the pig, there was a relatively large area of cortex devoted to the snout, with which the animal forages for food, and little, if any, for the remainder of the body (Fig. 13–2). In the horse, though, it was the nostrils that occupied half of the sensory area, while in the monkey substantial areas were given over to the face and paws. In each animal the part of the body most prominently represented was that with the greatest number of nerve fibers coming from the skin.[14] Adrian had even tried recording from the exposed human cortex, bringing equipment into the neurosurgical operating theater at the London Hospital.[15] Many years later, others would show that, in the early fetus, the cerebral cortex is a "tabula rasa," a blank slate waiting to be drawn upon by the nerve fibers growing toward it from the thalamus.[16] The more the fibers, the greater the cortical territory taken over. The larger the army, the greater the conquest.

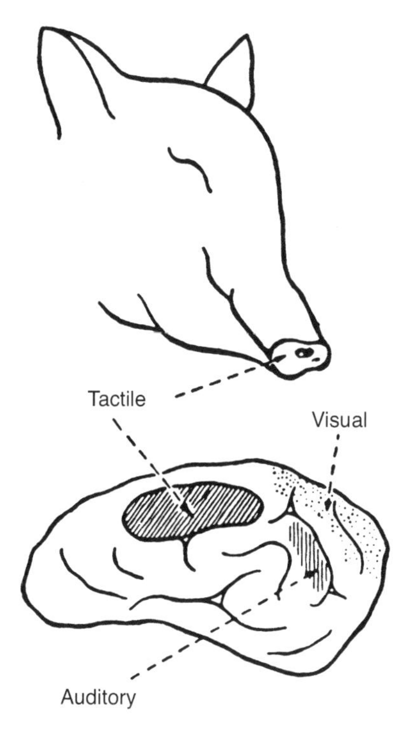

Figure 13–2 The sensory receiving areas in the pig brain. The entire area for the body surface has been devoted to the snout, which is especially important to the pig as it is used for touching, exploring, and digging. (From Adrian ED. Sensory areas of the brain. *Lancet* (July); 33–30, 1943. Courtesy of Elsevier.)

Having revisited his old department, Hodgkin, later in the same year, was finally able to write the first draft of the full paper describing the squid experiments, showing it to Andrew Huxley when the two met in London for lunch in January 1945. Almost 30 years earlier, during the 1914–18 war, Adrian and Keith Lucas had met under similar circumstances. The next few years would determine whether the reunion of the younger Cambridge neurophysiologists would have a more favorable outcome. Much would depend on what was happening on the other side of the Atlantic.

Although Hodgkin had been brought to the United States for a six-week period in 1944 to share his expertise with radar, he had not managed to see Kenneth Cole. Instead his visits to New York enabled him to rekindle a friendship with the daughter of Peyton Rous, the experimental pathologist who was to win a Nobel Prize at the age of 87, long after his demonstration that viruses were capable of inducing cancer. Within weeks, Marion ("Marni") Rous and Alan Hodgkin were married.

Had Hodgkin tried to contact Cole, he would, in any case, have found it difficult. After Hodgkin's visit to Woods Hole in 1938, Cole and Curtis had continued their rather leisurely experiments on the squid giant axon. During the next year they were, like Hodgkin and Huxley at approximately the same time, able to insert an electrode into a giant axon and to measure the full amplitude of the action potential. In two important respects, however, their experiment was inferior. First, they had used a metal wire as their internal electrode. Unlike the capillary electrode used by Hodgkin and Huxley, the metal wire produced a polarization potential as it contacted the electrolytes in the axoplasm, and this spurious potential made determination of the fiber resting membrane potential impossible. Even if they had used the correct type of electrode, the determination would still have been difficult as their amplifier was incapable of sustaining a steady (DC) signal, such as the resting potential. As a result, they were, unlike the Cambridge men, unable to conclude that the action potential exceeded the resting potential. In the following two years, 1940 and 1941, they worked with improved apparatus, but there was now a different type of error. Their new amplifier, although now suited for resting potential measurements, exaggerated the sizes of the action potentials. An important extension of their previous work was to determine the possible effects of replacing the sea water bathing the giant axons with a sugar solution (isotonic dextrose). The resting and action potentials, they reported, were "not appreciably affected."[17] This statement, wrong as it was, would have important consequences.

In 1942, following the entry of the United States into the war, Cole, like Hodgkin and Huxley, was obliged to discontinue his research on nerve fibers. Instead he became Principal Biophysicist at the Metallurgical Laboratory in the University of Chicago, studying the effects of radiation on biological tissues.[18] It was work vital to the Manhattan Project, the highly secret development of the atomic bomb in Los Alamos. As the first bomb neared completion, Cole, like some of the other scientists engaged in this work, voted against the decision to drop the bomb on a Japanese city.

Not only did Hodgkin not see Cole during his six-week stay in the United States, but there is no record of him visiting the Rockefeller Institute, where he had worked for a year in 1937–38. Had he gone, he would have found the situation very different to the one he had encountered before. Gasser, still the Director, had become the chair of a national committee supervising experiments dealing with chemical warfare. As the other scientists had been subsumed into the war effort, work in the neurophysiology laboratories had dwindled. There were two exceptions. David Lloyd continued to study reflex excitation and inhibition in the cat spinal cord, his recordings from the motor nerve fibers in the ventral roots providing a greater precision than had been possible with the muscle contractions of Eccles and Sherrington.[19]

The other occupant of the neurophysiology laboratories in the Institute was Lorente de Nó, toiling away at his measurements of demarcation potential and action potential in the bullfrog sciatic nerve, either unaware or unconcerned that more definitive results could be now be obtained in the squid axon. It was now five years since the note by Hodgkin and Huxley had been published in *Nature*.

Though the Institute was lacking many of its scientists, it had acquired a new glory. In October 1944 its director was awarded a Nobel Prize, Herbert Gasser sharing it with Joseph Erlanger for the work they had done together in St. Louis in the 1920s. It was an interesting choice for the Nobel Committee to have made, in that the two investigators had discovered little that was not previously known. The duration of the action potential, its conduction velocity along the fiber, the refractoriness during and immediately after the potential—all these had been determined before. The innovation of Gasser and Erlanger was to demonstrate these various features on the cathode ray oscilloscope, with a greater clarity and accuracy than had been possible before. They had also drawn attention to the different groups of fibers in a peripheral nerve, though, as Gasser would later show, the original observations were flawed since at least some of the notches on the oscilloscope display were artifacts, probably caused by the deteriorating condition of the continuously stimulated fibers. Worse, the original work on the C fibers, those extremely thin fibers without myelin sheaths, work that had been cited by the Nobel Committee, was not that of Gasser and Erlanger at all, but of Heinbecker and Bishop.[20] It was true, however, that Gasser had studied these fibers subsequently, and would continue to do so in his later years. It was also true that Erlanger, in collaboration with Edgar Blair, had done some later work on conduction in single frog A nerve fibers, including a detailed analysis of excitability changes during the passage of a polarizing current through the fibers; it was this work on which he dwelt in his Nobel lecture.[21]

But why was Bishop not a fellow recipient of the Prize? He had been involved in the use of the cathode ray oscilloscope almost from the beginning, and had been co-author of most of the influential early papers. In addition to the later study of the C fibers, he had taken neurophysiology in several new directions, examining the effects of pharmacological compounds on muscle excitability, the action potentials of skeletal muscles, and, above all, carrying out the first systematic survey of the electric responses of the different parts of the visual system.[22] In these last studies, which were the first evoked potential recordings from the brain, he had come to question whether the slow waves were due to the summated action potentials in groups of nerve fibers. Instead he suggested, correctly as it would turn out, that they emanated from the cell bodies and dendrites. Though not medically qualified himself, he had extended the grasp of neurophysiology to include topics of immediate clinical relevance such as nerve degeneration, poliomyelitis, spastic vascular disease, and pain. Some of these experiments had been done with Peter Heinbecker, the young Canadian surgeon, some with James O'Leary, another young man, this time from Texas, and some with others. In his range of enquiry, Bishop had shown himself to be the equal of Adrian, though without the other man's gift for exposition. During the pioneering days in St. Louis, there could be little doubt that, of the three neurophysiologists, Bishop was a more accomplished investigator than Gasser or Erlanger. It was an opinion that the modest and retiring Gasser would probably have agreed with.[23]

Most unusually for a potential Laureate, Gasser had declined to submit information about his work to the Nobel Committee when he and Erlanger had been nominated earlier, in 1938, feeling that it would have been self-promotion had he complied. On learning that he had won the Prize in 1944, his reaction was one of dismay, since he had long before put the St. Louis work out of his mind. It was necessary for him to read his old papers again "in order to regain touch," as he put it.[24]

Given Bishop's great contributions to neurophysiology, contributions he had somehow continued during the war when the practical difficulties were immense, how could he have been overlooked by the Nobel Committee? The answer may well have been Erlanger. The older man had never forgiven Bishop for his temerity, as Erlanger saw it, in carrying out, and submitting for publication, the work on the C fibers in the autonomic nervous system. That disagreement probably rested on a natural antipathy to Bishop—the big untidy man, unrestrained by authority, unafraid to speak out, and yet, unfairly as it must have seemed to Erlanger, blessed with a gift for experimentation and for attracting young scientists to him. Following their rift over the C fibers, Erlanger had minimized Bishop's contributions to the physiology of peripheral nerve in his writings and lectures.[25] Again, when Erlanger and Gasser had been nominated for the Nobel Prize in 1938, by Ranson, Erlanger would have made little mention of Bishop in submitting his own work for consideration by the Nobel Committee.[26] Yet Erlanger must have realized that an injustice had been done, for he later, with Gasser, gave Bishop a share of the award. Bishop, who had accepted his exclusion from the prize with his habitual good nature, put Erlanger's conscience money into the laboratory expenses.

There was one other aspect of the 1944 award that was unusual. The nomination of Gasser and Erlanger had been made not by another neuroscientist but by the Head of Surgery in St. Louis, a man who had probably never set foot in a neurophysiology laboratory and who was unlikely to have followed the work of Gasser and Erlanger in detail. There are two possible explanations, one of which lies in the fact that Evarts Graham, the surgeon, was the husband of Helen Graham, the accomplished neurophysiologist in the pharmacology department. Helen Graham may have persuaded her husband, the person with the necessary academic stature, to make the nomination. She would have done so without any regard for Erlanger, a man who had ignored her work and had excluded her from his brown-bag lunchtime discussions. In contrast, Gasser was a man she admired and had every reason to be grateful to. It was Gasser who, recognizing her potential and ignoring the prejudice of the time against female academics, had appointed her to the new Department of Pharmacology in St. Louis.[27] It was Gasser who would have explained the new neurophysiological equipment to her and introduced her to some of the experimental techniques. Together they had studied the after-potentials that followed the spike in the nerve compound action potentials,[28] and they had analyzed the slower responses that could be recorded from the surface of the spinal cord after the fibers in the sensory nerve root had been stimulated.[29] In keeping with the mandate of a pharmacology department, they had explored the effects of certain naturally occurring compounds on the nerve action potential. Long after Gasser had left St. Louis, Helen Graham would visit Gasser in New York to discuss her latest findings. It was during one of these visits, in 1937, that Alan Hodgkin had been introduced to her and had found her "very nice."[30]

The other possible influence on the Head of Surgery, in recommending Gasser and Erlanger for a Nobel Prize, was a colleague whom the surgeon would have met every week in faculty committee meetings in St. Louis. It was someone who would have made sure that Bishop was not a fellow nominee. . . Erlanger himself.[31]

Rockefeller Institute, New York, May 7, 1945

It had already started to rain, and was therefore a good day for an experiment. Lorente paused to watch the traffic crossing over the river on the Queensboro Bridge. It seemed as if the city never rested. How different it all was from Madrid. In this country, it seemed, everyone was in a rush, so that there was little time for contemplation and for observing the normal courtesies of life. It would be nice to go back when this study was finally complete, and to walk along the Paseo again, stopping for coffee under the shade of a tree at one of the sidewalk cafés. America though, America was the best place to do science. Oh yes, he had shown that in the past 12 years! And now that the war in Europe was over, and the Japanese were in their last, desperate retreat across the Pacific islands, it would not be long before the other neurophysiologists were back and he and David Lloyd would have someone else to talk to.

Rounding the corner, Lorente entered Smith Hall by its main door and turned left to his room. Putting on his white coat, he crossed to the aquarium in the corner of the room, slid the cover back, and reached for one of the bullfrogs. Another of Florida's finest. Much bigger than the leopard frogs, and therefore much longer sciatic nerves to work with.

Sitting at the bench, and still holding the frog round its middle, he picked up the needle by its wooden handle. Then, flexing the head of the frog down with the index finger of his left hand, he used his right hand to run the point of the needle down the back of the animal's skull until it found the small gap above the first cervical vertebra. Then, as he had done hundreds of times before, Lorente pushed the needle through the opening and into the brain, turning the point from side to side. The body of the frog, stiffened, its legs extending. It was now officially dead, incapable of feeling pain. Lorente laid the frog down on the dissecting board and used a pair of scissors to cut the skin round the animal's waist. Now for the "pantalones." Gripping the cut edge with a pair of blunt forceps, he pulled the skin down over the animal's legs. It had always struck him as a bizarre maneuver, and it was one that could only be carried out on frogs and toads. But it always worked, and it was so convenient, for the frog had been deprived of its "trousers." The muscles on the back of the thigh were now ready to be gently pulled apart, so as to expose the gleaming white sciatic nerve. Lorente continued the dissection, revealing more and more of the nerve and employing a fine pair of scissors to free it from the surrounding tissues, never touching the nerve except with a delicate glass hook. So long were these bullfrog nerves that he did not need to include the sacral plexus.

Instead, turning to the distal end of the sciatic nerve, he separated it from the back of the knee, cut the branches to the calf muscles, and found the slender continuation to the foot. Now, with more than 15 cm of nerve available to him, he tied a fine cotton thread to each end and cut the nerve free. Holding it by the threads at each end, and taking care not to stretch it, he laid the nerve in the Plexiglas chamber, maneuvering it so that it fitted into the slots at the sides of the three cylindrical compartments. Next, to make sure that the bathing fluid, the Ringer's solution, would not leak out through the slots, he carefully applied a small amount of Vaseline to each. Checking that there was water in the bottom of the chamber to help keep the nerve moist, he placed the cover on the chamber and sat back in his chair.

Still waiting for the nerve to equilibrate with its new surroundings inside the chamber, Lorente looked at the signed photograph of Cajal on the wall above the desk. What a man, what a Master, he had been—even though he, Lorente, had once proven him wrong when the Master had challenged him over the neurons in the mouse cortex! What would the Master think now, to see his protégé bringing the neurons to life, by recording their impulses?

Lorente turned his attention back to the nerve in the chamber, and took the first reading with the galvanometer. Good, no measurable potential between the segments of nerve in the first two compartments. With a dropping pipette, taking care not to touch the nerve, he removed the Ringer's fluid from inside one of the cylinders and replaced it

Theobald Smith Hall, at the Rockefeller Institute (now the Rockefeller University). It was in this building that Gasser, Lorente de Nó, and the other neurophysiologists had their laboratories.

with the potassium solution, starting his stopwatch at the same time. After 30 seconds he took his first reading. No difference in potential between the treated and untreated segments of nerve. Lorente made his first entry in his notebook. One minute, two minutes, and still no difference. One minute more and the first, small deflection of the galvanometer needle.[1] *By now the potassium had obviously penetrated the connective tissue sheath around the sciatic nerve and had begun to depolarize the nerve fibers. This experiment, this examination of the possible barrier presented by the sheath, was one that he really ought to have done at the outset, before he had started to examine the effects of the drugs. Still, everything seemed to be in order, and these last pieces of the huge study would fit nicely into place . . .*

14

Sodium Unmasked

The end of the war came in 1945, first Germany, and then Japan, accepting defeat. And so, one after another, the neurophysiologists were discharged from the Forces. In Sydney, Katz received, and accepted, an invitation from A. V. Hill to resume his work at University College London, this time as Assistant Director of Research in the Biophysics Unit. For the return to Britain, Katz and his new wife managed to find spartan accommodation in one of the liners that were still serving as troop ships. Kuffler had already left, also in the company of an Australian bride, but for Chicago. Eccles, meanwhile, was almost single-handedly teaching the whole physiology curriculum to the medical students in New Zealand. Although the deprivation was almost unbearable for him, there was little time for experiments.

In Chicago, and then in Oak Ridge, Tennessee, Cole had spent the war in charge of an increasingly large research unit, investigating the biomedical problems that might result from exposure to atomic radiation.[1] At the end of the war he was invited to return to Chicago, as the Head of the new Biophysics Institute of the University of Chicago. Although he no longer had Howard Curtis as a colleague, he was joined by George Marmont, who had been with him at Columbia University in New York. Together they had gone to Woods Hole in the summer of 1946, but the research, more experiments on longitudinal resistance, had gone badly.

In England, things were rather better. In the final year of the war, Hodgkin had completed the paper with Huxley, describing the 1939 experiments on the squid axon in Plymouth, experiments in which they had been the first to record the action potential with an intracellular electrode, and to observe the transient reversal of membrane polarity. Not long after the submission of the paper to the *Journal of Physiology*,[2] in February 1945, Hodgkin was released from service, Adrian having made a plea for help in the teaching of physiology at Cambridge. Although some of his, Hodgkin's, apparatus had been destroyed during the bombing of Plymouth, the most valuable part, the rack of electronic equipment, had been saved through the fortuitous intervention of one of his colleagues. After Huxley rejoined Hodgkin, at the end of 1945, the two used a rather indirect method, based on the capacitance of the membrane, to estimate the amount of potassium that might escape from a nerve fiber during the passage of a single action potential. The experiments, performed on single crab axons, indicated that about 10,000 potassium ions escaped through 1 cm^2 of membrane. Material for another paper in the *Journal of Physiology*.[3] Already, the contrast with the work in Cole's laboratory was striking. The Cambridge men knew exactly what they wanted to do, carried out the requisite experiments, published the work quickly, and moved smartly on to the next part of the nerve problem. If Cole had had either Hodgkin or Huxley as his partner, the story would have been different.

It was now time to start work on the key question that had arisen from the earlier work on the squid axon. What happened to the membrane to cause it to reverse its polarity during the action potential? The one person who had known, instantly, on reading Hodgkin and Huxley's 1939 letter to *Nature*, was Bernard Katz (Fig. 14–1). Giving a lecture to medical students in Sydney, one of whom was Phyllis Shewcraft, Stephen Kuffler's future wife, Katz had remarked that the membrane must have become permeable to sodium ions.[4] Had they scanned the early literature, Hodgkin and Huxley might have made this suggestion themselves. As long ago as 1902, Ernest Overton had published a very interesting paper in *Pflügers Archiv fur die Gesammte Physiologie des Menschen und der Thiere*.[5] Overton, born in England, had worked in Germany before moving to the small Swedish university town of Lund. While in Germany, at the University of Würzburg, he had observed that, if frog muscles were bathed in sodium-free solutions, they ceased to contract; they had become inexcitable. Restoring sodium to the bathing fluid brought back the excitability; the only ion that could substitute for sodium was lithium. So overcome was Overton by these dramatic results that he had rushed out into his garden to recover from his excitement. In reviewing his findings, he concluded—prophetically:

> "The role played by sodium or lithium ions in the transmission of excitability and in the contraction of muscle has not yet been explained; perhaps during these processes a certain exchange takes place between the potassium ions of the muscle fibres and the sodium ions of the solution flowing around them [extracellular], but this assumption is associated with considerable difficulties."

Overton was unable to demonstrate a similar dependency on sodium by nerve fibers, though he thought this might be due to the presence of sodium ions underneath the myelin sheaths surrounding the axis cylinders.

Instead of considering increased sodium permeability of the nerve membrane as a possible mechanism, Hodgkin and Huxley put forward four other explanations in their 1945 paper. The membrane might become selectively permeable to the anions present in the interior of the nerve fiber, as Adrian had speculated some years earlier in relation to the initiation of impulses in a sensory receptor. Alternatively, the electrically charged fatty acid molecules, which formed the two layers of the nerve fiber membrane, might momentarily change their orientation. The third and fourth possibilities were a change in the capacitance of the membrane, or the presence of an inductance. The fact that there were four explanations for the data was evidence in itself that the authors had little confidence in any of them. Indeed, all four would eventually be shown to be wrong.

There may have been another reason for Hodgkin and Huxley's failure to consider the possibility of an increase in the permeability of the nerve fiber membrane to sodium, though it was not mentioned in their 1945 paper. When Curtis and Cole had, in 1942, published their own findings with intracellular recording electrodes, they reported the sizes of the resting and action potentials of the squid giant axons.[6] Since they had taken particular care to minimize the unwanted junction potentials in the recording circuit, their values of resting potential were actually more reliable than Hodgkin and Huxley's. They had then described the effects of changing the composition of the fluid bathing the giant axons. In the first series of experiments, it was the concentration of potassium ions in the sea water that had been altered. They found that increasing the concentration caused the resting membrane potential to fall, as would be expected if this potential had been due to the membrane being selectively permeable to potassium. This result was not surprising, for it had long been known that raising the potassium concentration of the fluid surrounding frog sciatic nerve would produce a demarcation potential, as described earlier.[7] However, Curtis and Cole were able to show that the decrease in resting potential was proportional to the logarithm of the potassium gradient across the membrane, as predicted from the Nernst equation.[8] Thus, a tenfold increase in the external concentration of potassium reduced the resting potential by approximately 50 mV, which was close to the value of 58 mV calculated from the Nernst equation. This was a very important result, and one for which the authors were somehow not given full credit. However, it was the final part of Curtis and Cole's study that may have caused Hodgkin and Huxley to hesitate in attributing the action potential to a sodium effect. Curtis and Cole had changed the concentrations of other ions in the circulating fluid. Increasing the calcium ion concentration, they found, had little effect on the resting potential. Further, when all the ions were removed, by replacing the sea water with isosmotic dextrose, "the height of the action potential was not appreciably altered." Since sodium was by far the most prevalent cation in sea water, it appeared that it had little to do with the action potential.

It was Katz who provided Hodgkin with the stimulus to investigate the possible role of sodium. Since his arrival in London, in 1946, Katz had helped to get A. V. Hill's Biophysics Unit at University College operational again. The college had been damaged in the bombing, and parts of it were still occupied by the civil service. Not only was Katz prepared to put his wartime knowledge of electronics to the building of new equipment from surplus radar gear, but he even painted the laboratory walls.[9] Once installed again, he had spent a year investigating the local response in single crab nerve fibers, and then the electrical properties of muscle fiber membranes. In the crab axons, he had observed, among other things, that replacing the normal bathing fluid with a salt-free sugar solution resulted in inexcitability.[10]

Figure 14–1 Bernard Katz dissecting a squid axon at the Marine Biological Laboratory, Plymouth. (Photograph courtesy of Mrs. Ruth Weidmann)

The year that Katz published his findings, 1947, started off brutally cold in the United Kingdom. To make matters worse, there was a shortage of coal, so that many buildings were without heat, including those at Cambridge University. In his rooms at Trinity College, Andrew Huxley could be found with mittens on his hands, the slender fingers emerging to operate the levers of a Brunsviga calculating machine. He had begun to calculate the flow of currents through the squid axon membrane that would account for the action potentials observed by Hodgkin and himself in 1939.

But was the initial current really carried by sodium ions, as suggested by Overton's old work on muscle, and Katz's recent observations on single crab nerve fibers? Was sodium responsible for the dramatic decrease in membrane resistance found during the action potential by Cole and Curtis? By the middle of June, Hodgkin had settled into the Marine Biological Laboratory at Plymouth once more, having brought his electronic equipment with him, just as he had eight years previously.

Plymouth had changed since his previous visit, that with Huxley in 1939. Indeed, much of the city was barely recognizable.[11] In the center, the buildings lining the narrow streets, streets that had bustled with activity before the war, had been demolished during the German bombing raids. Elsewhere a few standing walls enabled a more substantial structure to be identified—the Guildhall here, St. Andrew's Church there. Nor was the city center the only casualty of the war. The Marine Biological Laboratory had also been hit by bombs and the repairs were still under way when Hodgkin took up residence once more. Elsewhere in the country, things were little better. A population that had celebrated the end of the war two years earlier still found itself saddled with many of the wartime restrictions—among them food rationing. The new Labour government, the government that had been elected to create a postwar Jerusalem, seemed curiously incapable of improving the lot of its people. Even worse was the situation in Germany, a country divided up among the victors, many of its combatants still unaccounted for, and with tens of thousands of civilians killed in the horrific bombing raids on Hamburg and Dresden in the last phases of the war.

Alone in his laboratory, Hodgkin had to learn again the techniques needed to impale the squid giant axons with glass microelectrodes. There was no Huxley to help him, since his former collaborator was about to get married. Once again, as always seemed to be the case with Hodgkin, the results came quickly. By September, when he was joined by Katz, the key evidence had already been obtained.[12] Both the amplitude of the reversed membrane potential, and the rate of rise of the action potential, were proportional to the concentration of sodium in the bathing fluid (Figs. 14–2, 14–3). Interestingly, the underswing following the spike of the action potential was sensitive to the potassium concentration, becoming larger if the concentration was lowered. To Hodgkin and Katz, this suggested that a phase of high potassium permeability followed the increase in sodium permeability. Thus, a sodium current was associated with the reversal of the membrane potential, and a potassium current with the return of the membrane to its resting state. Hodgkin and Katz were able to relate their results to an equation derived by Goldman for a membrane across which a constant voltage had been applied, with

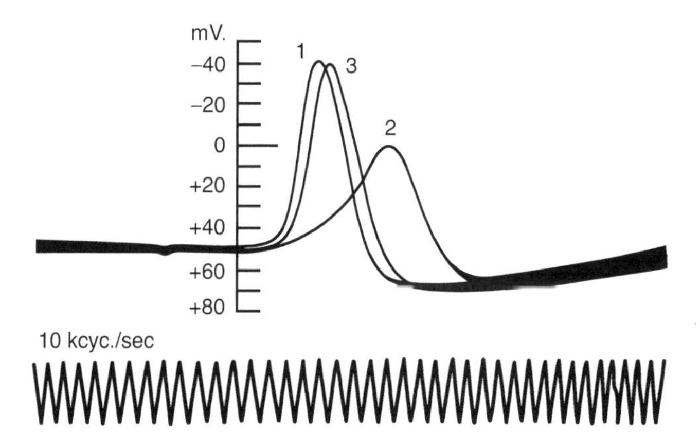

Figure 14–2 Action of sodium-deficient solution on the resting and action potentials of a squid giant axon. In *1* the axon is in sea water; in *2* the sea water has been replaced by a sugar solution (isotonic dextrose) and the action potential amplitude is halved; in *3* the axon is in sea water again, and the action potential has recovered its full size. Note that the absence of sodium does not affect the resting potential (horizontal traces to the left of the vertical calibration). (Modified from Hodgkin AL, Katz B. The effect of sodium ions on the electrical activity of the giant axon of the squid. *Journal of Physiology*; **108**: 37–77, 1949. Courtesy of John Wiley & Sons.)

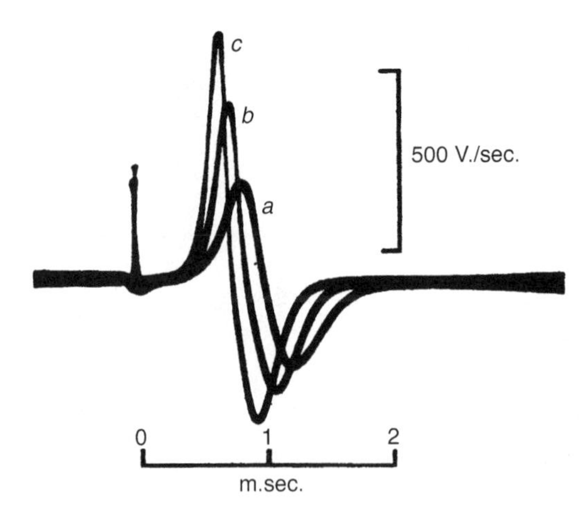

Figure 14–3 Effect of sodium concentration on the rate of rise of the action potential. The rise is greatest when the concentration is increased by rather more than 50% (*c*) and least when the concentration is halved (*a*), in comparison with that of normal sea water (*b*). (From Hodgkin AL, Katz B. The effect of sodium ions on the electrical activity of the giant axon of the squid. *Journal of Physiology*; **108**: 37–77, 1949. Courtesy of John Wiley & Sons.)

the ions on either side free to move through it under the influence of the electric field and their respective concentration gradients.[13] Thus, at room temperature:

$$V = 58 \log_{10} \frac{\{P_K[K]_o + P_{Na}[Na]_o + P_{Cl}[Cl]_i\}}{\{P_K[K]_i + P_{Na}[Na]_i + P_{Cl}[Cl]_o\}}$$

(Where V is the voltage across the membrane, and P_K, P_{Na}, and P_{Cl} are the permeabilities of the membrane to potassium, sodium, and chloride ions respectively. The values in brackets are the concentrations of the same ions.)

The equation could be made to fit the observed resting potential if the relative permeabilities for potassium, sodium, and chloride were taken as 1, 0.04, and 0.45 respectively. In contrast, the reversal of the membrane polarity during the action potential could be explained if the membrane had momentarily increased its permeability to sodium 500-fold.

How was it that Curtis and Cole had not seen the dramatic effects on the action potential of altering the sodium concentrations in the bathing fluid? The only possible explanation is that they had never carried out the experiment. This is not to suggest fraud, for in their 1942 paper, no numbers were given to back up their statement that, in the absence of sodium ions, "the height of the action potential was not appreciably altered." Most likely, they *thought* they had done the experiment, which is another thing entirely. It is something that can happen when the writing is done some time after the experiments. In this case, it is almost certain that the paper had been written by Curtis rather than Cole. Not only is it Curtis's name that appears as the leading author, but the discussion of the results contains none of the mathematics in which Cole delighted. There is another clue. At the time the paper was written, Curtis had already left Cole to take a position at Johns Hopkins, with the intention of becoming a physiologist. Another paper, this time with himself as first author, would have been a useful start to his new career. Writing the paper well after the end of the experiments, in a new department, and without Cole available for discussion, Curtis could have made a genuine mistake. As for Cole, he would later have no recollection of the dextrose experiments.[14] For Cole's reputation, however, the error was unfortunate.

If Curtis and Cole were short of data, at least in relation to the sodium experiments, there appeared, in 1947, another publication that was overflowing with it. Lorente de Nó's massive study of the frog sciatic nerve was published by the Rockefeller Institute. In one sense, it was a splendid achievement, the culmination of 10 years of intensive work by one man. With over a thousand pages, *A Study of Nerve Physiology* appeared in two volumes, each page filled with small print.[15] There were 453 figures, most of which showed the results that had been obtained in approximately 400 nerves and 300 experiments. In those figures in which the data were plotted, there could be as many as 18 curves, each with 50 or so points on it. In the illustrations prepared from oscilloscope recordings, as many as 40 separate panels were carefully arranged in montages. The traces in each panel were perfectly photographed, developed, and printed, so that no retouching was needed. Each of the panels was annotated by hand, with neat writing in white ink on the black backgrounds of the photographs (Fig. 14–4).

A Study of Nerve Physiology was magnificent, and it is doubtful if anyone else could have done it. It was the pinnacle of the classical approach, in which all the measurements had been made from large populations of nerve fibers with external electrodes. Had the work on the squid giant axon never taken place, Lorente's study would have become the standard reference, the electrophysiologist's bible.

Unfortunately, Lorente's main conclusions were wrong. Influenced by the long times taken for changes in the nerve bathing solution to achieve their maximal effect, he argued that the resting membrane potential was not determined by the concentrations of potassium ions inside and outside the nerve fibers. Nor, he stated, was the action potential dependent on a high concentration of sodium ions in the external fluid, for excitability was maintained in the presence of a small fraction of the normal sodium value. He also concluded that, in a nerve fiber with a myelin sheath (that is, the fatty covering around the core of the fiber), the impulse did not jump from one interruption in the sheath to another ("saltatory" conduction) but traveled continuously. Another error. The membrane potential itself, he postulated, though again incorrectly, was composed of three fractions, each of which corresponded to a double layer in the membrane.

How had these mistakes come about? It was not that the experimental observations were faulty; indeed, there was abundant evidence that the studies had

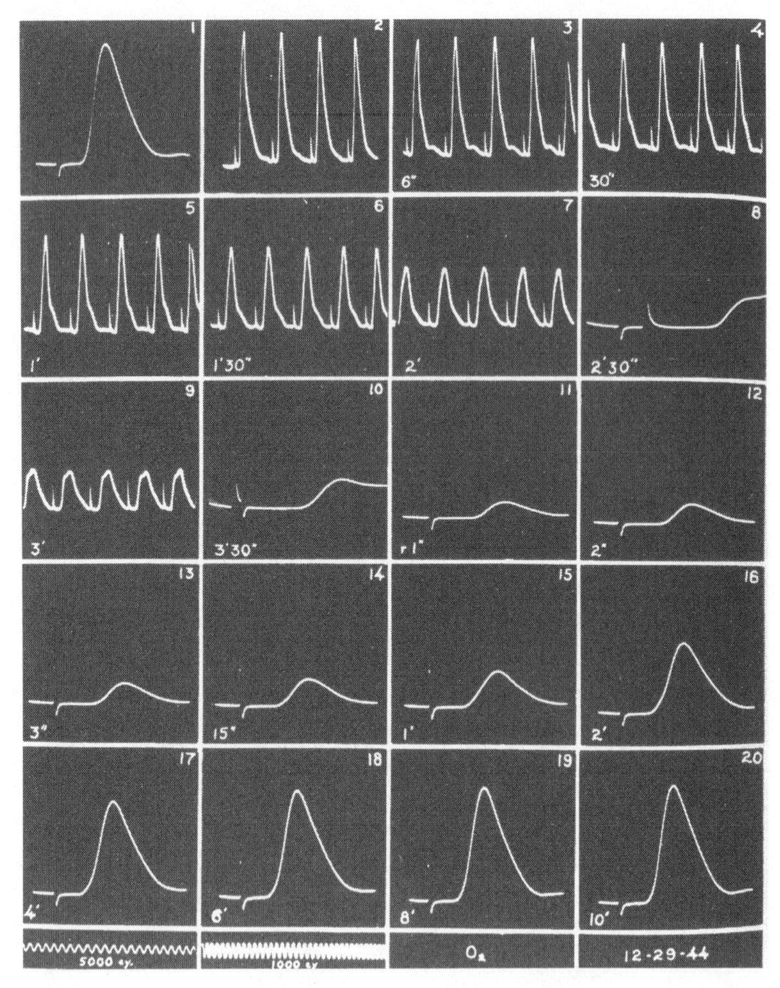

Figure 14–4 A typical illustration of one of Lorente de Nó's experiments. The oscilloscope traces, their layout, and their annotation are exquisite. In this experiment Lorente was investigating the effects of prolonged stimulation on the action potential of frog sciatic nerve (*panels 2 to 10*) and also the recovery of the nerve (*panels 11 to 20*). It is interesting that the notches and bumps so important to Erlanger are absent in the control action potential (*panel 1*) and appear only as the nerve action potential starts to fail during the repetitive stimulation (*panels 2 to 7*). (From Lorente de Nó R, 1947. Courtesy of the Rockefeller Archive Center.)

been very carefully carried out. The fundamental problem was that Lorente underestimated the influence of the sheath of connective tissue surrounding the thousands of fibers in the sciatic nerve, and especially of the thin layer of flattened epithelial cells lining the inner surface of the sheath. He was aware that the sheath might act as a diffusion barrier and his own experiments with potassium would show that, having changed the composition of the bathing fluid, it could

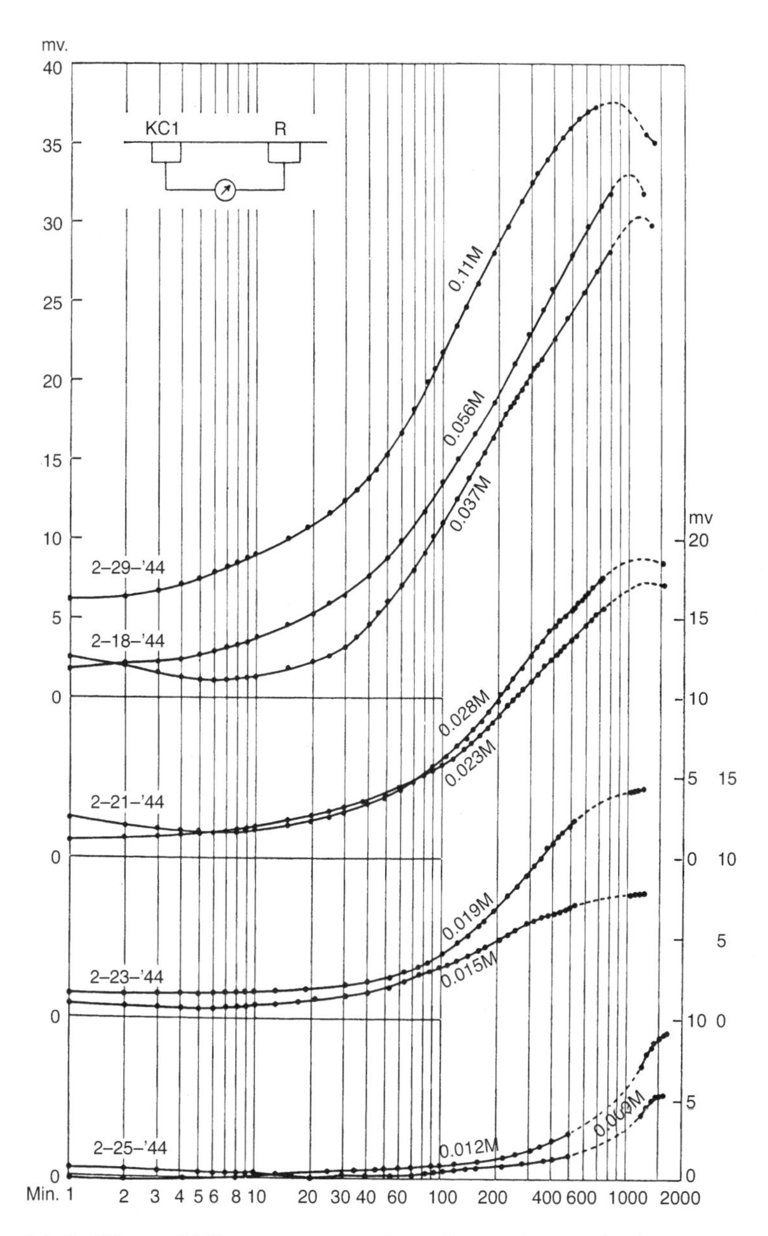

Figure 14–5 Effects of different concentrations of potassium on the demarcation potential of frog sciatic nerve (the greater the demarcation potential, the smaller the resting potential). Note the logarithmic time scale, with experiments lasting 24 hours, and the long times taken for the demarcation potentials to become maximal. To Lorente de Nó these long delays suggested (wrongly) that potassium concentrations did not determine the resting membrane potential. (From Lorente de Nó, 1947. Courtesy of the Rockefeller Archive Center.)

take hours before the maximal effect on the demarcation potential was achieved (Fig. 14–5). However, rather than attribute the result to the slow diffusion of potassium ions across the sheath, Lorente decided that potassium diffused readily but was not responsible for the resting potential.

The faulty conclusion would never have been reached if he had made a single cut in the connective tissue sheath with a pair of fine scissors: events that had taken minutes or hours would have appeared within seconds. Perhaps Lorente's disasters would have been avoided if he had called in to see Hodgkin, at the time when the young research fellow was just a few doors away in the Rockefeller Institute, working on crab nerves. Had he visited, he would have seen what could be done with a single fiber, and might have sensed that the study of the nerve impulse was about to enter a new era.

There was one conclusion of Lorente's that *was* important, however. He found that the demarcation potential decreased if the supply of oxygen to the nerve was removed, or if a very small amount of cyanide was added to the bathing fluid. Therefore, he argued, correctly in this case, the resting membrane potential depended on oxidative metabolism.

Why had Lorente done all this work, so much of which had been invalidated by a few quick experiments on the squid axon? The official reason was that the analysis was "only a preliminary step in the study of the physiology of nerve, for an understanding of nerve function cannot be reached until those electrochemical reactions which establish the resting membrane potential and those which produce the alteration have become known in detail."[15] The deeper reason, surely, was that the study might have given him insights into the way that the synapses operated in the central nervous system. It was the approach that had been favored by Gasser, on taking up his appointment as Director of the Rockefeller Institute. Lorente was more familiar than anyone else with the anatomical arrangements of the nerve fibers in the cerebral cortex and brain stem, and he had already made his own measurements of the synaptic delays involved, when the motoneurons of the eye muscles were excited in the brain stem. Further, in *A Study of Nerve Physiology*, he had carried out many experiments in which the membrane potential had been changed by applying long pulses of current to the nerve. As would emerge later, he must have been thinking, like Eccles, that excitation or inhibition in the central nervous system was brought about by currents flowing across the synapses from one nerve cell through the membrane of the other.

It was a concept that Eccles himself would soon put to the test. Before that would happen, however, the research on peripheral nerve would turn in a new direction.

15

The Voltage Clamp

For most people, understandably concerned with matters able to affect their lifespans, interest in the nervous system is confined to its diseases—amyotrophic lateral sclerosis (ALS), stroke, multiple sclerosis, and, especially in the past decade, Alzheimer's disease. And others. Only occasionally is there an advance in "pure" neuroscience sufficiently spectacular to capture the public's imagination. Penfield's demonstration, more than 50 years ago, of the recall of memories, by electrical stimulation of the exposed human brain, was one such event.[1] Another, at about the same time, was the demonstration by Adrian and Matthews of the electrical activity in the human brain, the waves that they were able to record with electrodes attached to the scalp.[2] A later example was the discovery by Hans Kosterlitz and John Hughes that the brain manufactures its own morphine-like compounds, the endorphins[3]—a discovery all the more surprising for having been made by an elderly professor and his young colleague in a remote Scottish university, using antiquated equipment. Then, rather earlier but unsurpassed for sheer drama, was Delgado's ability to stop the charge of a bull by radiofrequency stimulation of a small nucleus deep in the animal's brain.[4]

For the neuroscientists themselves, the situation is not dissimilar. So broad has the field become, with so many subspecialties, that most investigators are unaware of the bulk of the research published. It is also true that most of the research consists of incremental advances, or sometimes the mere confirmation or limited extension of something already known. But every now and again, something rather special will be reported that has a much wider impact in the neuroscientific community. An early example would have been the first report of chemical transmission at a synapse—Otto Loewi's demonstration, in 1921, of a substance, later shown to be acetylcholine, that was released by the endings of the vagus nerve on the heart.[5] In the past few decades, there have been the reports of genetically programmed cell death in the nervous system, of trophins—substances manufactured by some cells for the maintenance of others, of serial and parallel analysis of the visual picture by different regions of the brain, and of brains that can learn how to operate prostheses. And there are others. But aside from the spectacular ones, there is a different kind of publication that will also command attention, one in which the work described is both exceptionally elegant and thorough, and at a level few could have envisaged. The analogy with a long-awaited mathematical proof is not out of place. Thus it was that in 1952, a series of five papers on the ionic mechanisms responsible for the nerve impulse appeared in the *Journal of Physiology*.[6]

The authors of the five papers were Alan Hodgkin and Andrew Huxley, with Bernard Katz as co-author in the first in the series. The papers, the most important to have been written on peripheral nerve, described the amplitudes and time-courses of the ionic currents responsible for the action potential, the identification of sodium and potassium as the ions carrying the currents, the likely physical attributes of the structures in the membrane that made the currents possible, and a mathematical model of the propagating action potential. There was also a description of the novel methodology that had enabled the relevant observations to be made. The quality of the writing, the originality of the approach, the comprehensive nature of the results, and the depth of the mathematical analysis transformed the study of the nerve impulse. Yet, as sometimes happens with success, controversy and resentment followed.

The new advance started not in Cambridge, from whence the five papers had been submitted, but in Chicago. It was to Chicago that Kenneth Cole had moved at the end of the war, taking up the position of Director of Biophysics in the newly created Institute of Radiobiology and Biophysics. He no longer had Howard Curtis with him, but had instead been joined by George Marmont, a former colleague at Columbia University who had also done some work on the squid giant axon. Following the suggestion of a research fellow, Marmont devised a new method for studying the giant axon. A long electrode would be inserted into the interior of the axon and used to pass current through the membrane of an appreciable length of fiber. The size of the current would be controlled electronically, regardless of the amplitude of the potential that developed across the membrane. For his internal electrode, Marmont coated a thin glass capillary with a layer of silver. For the external electrode, he used a spiral of silver wire around the axon. Similar spirals on either side served as "guard" electrodes, protecting the central electrode from any artifacts. For the control of the current, Marmont used a feedback system, applying a command signal to a differential amplifier and taking the output from the amplifier to the electrodes. The same electrodes were used for passing current and for measuring the resultant voltage change.[7]

Marmont's interest was in determining the current threshold for initiating an impulse, if there was a threshold. In addition, the current/voltage plots enabled the resistance and capacitance of the axon membrane to be determined. Cole, however, could see that more useful results could be obtained if the apparatus was employed in the reverse mode—instead of "clamping" the current through the membrane, it was the voltage across the membrane that should be controlled. The current needed to "clamp" the voltage would then be a measure of the current normally flowing through the membrane at a particular instant during the action potential. Unfortunately for Cole, Marmont, having thought out the design of the apparatus and having built it, had proprietary rights to it. Nevertheless, on four occasions Cole was able to borrow it and to try voltage clamping. The experiments worked immediately.[8] When the membrane was made to depolarize, there was an initial inward current followed by an outward current. The outward current, Cole thought, was carried by potassium ions. For the inward current Cole had no suggestions. Sodium was not considered.

The next event was a fateful exchange of letters with Hodgkin. The Cambridge man had been the first to write, at the end of August 1947, telling Cole of his interest in stimulating the giant axon diffusely over a considerable length, so as to avoid the complication of dealing with a propagated action potential. Preventing the impulse from traveling was, of course, what Cole and Marmont had just accomplished by clamping either the voltage or the current. Cole wrote back, informing Hodgkin of the new experiments at Woods Hole. The correspondence was followed by Hodgkin's visit to Cole's laboratory in Chicago the following spring.

Hodgkin, in his autobiography, downplayed the importance of the visit.[9] Thus one of the main purposes of the trip to the United States had been to enable his wife, Marni, the daughter of Peyton Rous, to see her parents in New York. The other main reason for traveling was to give a lecture in Chicago: he would also exchange information with Cole and Marmont while he was there. Cole placed the emphasis rather differently.[10] On learning of the voltage and current clamp experiments, Hodgkin immediately asked to visit Cole's laboratory to see the results for himself. It was Cole who engineered an official invitation so that Hodgkin could use some of the Rockefeller Foundation grant, and it was Cole who arranged for Hodgkin to give a lecture while in Chicago. As to the laboratory visit, Hodgkin seemed uninterested in the reasoning that led Cole to the voltage clamp experiments. Instead, "he was already thinking ahead about using it." Further, in Hodgkin's opinion, the non-instantaneous rise in inward current, which Cole had found, must have been an artifact, a fault of the equipment. Notwithstanding this disagreement, the ever-generous Cole allowed Hodgkin to borrow "all of the various pieces of apparatus to repeat my experiments."

It is not known whether Cole also gave Hodgkin the circuit diagrams for the feedback system. It probably does not matter, for Hodgkin, with his wartime expertise in radar, could easily design his own system, and proceeded to do so. No sooner had Hodgkin returned to Cambridge than he put the departmental instrument maker onto building a feedback amplifier. It had to be ready for that summer's experiments in Plymouth—only two months away. For these experiments, conducted in 1948, Hodgkin once again had Huxley, newly married, as a partner. There was also Bernard Katz, who had worked with Hodgkin the previous year, and had helped demonstrate that the action potential required the presence of sodium ions.

The Plymouth experiments differed from those of Cole and Marmont at Woods Hole in several respects. One of these was the use of separate silver wires for stimulating and recording inside the axon, so as to avoid the complication of surface polarization that occurs when a metal passes current in solution, or, in this case, axoplasm (Fig. 15–1). Hodgkin and Katz made the first pairs of internal electrodes themselves, twisting two fine silver wires around a glass capillary. Inevitably, there were teething problems, some of which were overcome when Andrew Huxley arrived and was able to get the feedback system working properly. The speed was impressive—within a bare three months of returning from the United States, where he had seen Cole and Marmont's voltage clamp results, Hodgkin managed to get his own voltage clamp apparatus built and operational (Fig. 15–2).

With their own system, the three investigators in Plymouth were able to obtain successful recordings of the membrane currents corresponding to various programmed depolarizations. As Cole had already observed, the initial current, the inward one, was not instantaneous, but built up to a peak within half a millisecond

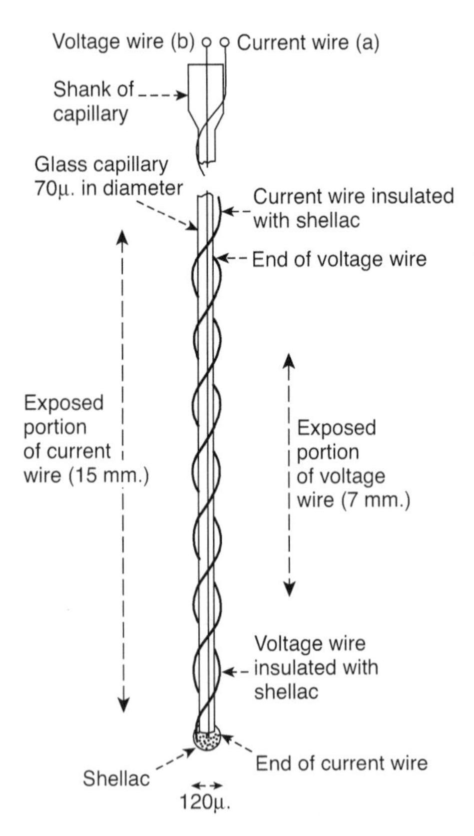

Figure 15–1 The electrode that was inserted into the squid giant axon for the voltage clamp experiments. Two fine silver wires, one for recording membrane potential and the other for passing current, were wound around a glass capillary, 70 μm in diameter. (From Hodgkin AL, Huxley AF, Katz B. Measurement of current-voltage relations in the giant axon of *Loligo*. *Journal of Physiology*; **116**: 442–449, 1952. Courtesy of John Wiley & Sons.)

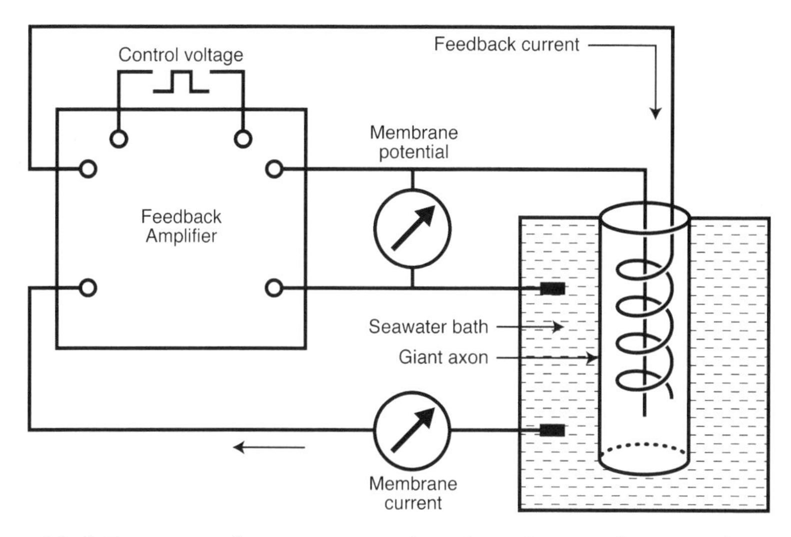

Figure 15–2 Experimental arrangement in the voltage clamp studies. The voltage between an intracellular and an extracellular electrode is measured and compared with a controlling voltage; any difference is automatically eliminated by passing a feedback current through another pair of electrodes on the two sides of the membrane.

or so, before decaying and giving way to a slower outward current. The size of the inward current, the Plymouth investigators showed, was affected by the concentration of sodium ions in the bathing fluid, strong evidence that it was sodium ions that formed the current. The results were sufficiently clear-cut that Hodgkin and Huxley were able to present the work at a symposium in Paris in April 1949, held to honor Louis Lapicque, the neurophysiologist who had so extensively explored the relationship between the intensity and the duration of a stimulus just sufficient to excite a nerve or muscle. Cole, who was also present at the symposium, was surprised at Hodgkin's presentation. As he was to remark later, "In a little over a year they had corrected most of my difficulties, caught up and run past me."[11]

It was indeed an impressive achievement, but much more was to come in the summer of the same year. This time Hodgkin and Huxley were the first to arrive in Plymouth and to start the experiments, with Katz joining them the following month. By this time the supply of squid had improved, and in a single month the three investigators were able to carry out the experiments that formed the basis of the 1952 papers. One of the key experiments was to replace the sodium in the bathing fluid with choline. Since choline was unable to move through the membrane, a depolarization of the membrane could no longer induce an inward current, but only the later, potassium, current. When the latter was subtracted from the total current carried by the two types of ion, the full form of the inward, sodium, current was seen (Fig. 15–3). As Hodgkin and Huxley pointed out, the sodium permeability mechanism was regenerative. Thus, the entry of the positively charged

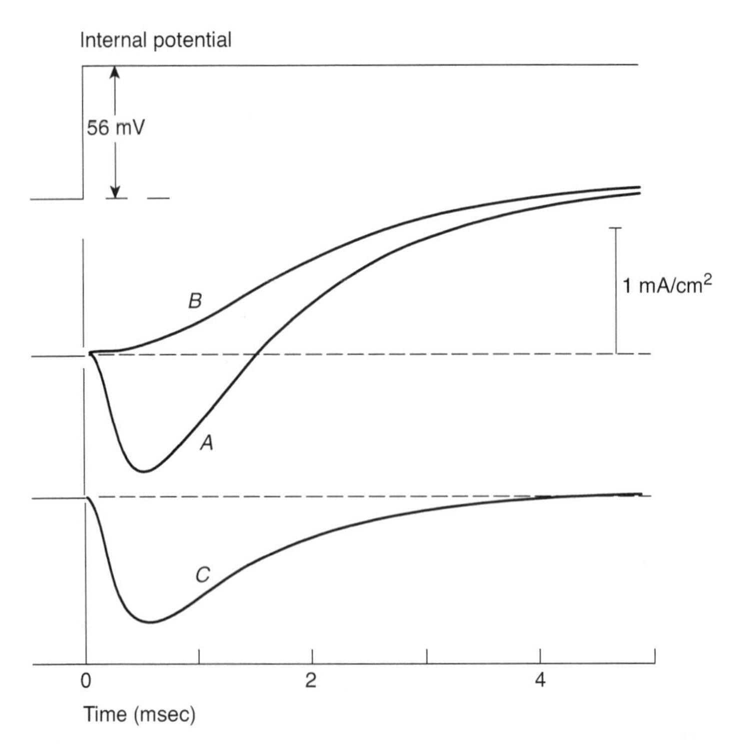

Figure 15–3 Separation of sodium and potassium currents following a voltage step (depolarization) of 56 mV. *A* shows combined currents with the giant axon in sea water. In *B* most of the sodium in the bathing solution has been replaced by choline, so that only the potassium current is left. *C* shows the sodium current, obtained by subtracting the curve in *B* from that in *A*. (From Hodgkin, Sir Alan. *Chance & Design. Reminiscences of Science in Peace and War*, 1992. Courtesy of Lady Marion Hodgkin and Cambridge University Press.)

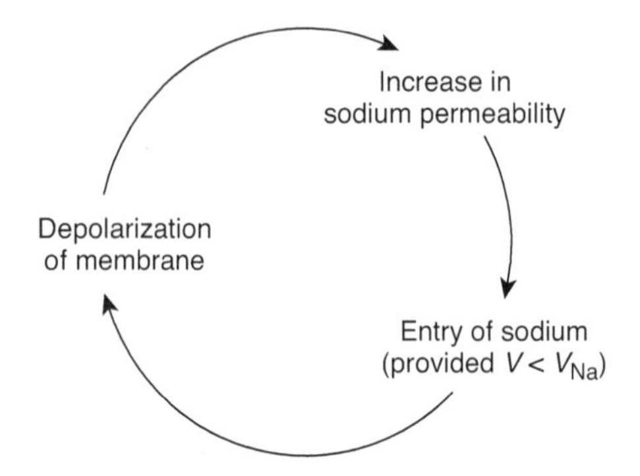

Figure 15–4 Regenerative (positive feedback) cycle of sodium conductance in squid axon membrane.

sodium ions into the nerve fiber added to the depolarization of the membrane and this, in turn, increased the its sodium permeability (Fig. 15–4). It was a rare example of positive feedback in biology.

The summer of 1949 was the last time the three would work together, for at its end Katz withdrew. He realized that, with Huxley present, there was little of his own that he could contribute to Hodgkin. He was well aware of the importance of the voltage clamp work, and the likely consequences of its success for the investigators.[12] A lesser person would have clung on, trying, at all costs, to remain part of the team. Katz, however, exhibiting the same nobility that had caused him, in 1939, to leave a safe academic position in England for an unknown one in Australia, did the opposite.

The experiments completed, the next step was to write them up and, from the data, find equations that would describe the relationships between the various properties of the active membrane—the current, the capacitance, the potential difference, the equilibrium potentials for the different types of ion, and, above all, the conductance (permeability) of the membrane to those ions (Fig. 15–5). Andrew Huxley had already spent considerable time on this problem, making many calculations during the cold winter of 1947, and had eventually been able to reproduce the general form of the action potential. But the assumptions were too many, and a mathematical success did not necessarily imply the correctness of the biological

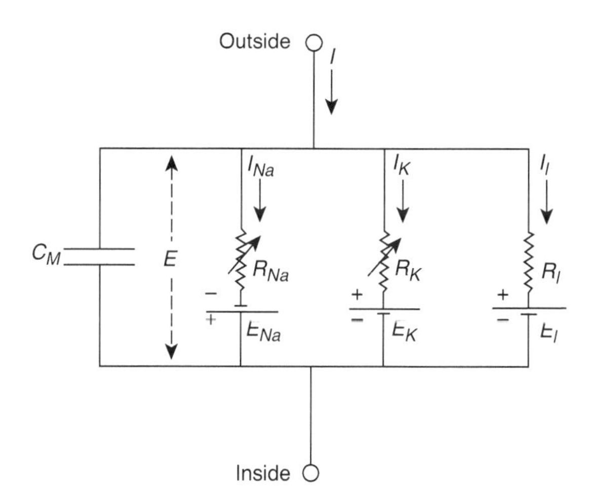

Figure 15–5 Electric analogue of nerve fiber membrane. The membrane potential, E, and the current, I, flowing through the membrane, are determined by the equilibrium potentials for sodium (E_{Na}), potassium (E_K), and "leakage" ions such as chloride (E_l), together with their respective variable resistances (R_{Na}, R_K, R_l). The membrane also has a capacitance (C_M). (From Hodgkin AL, Huxley AF. Currents carried by sodium and potassium ions through the membrane of the giant axon of *Loligo*. *Journal of Physiology*; **116**: 449–472, 1952. Courtesy of John Wiley & Sons.)

model on which it was based. Now, with the new data from the voltage clamp experiments available, there would be more precision in constructing the equations. Because long internal electrodes had been used in the voltage clamp experiments, the longitudinal resistance of the axoplasm did not have to be considered. Nor, since the membrane potential had been held steady by the voltage clamp, did the propagation of the impulse intrude into the calculations. The formula that Hodgkin and Huxley eventually came up with, to describe the flow of current through the membrane at any instant, was:

$$I = \frac{Cdv}{dt} + (V - V_k) g_k n^4 + (V - V_{Na}) g_{Na} m^3 h + (V - V_L) g_L$$

In this equation, I, the density of the current flowing through the membrane, was equal to the sum of the capacitance current CdV/dt, and the currents carried by potassium ions (second term in the equation), sodium ions (third term), and all other ions (fourth term). For each of the ion species, the current was the product of the driving potential (the term in parentheses) and the conductance (or permeability), g, for that ion. To make the equation account for the steep rise in conductance as the membrane was depolarized, Huxley had, with Hodgkin's agreement, introduced the additional terms n^4 for the potassium current and m^3h for the sodium current. In introducing these terms, Huxley had envisaged each current pathway as having several particles that would have to move to critical sites before the ions could start moving through the membrane. In the case of potassium, there would be four such particles (hence the term n^4), while for sodium there would be three particles (hence the term m^3). However, the sodium pathway had an additional particle that served to block the movement of ions, and that was responsible for the later decline in the sodium current observed in the voltage clamp experiments. The term h signified that this fourth particle had yet to move to its blocking position.

The next step for Hodgkin and Huxley was to compute the form and propagation velocity of the action potential from the equations they had developed. This involved combining two equations for the membrane current, the one given above, with its n^4 and m^3h terms, and a second one, with terms for the radius of the axon (a), the resistivity of the axoplasm (R), the conduction velocity of the impulse ($Ø^2$), and the membrane potential (V):

$$I = \frac{a}{2R\phi^2} \times \frac{d^2V}{dt^2}$$

In this equation, a "second-order" one, the velocity was unknown at the start of the computation but could found by inserting a trial value and solving the equation. If the velocity value was too high or too low, the term V went to plus or minus infinity. When the value chosen was correct, however, the equation brought V, the membrane potential, back to its resting value.

It was the intention of Hodgkin and Huxley to use the Cambridge University computer—the only computer in the entire university—to carry out the formidable amount of calculation involved, but the machine was undergoing major modifications at the time and would not be available for six months. Huxley then suggested to Hodgkin that he, Huxley, attempt to solve them himself, with the aid of his hand-operated Brunsviga calculating machine. It was an extraordinarily ambitious undertaking. The calculator, rather like an old-fashioned cash register, required that the data were entered by moving small levers in slots to appropriate positions beside numerically inscribed wheels. A handle at the side of the machine would then be turned so many times in one direction or another, and the results read off on the numbered wheels. These results would then have to be written down on paper, before proceeding to the next stage of the calculation. And these steps had to be repeated over and over again. The reconstruction of the action potential required numerical integration, and a complete set of data had to be produced for each small time interval. To calculate a complete "run" required 8 hours of intense mental and physical activity. It has been said that, in all the calculations, more than a million separate steps were involved. It is doubtful if anyone other than Huxley could have brought it off (Fig. 15–6). Hodgkin, who was a fine mathematician himself, albeit a largely self-taught one, would later admit that it would have been beyond him.[13] The derivation and solution of the equations, and the suggestion that voltage-sensitive particles had to take up new positions in the membrane before ions could pass through, would prove to be the crowning glory of the voltage clamp work by the British group. It was this

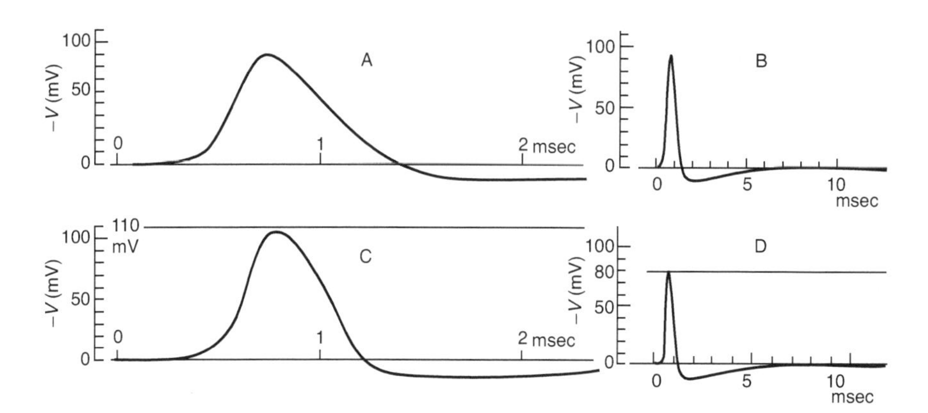

Figure 15–6 Comparison of computed propagated action potentials on fast and slow time scales (*A* and *B* respectively) with actual recordings on same time scales (*C* and *D*). (From Hodgkin AL, Huxley AF. A quantitative description of membrane current and its application to conduction and excitation in nerve. *Journal of Physiology*; **116**: 500–544, 1952. Courtesy of John Wiley & Sons.)

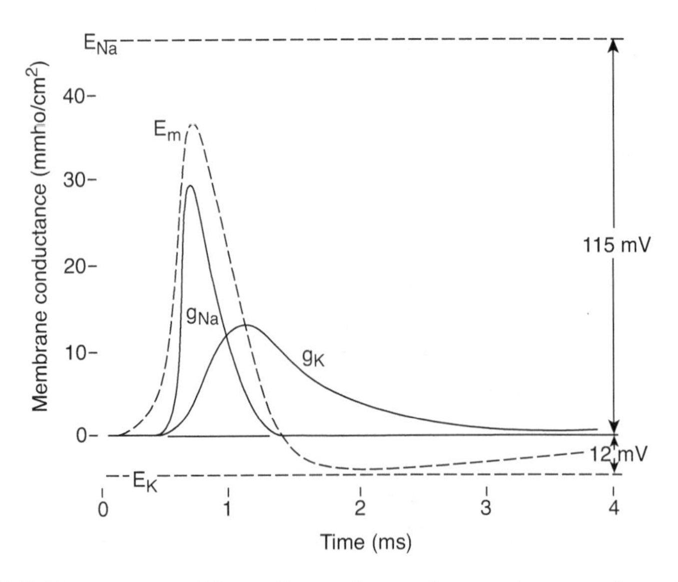

Figure 15–7 Time-courses of the sodium and potassium conductances (g_{Na}, g_K) during an action potential (E_m). (From Hodgkin AL, Huxley AF. A quantitative description of membrane current and its application to conduction and excitation in nerve. *Journal of Physiology*; **116**: 500–544, 1952. Courtesy of John Wiley & Sons.)

theoretical treatment that formed the substance of the last of the five papers in the *Journal of Physiology*. The figure that summarized the voltage clamp work would, like Cole and Curtis's impedance picture, become a classic (Fig. 15–7).

Once again, however, there was one omission in the published work. Just as with the letter to *Nature* in 1939 and the 1945 paper in the *Journal of Physiology*, there was no recognition of Cole's help. It is true that, in the first paper of the 1952 series, there was an acknowledgement that "the experimental method was based on that of Cole (1949) and Marmont (1949)," but that was all.[14] There was no mention of the fact that, during an induced depolarization of the axon membrane, Cole had been the first to detect an inward current followed by an outward one, correctly attributing the latter to potassium ions. A figure showing the inward and outward currents had been included in Cole's brief 1949 paper and Cole had shown his results to Hodgkin during the latter's visit to Chicago in the previous year. At that visit Cole had loaned him "all of the various pieces of apparatus to repeat my experiments."[15] Later, Hodgkin would freely acknowledge Cole's lead in the voltage clamp work, and at least some of the help that the American had given, but the damage had been done. The omission rankled, and in his later years, Cole was to become bitter about it.

Andrew Huxley was also concerned about the issue of priority. Although he had already displayed some of the family genius, both in the laboratory and, especially, in the mathematical and theoretical treatment of the results, he still saw himself as Hodgkin's student. He also had the family trait of loyalty. Like his illustrious grandfather, who had defended Darwin, so Huxley defended Hodgkin. Having read the draft of one of Hodgkin's autobiographical chapters, and reflecting on events of 30 years previously, he would write to Hodgkin: "This reads as if you hadn't thought of using feedback till it was suggested by Kacy's [Cole's] letter of October '47. I think I have a clear memory of your talking about a feedback system to do a voltage clamp before the end of the war."

It is certainly true that Hodgkin had been thinking of stimulating a length of axon uniformly, for this was stated in his own letter to Cole, in August 1947. In the previous year, Huxley had, in fact, made a Perspex chamber that allowed a single crab fiber to be stimulated simultaneously at ten different sites. The result had been disappointing, the action potential looking very much like one initiated in the normal way. If feedback had been discussed a year or two earlier, as Huxley thought, why was it not remembered then? As events had proved, it would have been a much more elegant and powerful method.

Youth is ever ambitious and, in Hodgkin's defense, it must be remembered that, despite his formidable achievements, he was a young man. When his and Huxley's first results on the squid axon had appeared in *Nature*, he had been 25, and the voltage clamp papers were published while he was still in his thirties. There is also this to be said. Although Hodgkin had been helped by Cole, helped very considerably, no one could have done the work better than he and Huxley had. Cole himself acknowledged that. Further, Hodgkin's credentials, prior to the 1952 papers, were the equal of the older man's, and he had done considerably more work on the nerve fiber. Even if Cole had not shown him his squid work in 1938, Hodgkin would have been bound to move to the squid as his preparation through his subsequent contact with J. Z. Young at the Plymouth Marine Laboratory. It is likely that Hodgkin and Huxley would have eventually thought of using a voltage clamp themselves, if Cole and Marmont had not already done so. Cole's main contribution was to give them the additional time; without his help and pioneering observations, the British work would have been delayed, possibly for several years. Finally, Cole could have done more to strengthen his own claims to the innovation of voltage clamping. Three years had elapsed between the publication of his own short paper in 1949—a paper with only two pages devoted to voltage clamping—and the appearance of the Hodgkin–Huxley series in 1952. It was ample time for additional experiments to have been done and for a more substantial account to have been written.

If, in 1963, the Nobel Committee had acted differently, the argument over priority in the voltage clamp work would not have mattered so much, for both Hodgkin and Huxley would have welcomed Cole as a fellow winner.[16] Instead, just as with Bishop in 1944, the third man was left out.

16

Aftermath

Lorente de Nó went to Spain for a long holiday. The publication of the Hodgkin–Huxley papers had been a bitter blow. Having labored for 10 years on his monumental study of peripheral nerve, Lorente now found that it was largely irrelevant, or, even worse, wrong in its main conclusions. What should have been his greatest achievement was likely to prove an embarrassment and, at the age of 50, his best years for experimentation were probably behind him. All this he must have known, and yet he refused to capitulate, let alone to walk away from a battle that only he wished to fight. He would appear at international meetings, rejecting the general applicability of the Hodgkin–Huxley findings, and referring dismissively to the "so-called sodium hypothesis." It was a sad end to a career that had been so full of promise.

Another person who would wage a futile struggle against the thoroughness of the squid work was a professor at Columbia University. David Nachmanson was an able biochemist who had done useful work on the synthesis and degradation of the chemical transmitter, acetylcholine, at the neuromuscular junction. Later, he had noted the presence of the degrading enzyme, acetylcholinesterase, in the axon itself, where it could be demonstrated close to the membrane, especially at the nodes of Ranvier—the interruptions of the myelin sheath. To account for the presence of the enzyme, he postulated that the action potential, in addition to its dependence on sodium ions, as Hodgkin, Huxley, and Katz had so convincingly shown, required the combination of acetylcholine with a membrane protein.[1] As Katz would remark of another scientist, it was "possible to entertain the hypothesis if, and only if, one is prepared to ignore everybody else's work on the subject."[2]

Hodgkin himself knew that, while the sodium and potassium story was almost certainly true, there was still more that should be done. Yes, he and his colleagues had been able to show that the action potential failed unless sodium ions were present outside the nerve fiber in sufficient concentration. And yes, if, during the action potential, the membrane became highly permeable to sodium ions, the latter should move through the membrane to the inside of the fiber, driven by the combined electrical and chemical gradients. But could it actually be proved that they had entered the fiber?

There was another unresolved issue. Hodgkin, with Katz and Huxley, had estimated that the resting nerve fiber membrane was only sparingly permeable to sodium ions. Nevertheless, even when the fiber was quiescent—not conducting impulses—there should be a small flow of these ions into the fiber and, since their concentration remained low, the fiber must have some means of continually expelling them. Might there be a pump in the membrane that would serve this function? The idea of sodium excretion had occurred to Overton half a century earlier, at the time that he was demonstrating the necessity of sodium for muscle excitability. If a small amount of sodium were to enter a cardiac muscle fiber with each heartbeat, he reasoned, then one might expect the concentrations inside the fibers to go on increasing during life. Despite the huge number of heartbeats involved, this did not happen—clearly, then, there had to be some mechanism at work that expelled sodium from the fibers.[3]

Following Overton's remarkable insights, the idea of sodium excretion had then lain dormant. Forty years later it was raised again, this time by a young physiologist in Rochester, New York. Taking part in a symposium on membrane permeability in Cold Spring Harbor, Long Island, in 1940, Robert Dean had stated, in the discussion following one of the papers, "The picture you showed of the potassium saturation of frog muscle as a function of the potassium in the external solution can be explained if you assume a mechanism constantly excreting sodium."[4] Dean himself did not take up the challenge of demonstrating the existence of such a pump, and the situation was to remain uncertain for several more years. Then, in 1947, buried in the thousand pages of *A Study of Nerve Physiology*, had come Lorente de Nó's statement that the demarcation potential, and therefore the resting membrane potential, declined if a nerve was deprived of oxygen.[5] It was one of the few findings of lasting value in the entire study. If Lorente was right, perhaps oxygen was needed for the metabolic machinery in the fibers that kept membrane pumps supplied with ATP? To find the answers to some of these problems, Hodgkin enlisted the help of another young Cambridge man.

Before the war, Richard Keynes (Fig. 16–1) had, like Andrew Huxley, been one of Hodgkin's students. He was also, like Huxley, a member of the Cambridge intellectual aristocracy. His father, Sir Geoffrey Keynes,[6] was a distinguished surgeon, and had been the first to show, convincingly, that removal of the thymus gland from the chest was curative for some of the patients with muscle weakness due to myasthenia gravis. He had also pioneered the implantation of radium needles in the tumor tissue of patients with advanced breast cancer, achieving astonishingly good results.[7] As a Cambridge undergraduate, Geoffrey Keynes' close friends had included Rupert Brooke, the poet, and the young mountaineer, George Mallory, later to die on Everest. Geoffrey's brother was Maynard Keynes, the most influential economist of his generation, and Geoffrey himself had married a descendant of Charles Darwin. As if these connections were not strong enough, he had, through the marriage of his sister, acquired A. V. Hill, the muscle physiologist and Nobel laureate, as a brother-in-law. To round it off, Richard Keynes, his son, had married Adrian's eldest daughter. How carefully tended was the genetic garden in Cambridge!

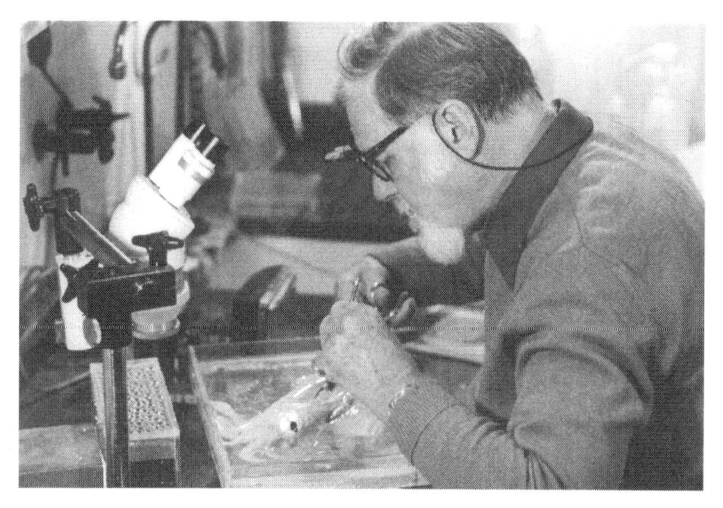

Figure 16–1 Richard Darwin Keynes dissecting a cuttlefish (or was it a squid?). (Courtesy of Professor Simon Keynes)

Given the expectations inherent in membership of such a family, Richard Keynes was to acquit himself well. Starting his work before the publication of the Hodgkin–Huxley papers, he was to provide much of the information missing from the sodium–potassium story. Like Hodgkin and Katz, he had specialized in radar during the war, returning to the Cambridge department of physiology at its conclusion to finish his undergraduate studies and to work for a Ph.D.[8] With radioactive tracers, Keynes reasoned that it should be possible to obtain very accurate information about the movements of sodium and potassium in and out of a nerve fiber. It was not an easy undertaking, however. For one thing, there was no supply of isotope commercially available for the first studies. Undaunted, Keynes made his own neutron bombardment of sodium and potassium, using the Cambridge University cyclotron. He also made an observation chamber out of Perspex and brass, with a narrow slot for the nerve fiber; the nerve fiber was held at each end by forceps and could be quickly raised and transferred to a second chamber containing radioactive sodium or potassium (Fig. 16–2). The radioactivity gained by the nerve fiber could be measured with a Geiger counter mounted immediately below the observation chamber; as ions either diffused or were pumped out of the fiber they were carried away in the sea water perfusate.

The fiber was first primed with radioactive sodium or potassium by immersing it in radioactive sea water. The fiber was then left in the resting condition for a number of minutes or else it was stimulated, so as to make it fire a given number of action potentials (typically 30,000). The fiber was then transferred back to the observation chamber and its radioactivity measured again. The experiments, carried out first on crab nerve and then on single cuttlefish axons, showed that there was a steady leakage of potassium from the resting fiber and that the loss was

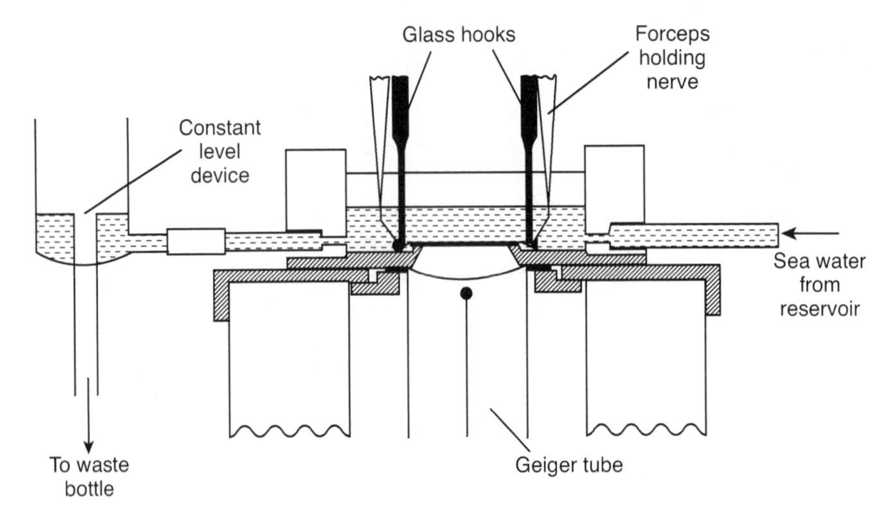

Figure 16–2 Observation chamber for nerve isotope experiments, designed and built by Keynes. (From Keynes RD. The leakage of radioactive potassium from stimulated nerve. *Journal of Physiology*; **113**: 99–114, 1951. Courtesy of John Wiley & Sons.)

enhanced when the nerve was made to fire impulses—just as Hodgkin and Huxley, and for that matter, Cole, had thought. For each impulse, he was able to calculate the number of potassium ions passing through one square centimeter of axon membrane, obtaining a value close to that obtained indirectly by Hodgkin and Katz on the basis of measurements of membrane capacitance.[9]

By determining the sodium radioactivity of the nerve fibers, Keynes found that, with the fiber at rest, there was, as predicted, a small but steady entry of sodium ions. Nevertheless, it was clear that the resting membrane was much less permeable to sodium than to potassium, just as Bernstein had postulated at the turn of the century, and as Hodgkin and Katz had estimated on the basis of their intracellular recordings of membrane potentials in the squid axon. When the nerve fibers were stimulated, however, Keynes found that the rate of sodium entry increased 18-fold,[10] the sort of large increase indicated by the squid experiments of Hodgkin, Huxley, and Katz (Fig. 16–3).

With the aid of a simple mechanical model, Keynes, with Hodgkin, went a step further, and deduced something about the nature of a potassium channel. When unequal numbers of steel balls, representing potassium ions, were placed in two connecting chambers and shaken, Hodgkin and Keynes found that ratio of balls moving in opposite directions through the opening could be explained only if the passage was sufficiently long for several balls to line up in single file. Here, then, was another prediction about the structure of an ion channel in the membrane: it had to be long enough to accommodate several ions (perhaps three) at the same time.[11] This information could be added to what was already suspected. Thus, Huxley, in finding equations that fitted the experimental data in the squid axon,

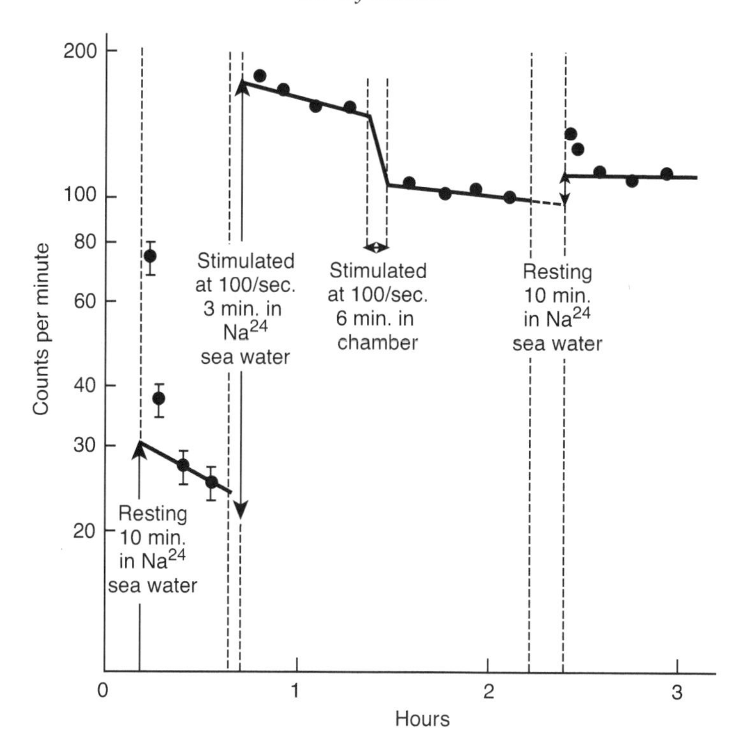

Figure 16–3 Movements of radioactive sodium in a *Sepia* (cuttlefish) axon. The decline in the first segment is due to extrusion from the fiber, brought about by the sodium pump. Repeated stimulation then causes sodium entry with each action potential, followed by further pump-mediated extrusion. (From Keynes RD. The ionic movements during nervous activity. *Journal of Physiology*; **114**: 119–150, 1951 Courtesy of John Wiley & Sons.)

had already postulated that each channel had several parts ("particles") that moved to new positions in the membrane under the influence of a voltage change. It was the displacement of the particles that was thought to open up the channel, and to allow ions to move through the membrane.

So far, so good, but what about the putative sodium pump? Keynes and Hodgkin found that, in the presence of a poison such as DNP (dinitrophenol) or cyanide, the extrusion of sodium from a cuttlefish axon was eventually abolished. Further, in an untreated axon, the extrusion of sodium was linked to the entry of potassium. Since DNP acts by blocking the synthesis of ATP through oxidative metabolism, the result strongly suggested that sodium was being pumped out (in exchange for potassium) by an energy-dependent process.[12]

Hodgkin was still not satisfied with the level of proof. The next stage in the investigation must be to see if the pumping activity was restored when ATP was injected into the squid giant axon. For this part of the project, Hodgkin, continuing to display admirable leadership qualities, recruited Peter Caldwell, an Oxford-trained biochemist. Caldwell, with Keynes, showed that, indeed, ATP was able to restore pumping in a giant axon poisoned with cyanide, as were certain other types of energy molecule normally found in the axoplasm.[13] This part of the research on the nerve fiber had now been taken as far as it could go. The membrane evidently contained a pump that hydrolyzed ATP as it extruded sodium in exchange for potassium. It was this pumping process that was ultimately responsible for creating the marked concentration differences across the membrane for sodium and potassium, upon which the action and resting potentials depended.

As the Cambridge investigators collected their evidence for a sodium pump, others were doing the same in other tissues—the gastric mucosa, frog skin, and red blood cells. Finally, in 1957, came a piece of crucial evidence from Denmark. Jens Skou, a biochemist, succeeded in isolating a membrane enzyme from fragments of crab nerve, an enzyme that split ATP and that required the presence of sodium and potassium ions to do so. It was the sodium pump.[14]

More than any other tissue, the giant axon of the squid had provided the means to understand the nature of the nerve impulse. In the hands of ingenious investigators, however, the squid giant axon could be made to do still more tricks. Two of Hodgkin's later recruits, Peter Baker and Trevor Shaw, discovered that, by pushing a small rubber-covered roller along the surface of an axon, they could squeeze the axoplasm out from the cut end, leaving only the membrane with its supporting sheath (Fig. 16–4). Almost miraculously, if the empty sleeve of membrane was perfused with a solution containing potassium and a little sodium (Fig. 16–5), it not only exhibited a resting potential but became capable of conducting impulses once more—almost a million of them (Fig. 16–6). Further, if the normal potassium concentration gradient across the membrane was reversed, by perfusing the interior of the axon with a potassium solution weaker than the bathing fluid, the resting membrane potential also reversed, just as the Nernst equation predicted.[15] These dramatic experiments proved, beyond all doubt, that the resting and action potentials depended on the relative permeabilities of the membrane to potassium and sodium, and on the availability of those ions on the two sides of the membrane.[16,17]

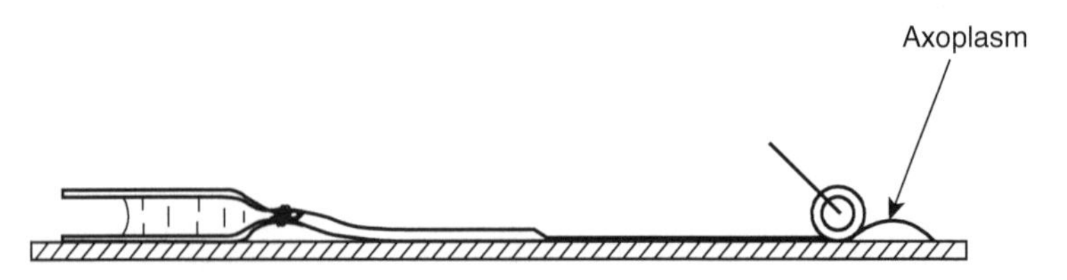

Figure 16–4 Expulsion of axoplasm from a squid giant axon, using a miniature rubber-covered roller. (From Baker PF, Hodgkin AL, Shaw TI. The effects of changes in internal ion concentration on the electrical properties of perfused giant nerve fibres. *Journal of Physiology*; **164**: 355–374, 1962. Courtesy of John Wiley & Sons.)

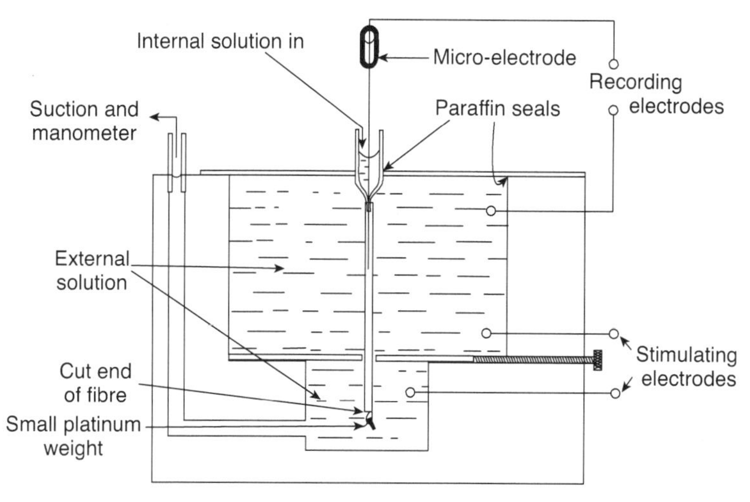

Figure 16–5 Arrangement for internally perfusing a squid giant axon, and for stimulating and recording from it. Negative pressure applied to the experimental chamber ensures the flow of fluid through the fiber. In some experiments a different arrangement was used, in which the axon was perfused through a cannula inserted into its lower end. (From Baker PF, Hodgkin AL, Shaw TI. The effects of changes in internal ion concentration on the electrical properties of perfused giant nerve fibres. *Journal of Physiology*; **164**: 355–374, 1962. Courtesy of John Wiley & Sons.)

Andrew Huxley also made further contributions to nerve fiber physiology. While he was still working with Hodgkin on the voltage clamp he had been joined in Cambridge by a young Swiss. Robert Stämpfli had taught himself to dissect out single myelinated fibers from frog nerve. Since the diameter of the largest fibers was only 18 to 20 μm, this was a considerable feat, though it was one that several Japanese workers had mastered some years before. Acting on a suggestion of Hodgkin's, Huxley and Stämpfli set out to determine whether, as seemed likely, the action potential traveled down such a nerve fiber by a series of jumps, the current flowing in and out of the fiber at the interruptions in the myelin sheath (the nodes of Ranvier). This seemed likely, since the myelin sheath, being largely composed of fat, would act as an insulator. Such a mode of impulse conduction would be in contrast to that demonstrated for the giant axons of the squid, in which, since there was no myelin sheaths, the impulses flowed smoothly along the fibers.

Huxley and Stämpfli tackled the problem by slowly drawing a single nerve fiber through a short glass capillary and simultaneously measuring the current flowing around the outside of the fiber during the passage of an action potential. As expected of an impulse that skipped from node to node, the action current had a constant latency in between nodes and increased only when it reached the next node (Fig. 16–7).[18] One advantage of this "saltatory" conduction over continuous

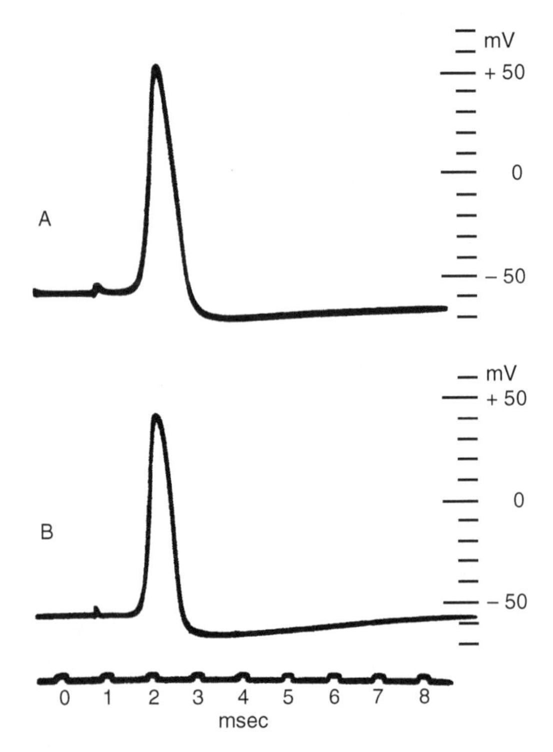

Figure 16–6 *A*, action potential recorded with an internal electrode in a squid giant axon filled with isotonic potassium sulfate in place of axoplasm. The action potential is virtually identical with that recorded from an intact axon (*B*). (From Baker PF, Hodgkin AL, Shaw TI. The effects of changes in internal ion concentration on the electrical properties of perfused giant nerve fibres. *Journal of Physiology*; **164**: 355–374. Courtesy of John Wiley & Sons).

conduction was that the velocity of the action potential was considerably increased (by the square root of the fiber diameter). In addition, ion exchange across the membrane was limited to the nodes, so that less pumping was needed to restore the internal concentrations of sodium and potassium. Finally, the fatty myelin sheath, through its insulating properties, prevented action currents flowing between one nerve fiber and another, and generating spurious action potentials in the second fiber.

Having proved the existence of saltatory conduction in myelinated fibers, Huxley went on to devise a special method for measuring their resting and action potentials. Since the fibers were 20 μm or less in diameter, they were too small to allow penetration with a microelectrode, and the measurements were made with external electrodes instead, using a special circuit in which the resistance of the fluid around the fibers was made very large.[19] The experiments, undertaken with Stämpfli, were extended to show that the resting and action potentials of single myelinated nerve fibers in a vertebrate (the frog) were dependent on potassium

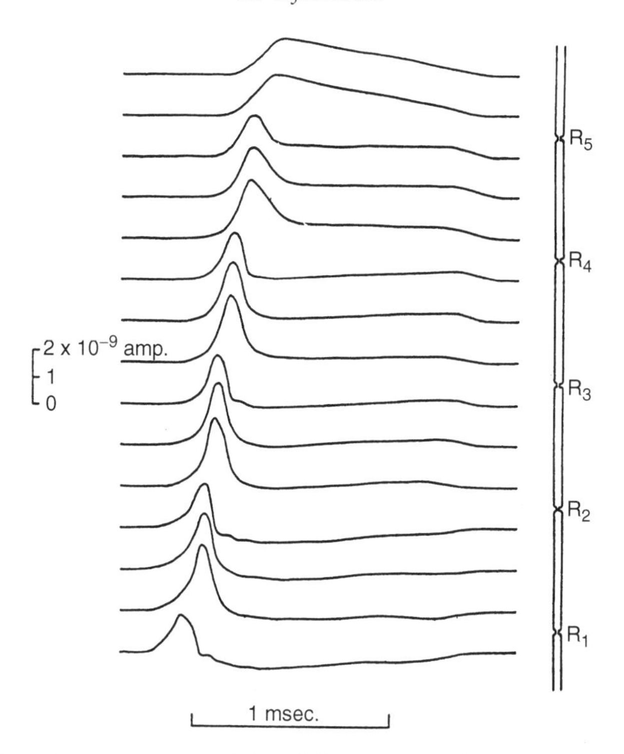

Figure 16–7 Action currents recorded at different points along a single myelinated nerve fiber. The nerve fiber is shown on the right, with the positions of the nodes of Ranvier indicated (R_1, R_2, R_3, R_4, R_5). It can be seen that, although their amplitudes and shapes differ slightly, the currents in the three records taken from each "internode" are synchronous—the delays occur only at the nodes. This pattern indicates that the impulse is jumping from one node to the next. (From Huxley AF, Stämpfli R. Evidence for saltatory conduction in peripheral myelinated nerve fibres. *Journal of Physiology*; **108**: 315–339, 1949. Courtesy of John Wiley & Sons.)

and sodium permeabilities of the membrane—just as in the squid. Lorente de Nó's last defense of his work, that the findings in the squid were not relevant to vertebrate nerves, had been demolished.

Hodgkin and Huxley were not the only ones working on nerve. Others had been quick to grasp the significance of the 1952 papers and had accepted the impossibility of improving on the squid experiments or, especially, on the theoretical treatment of the results. Nevertheless, other things could be done. One of these was to estimate the density of the sodium and potassium channels in the nerve fiber membrane. It was an interesting challenge for, even if one had known what to look for, the channels were far too small to be seen with a microscope, even an electron microscope. A possible answer to the problem would be to use a toxin, one that had recently been found to produce its effects by combining with the sodium channels and paralyzing the muscles.

It would not be the first time that neuroscientists had used naturally occurring poisons. As described previously, curare, found in certain plant species of the Amazon rain forests, had been used to block transmission at the neuromuscular junction by Eccles and his team in Australia, though it had, in fact, been employed for the same purpose much earlier by the famed French scientist Claude Bernard, in 1844.[20] Then there were compounds such as atropine and ergot, produced by the deadly nightshade plant and by a bread mold respectively, that had powerful effects on the autonomic nervous system—that is, the nerves that control such involuntary actions as the heart rate, sweating, and the size of the pupil. So far as sodium channel poisons were concerned, there were two possible candidates. One was produced by a curious fish found in the seas around China and Japan, a fish that could inflate itself so as to appear more formidable, the puffer fish. It was a fish that was well known to the ancient Chinese, having been described in the first Chinese pharmacopeia, *Pen-T'so Chin* (the Book of Herbs), written during the reign of Emperor Shun Nung (2838–2698 BC).[21] It was the poison of this fish, tetrodotoxin, that had almost killed Captain James Cook, the British explorer and navigator, during his voyage around the world in 1777. Cook's account of the incident is available to us. Referring to the puffer fish, he wrote:

> "Having no suspicion of its being of a poisonous nature we ordered it to be dressed for supper; but very luckily, the operation of drawing and describing took up so much time that it was too late, so that only the liver and roe were dressed, of which the two Mr Forsters and myself did but taste. About three o'clock in the morning, we found ourselves seized with an extraordinary weakness and numbness all over our limbs. I had almost lost the sense of feeling, nor could I distinguish between light and heavy bodies of such as I had the strength to move, a quart pot, full of water, and a feather being the same in my hand."[22]

The other poison, saxitonin, is synthesized by certain algae that periodically infect shellfish and that are responsible for the appearance of a red tide in coastal waters. Like tetrodotoxin, saxitonin is quite capable of causing death through paralysis. Both toxins, in dilute solution, can be made to block impulse conduction and the number of molecules bound to the membrane can be determined; this value will be the same as the number of sodium channels, assuming that each channel is inactivated by a single molecule of toxin. An alternative method, developed at about the same time, enables the number of channels to be found by a statistical analysis of the fluctuations in current flow through the membrane. Using the latter method to study unmyelinated nerve fibers in the lobster, Trevor Shaw, working with John Moore and Toshio Narahashi in the Physiological Laboratory in Cambridge, found that, on average, there were no more than 13 channels in one square micrometer of nerve membrane.[23] In another study from the same laboratory, Richard Keynes, Murdoch Ritchie, and Eduardo Rojas found similarly low values for unmyelinated fibers in crab leg nerve and in the rabbit vagus nerve.[24] In nerve fibers with myelin sheaths, such as the largest fibers found in our own peripheral nerves, it is only the membranes at the nodes of Ranvier that possess sodium channels. This makes good sense, for this is the only part of the membrane that is required to generate an action potential. In contrast to membranes of unmyelinated fibers, the membrane at a node is replete with sodium channels, one estimate being 21,300 channels for a single node in a rat nerve,[25] corresponding to a density of 700 channels per square micrometer.[26]

The efficiency of the channels in moving sodium ions across the membrane is surprising, as an analogy shows. Imagine a person standing in a large garden enclosed by a high wall in which there are a number of doors, all of them closed. While looking at the wall, he (or she) suddenly finds himself surrounded by a hundred other people. This is strange, for none of the doors appeared to have opened. In fact, one of the doors *did* open, but only for a fraction of a millisecond, too short a time for it to have been registered by the human visual system. In this analogy, each of the hundred thousand persons represents a sodium ion that has traveled though a single channel in the nerve fiber membrane in that extraordinarily brief time. To extend the analogy, suppose now that the door in the garden wall is rather stiff, and this time stays open for a full second. This time, still brief, would nevertheless be long enough for the entire populations of New York City or London to rush through![27] Although, as described above, there had already been speculation about some of the properties of an ion channel, the remarkable efficiency came as a revelation. There was also the specificity of the channel to explain. How could a channel admit sodium ions and not potassium ions, or vice versa? Returning to the analogy of the door in the garden wall, it was as if the door would allow either men or women to pass through, but not both.

The work on the densities of the ion channels formed a natural conclusion to the two decades of research following the publication of the Hodgkin–Huxley papers in 1952. A great deal had been found out about the nature of the nerve impulse. But what about other tissues, especially muscle, whose functioning also depended on the ability to generate and propagate action potentials?

17

Muscle: The New Physiology

Muscle—that is, skeletal muscle—is a very desirable tissue for a physiologist to study. Not only do its fibers fire impulses, but they also contract—two features that, in most experiments, can be easily measured. Then, for the biochemist, there is all the metabolic machinery to be found in other cells—the enzymes and the proteins, lipids, and sugars on which they act. There are also great practical advantages in working on muscle. For one thing, there is so much of it. For most experiments, only a simple dissection is required and, particularly in human subjects, pieces can be removed for further study without any ill effects. Moreover, unlike the spleen, say, or the pancreas or the thymus, muscle is a tissue that is readily identified. People can recognize the main muscles in themselves, they are aware of the contractions that are needed, for example, to lift a heavy load or to climb a ladder, and, on television, they can see the bulging muscle masses of professional wrestlers. Further, for most people, muscle is the main source of dietary protein. As the Thanksgiving turkey is carved, the different muscles reveal themselves, most of them white but some red, especially in the leg, and each with its bundles of fibers running through the belly. It is not surprising that muscle had already been well studied prior to the 1950s. But there were still large gaps in knowledge. Not only were the ionic mechanisms responsible for the muscle action potential not known, but it was still a mystery as to how the muscle fibers contracted. And how did the excitation spread into the muscle fibers from the nerve?

In the light of their previous achievements, perhaps it was not altogether surprising that three of the people who were to make the greatest advances in muscle physiology, in the second half of the 20th century, were Alan Hodgkin (Fig. 17–1), Andrew Huxley, and Bernard Katz. Oddly enough, the fourth person was another Huxley, though not a relative of Andrew Huxley—this was Hugh Esmor Huxley.

Hodgkin, despite the intensity of his work on the squid giant axon, was the first to turn his attention to muscle. As with the squid axon, he obtained inspiration from the United States. During his trip to that country in 1948, he had not only had rewarding discussions with Cole in Chicago, discussions that had led to the Hodgkin–Huxley voltage clamp experiments, but he had also visited the physiology laboratories of Ralph Gerard and Gilbert Ling in the same university. Gerard, a well-established neuroscientist, had found an easy way of measuring the resting membrane potentials of skeletal muscle fibers in the frog. He had made

Figure 17–1 Sir Alan Hodgkin. (From Hodgkin, Sir Alan. *Chance & Design. Reminiscences of Science in Peace and War.* Cambridge University Press, 1992. Courtesy of Lady Marion Hodgkin and Cambridge University Press.)

microelectrodes out of glass, heating a piece of fine tubing over a flame until it began to melt, and then drawing the two ends rapidly apart. Each of the two pieces of glass now had a fine taper, ending in a tip less than 1 μm in diameter. The tubing could then be filled down to the tip with a salt solution, so that, now being able to conduct electricity, it had become an electrode. Gerard had found that, with a micromanipulator, the fine tip could be inserted perpendicularly through the membrane of a muscle fiber and the full resting potential measured with a galvanometer. He had started these experiments in 1940 with a graduate student, Judith Graham, and had continued them with his colleague Gilbert Ling.[1] Interestingly, Ling was, like Lorente de Nó and David Nachmanson, one of the few people who would be unable to accept the Hodgkin–Huxley account of the nerve impulse. The essence of his disagreement would be his belief that the potassium inside the nerve fiber was firmly bound to the proteins in the axoplasm. Later, despite overwhelming evidence to the contrary, he would deny the existence of the sodium pump. In 1948, however, he did not hesitate to show Hodgkin the tricks in recording muscle fiber membrane potentials.

Inevitably, and it seemed to be a characteristic of the Cambridge physiologists, the topic was not only taken up, but taken over, and in a very short period of time. Within only two years, a full paper, "The Electrical Activity of Single Muscle Fibers," would appear in the *Journal of Cellular and Comparative Physiology*.[2]

Written jointly by Hodgkin and William Nastuk, a visiting fellow from Columbia University, New York, the paper gave values for the resting membrane potentials of muscle fibers. As in the squid giant axon, and in the myelinated fibers of the frog sciatic nerve, the inside of the muscle fiber was negative with respect to the outside, in this case by some 80 mV. Unlike Ling and Gerard, who had published their own results the previous year, Hodgkin and Nastuk had employed a cathode ray oscilloscope for their measurements, and could therefore record the action potentials as well, and give a mean value for the overshoot. In keeping with the sodium hypothesis for the action potential, they found that the amplitude of the action potential was markedly affected by the concentration of sodium in the fluid bathing the muscle fibers. Unlike the papers that Hodgkin, with Huxley, had written on the squid giant axon, there was, at the conclusion, a full acknowledgement of the American help. Just as with the squid axon, however, Hodgkin had done the experiments quickly and thoroughly, and once again the American pioneers had lagged behind.

The paper with Nastuk was to prove extremely influential, and would be the inspiration for many other studies of muscle, including some on the special conducting fibers in the heart (the Purkinje fibers). In these fibers Silvio Weidmann, a Swiss who had worked with Hodgkin in Cambridge, showed that the action potential, like those in nerve and muscle fibers, had a large inward sodium current; however, there was a subsequent plateau before the current was turned off, giving the potentials much longer durations (Fig. 17–2).[3] It was from investigations such as these that the true basis of the electrocardiogram (EKG) could be found, as well as the mechanisms of certain irregularities of the heartbeat.

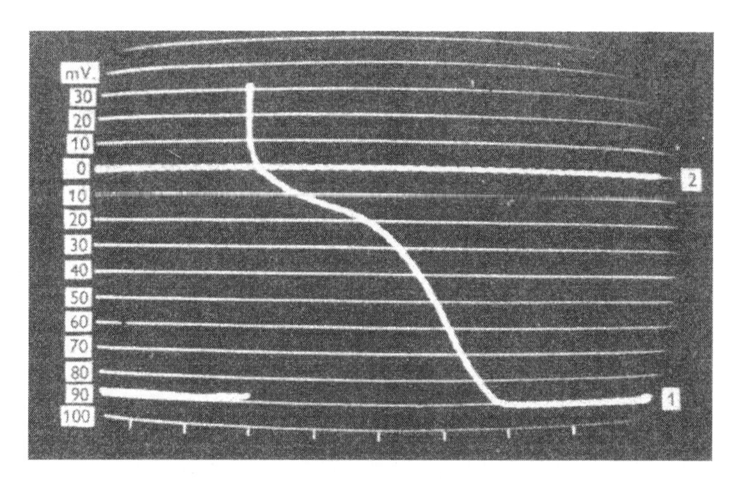

Figure 17–2 Action potential of a single cardiac muscle fiber, showing the characteristic long duration. (From Draper MH, Weidmann S. Cardiac resting and action potentials recorded with an intracellular electrode. *Journal of Physiology*; **115**: 74–94, 1951. Courtesy of John Wiley & Sons.)

From Weidmann's microelectrode studies came also the first evidence that low-resistance pathways might exist between cells in the same tissue, an important conclusion verified by the demonstration of gap junctions in heart muscle a few years later.[4] The Hodgkin and Nastuk paper would also launch a host of studies on the nerve cells in the brain and spinal cord.

One of the persons to read the paper was Bernard Katz, still the assistant director of the Biophysics Unit at University College London, and the third man in the squid axon experiments at Plymouth. Had he wanted, he could have been a full partner in those studies, but he had quietly withdrawn, sensing that Hodgkin and Huxley could manage perfectly well without him. Katz had once described his British mentor, A. V. Hill, as "the most upright person" he had ever known. It was a description that could have applied equally well to Katz. Well-built, tidy in appearance, his wavy brown hair neatly combed, and habitually bespectacled, Katz was already widely respected in physiological circles. With more than a hint of a German accent in his thick voice, he would, when lecturing or giving a communication to the Physiological Society, impose the same clarity and authority that characterized his writing.

After leaving the squid axon, and with it the ionic nature of the nerve impulse, Katz had turned his attention to muscle. He had already, in his brief time at University College London, made one discovery that, had he pursued it, would have assured him of a prominent place in international physiology. He had been able to record small potentials generated in muscle spindles, the elongated sensory structures in muscle that respond to stretch.[5] As Sherrington had shown many years before, the spindles are necessary not only for the tendon reflexes elicited by a physician, but also for the normal maintenance of posture. Were it not for the presence of the muscle spindles, we could neither stand nor hold our heads up. The small electrical signals that Katz had recorded from the spindles were the first example of "receptor potentials"—that is, depolarizations of the sensory nerve fiber membrane brought about by a natural stimulus, in this case a mechanical deformation. By changing one form of energy into another, the muscle spindle was acting as a transducer, in much the same way that a microphone can turn a mechanical signal, the sound waves, into an electrical one. It was the sort of response in sensory nerve endings that Adrian had speculated upon but had never been able to record. Nor, until Katz, had anyone else. Later, Katz was to make another important discovery in muscle, which he again chose not to follow up. This was the recognition of some peculiar cells at the peripheries of the muscle fibers. Few in number, and consisting mostly of nucleus, these were the "satellite" cells, and their importance, demonstrated later by others, was that they could multiply and repair any damage to the muscle.[6,7]

The reason why Katz disregarded these two discoveries was that they were unrelated to his main project. Katz wanted to know what happened when an impulse reached the end of a nerve fiber. During his time in Sydney with Eccles and Kuffler he had, with them, made recordings of the electrical response, the "end-plate potential," that developed in the muscle when it was stimulated through its nerve.[8] Well before that, in the 1930s, Brown, Dale, and Feldberg, working together in the Medical Research Council laboratories in Hampstead, London, had provided convincing evidence that the chemical acetylcholine was released by the motor nerve endings and acted in some way on the muscle fibers so as to excite them.[9] Eccles, on the basis of the work in Sydney, had reluctantly agreed that the transmission of excitation at the neuromuscular junction was, contrary to his earlier opinion, chemical and not electrical.

In truth, the papers from Sydney, published in the *Journal of Neurophysiology*,[10] had none of the elegance that would later characterize the work of their three authors. Instead of concentrating on a single tissue and investigating it exhaustively, there were some observations on the cat soleus muscle, others on the frog sartorius, and yet others on Kuffler's single nerve fiber–muscle fiber preparation. The work did not seem to hang together. Further, the papers themselves, probably written mostly by Eccles, were unnecessarily long. Yet, the work itself had important, and very different, consequences for all three investigators. For Eccles, it had been the means to establish himself as a medical scientist in Australia. For Kuffler, it had been the first exposure to physiological research and the opportunity to prove himself capable of intricate and beautiful dissections. For Katz, it was an introduction to the neuromuscular junction, a subject that would occupy him for the remainder of his career, and that would ultimately reward him with a Nobel Prize.[11]

On reading the Hodgkin and Nastuk paper, Katz realized that, instead of recording end-plate responses with wire electrodes on the surface of the muscle, as had been the custom in Australia, he would probably be able to record much larger signals with the kind of glass microelectrode that Hodgkin and Nastuk had employed. Also, instead of having to interpret the summed activity of an unknown number of muscle fibers, he would only have to deal with the responses of the single muscle fiber penetrated by the microelectrode. Best of all, the responses recorded would be the actual change in the membrane potential of the muscle fiber. Katz's expectations concerning the power of the new technique were confirmed when he visited Hodgkin in Cambridge and saw him demonstrate it.

And so Katz began his new project. Over the years he would be assisted in his experiments by outstanding young scientists—especially Paul Fatt, Jose Castillo, and Ricardo Miledi—and there were others who would come to work under his general direction in the biophysics laboratories at University College. In 1952, the Unit would become a full university department and, with A. V. Hill's retirement, Katz would become its Chair. It was the reward for some outstanding research.

At the beginning there was the luck of an unexpected finding. Just as Adrian, when recording from a muscle nerve, had noticed impulse activity from the muscle spindles going on in the background, so Fatt and Katz had a similar experience with the muscle fiber. It was another kind of spontaneous activity, and it was found only when the exploring microelectrode was in that region of the muscle

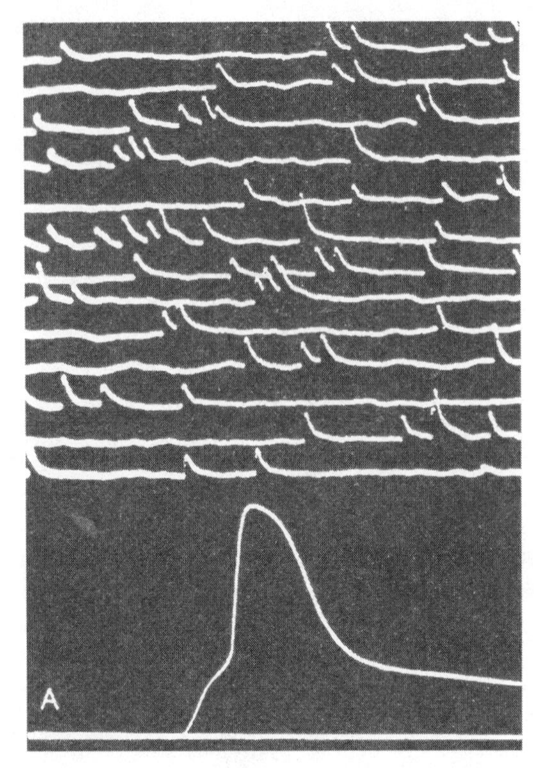

Figure 17–3 *Top*: Spontaneous "miniature end-plate potentials" recorded with a microelectrode inserted into a frog muscle fiber at the neuromuscular junction. *Bottom*: Response at the same location to stimulation of the nerve—the muscle fiber develops an end-plate potential, which, in turn, triggers an action potential (the transition occurring at the "step" in the upstroke). (From Fatt P, Katz B. Spontaneous subthreshold activity of motor nerve endings. *Journal of Physiology*; **117**: 109–128, 1952. Courtesy of John Wiley & Sons.)

fiber lying under the fine nerve endings. At the amplification normally used for recording the resting potential, the deflections were barely visible. But when the amplification was increased, there they were—small (typically 0.5 mV), short-lived depolarizations of the muscle fiber membrane (Fig. 17–3, *top*).[12] Though very much smaller, the depolarizations had the same shape as the end-plate potential, evoked by stimulation of the muscle nerve. Accordingly, Fatt and Katz called them the miniature endplate potentials (m.e.p.p.'s). The depolarizations also resembled the end-plate potential in that they were affected by drugs in the same way, their amplitudes being diminished by tubocurarine (curare) and increased by eserine. Since the end-plate potential was known to be a response of the muscle fiber to the chemical transmitter acetylcholine, and since tubocurarine and eserine produced their effects by modifying the action of acetylcholine, it was logical to suppose that the small depolarizations, the m.e.p.p.'s, were also due to acetylcholine. But how did the acetylcholine come to be spontaneously released from the nerve endings, and, as shown later, in packets of several thousand molecules?

The next clue came not from neurophysiology, but from microanatomy. The electron microscope had been invented in the early 1930s by Max Knoll and Ernst Ruska in Germany, but it was only after the 1939–45 war that these large devices became commercially available and began to appear in universities.[13] Whereas the resolving power of a light microscope, the sort of instrument used by high-school students to study the general structure of plant and animal tissues, is of the order of 1 µm, the electron microscope can distinguish objects a thousand times smaller. A single red blood cell can appear like the surface of the moon. A new universe, the universe of the very small, waited to be explored, and some of the first targets were the nerve cells, their fibers and their endings. Even in the earliest pictures it was possible, for the first time, to see the cell membrane—a structure approximately 50 Å wide.[14]

When Katz had his own electron micrographs of the neuromuscular junction, work for which he enlisted the help of Hugh Huxley and Richard Birks, he noted, as others had, that there was a very narrow space (the synaptic cleft) separating the membrane of the nerve terminal from that of the muscle fiber.[15] Within the nerve terminal were a number of small rounded structures, the "synaptic vesicles," and Katz and his colleagues thought it likely that each of these consisted of a packet of acetylcholine (Fig. 17–4). They noted that the vesicles tended to collect near regions of membrane facing folds in the surface of the adjacent muscle fiber. It looked as if the vesicles were clustering at the sites where the acetylcholine would be released, the membrane of each vesicle momentarily fusing with that of the nerve ending, and the vesicle then bursting open and discharging its acetylcholine. Katz supposed that each m.e.p.p. resulted from the spontaneous rupture of a single vesicle and that, when an impulse arrived at the junction, instead of a single vesicle, a hundred or so would liberate their packets of acetylcholine. The combination of the acetylcholine with receptors in the muscle fiber membrane would produce an end-plate potential—the response that he had studied in Sydney with Eccles and Kuffler. When the end-plate potential was sufficiently large, it would trigger an action potential in the muscle fiber (Fig. 17–3, *bottom*).[16]

It was a nice story, and, such is evolution, one that made sense. For example, having the acetylcholine in packets enabled the nerve terminal to release much more transmitter from the restricted sites in the membrane than would otherwise have been possible. But there were still unanswered questions, especially to do with the nerve terminal. Did the impulses travel to the endings of the fine nerve branches on the muscle fibers, or did they stop some distance before that? And how did the impulses mobilize the synaptic vesicles within the terminals? Katz and Miledi answered the first question by stimulating a nerve fiber and moving a recording microelectrode along the fine endings on the muscle fiber. They found

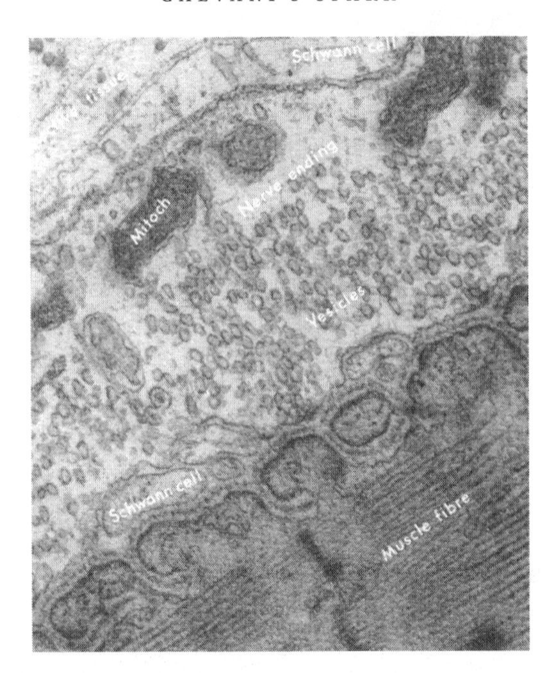

Figure 17–4 Electron micrograph of part of a frog neuromuscular junction. The lower, darker part of the figure is the muscle fiber and shows the actin and myosin filaments. The upper, lighter part of the figure is the motor nerve fiber ending, and contains many small, circular bodies—the synaptic vesicles. At the interface between the nerve and muscle fibers, the muscle fiber membrane is repeatedly invaginated. (From Birks R, Huxley HE, Katz, B. The fine structure of the neuromuscular junction of the frog. *Journal of Physiology*; **150**: 134–144, 1960. Courtesy of John Wiley & Sons.)

that the impulse, though it was now traveling more slowly, did indeed reach the nerve endings.[17] As to the second question, they showed that there was a chemical link between the impulse and the rupture of the synaptic vesicles, and that the link was calcium.[18] Thus, as the impulse invaded the nerve terminal, not only did sodium ions enter through the membrane, as Katz and Hodgkin had concluded from their squid axon experiments, but calcium ions also. Once inside the terminal, the calcium ions appeared to aid the momentary fusion of the vesicle membrane and the nerve fiber membrane, prior to the bursting open of the vesicle. By acting in this way, calcium was serving as a messenger between the nerve fiber membrane and the synaptic vesicles. The nerve impulse could be regarded as the "first" messenger, and calcium as the "second" messenger. In time, other second messengers would be discovered in a variety of cells, with different types of organelle as their targets.

As he uncovered the details of transmission at the neuromuscular junction, Katz must have wondered to what extent similar mechanisms might operate at the synapses between one nerve cell and another in the brain and spinal cord. Yet it was typical of him not to speculate in print and, in any case, the answers were to come quickly from laboratories elsewhere. The presence of vesicles had already been noted in early electron micrographs of brain tissue by other investigators, and it was shown that specimens of nervous tissue, rich in vesicles, were also rich in transmitter—suggesting that, just as at the neuromuscular junction, the vesicles contained the transmitter.[19] This suggestion was strengthened when spontaneous miniature depolarizations were observed during intracellular recordings from nerve cells in the sympathetic ganglion, and later, from nerve cells in various parts of the brain and spinal cord.[20] It would also be shown that the chemical transmitter was not always acetylcholine. Indeed, a host of chemicals would be discovered, and their recognition would lead to the research and manufacture of an astonishing variety of drugs for the treatment of neurological and psychiatric disorders. But it was the neuromuscular junction, and the work of Bernard Katz, that laid the groundwork for so many of the later advances.

Andrew Huxley (Fig. 17–5) came to his study of muscle by a different route. With the publication of the five papers on the squid giant axon in the *Journal of Physiology* in 1952, he realized there was little more that could be done on the nerve impulse other than to fill in some of the details, as Hodgkin and Keynes were already doing with their radioactive tracer studies on the cuttlefish axon. Further understanding of the nature of the channels in the membrane, if it could be achieved at all, would require a different approach to the neurophysiological

Figure 17–5 Sir Andrew Huxley in his laboratory at University College London.

and mathematical one that he and Hodgkin had employed. In contrast, the muscle fiber, and the way in which it was able to shorten, represented an interesting challenge. Instinctively, Huxley felt that the best, and certainly the most direct, line of attack was simply to look down a microscope and observe how a muscle fiber might change its appearance as it contracted. As a boy, he was given his first light microscope by his mother, and not only had he studied plant and animal tissues with it, but he had become interested in the physics of the optical system.[21] He also knew that, while the light microscope was very good for looking at tissues that have been taken from the body, cut into thin sections and then stained with a dye, it could not be used satisfactorily for living cells. Although earlier investigators had tried, it was too difficult to get a clear image of a muscle fiber as it contracted. As it happened, Huxley had a book on interferometry, a school prize, and, with his knowledge of optics, he realized that it might be possible to design an interference microscope that would be suitable for living cells. A prism would split a source of light into two beams, only one of which would pass through the tissue, and then a second prism would combine the beams again in the viewing system. The difference in wavelength between the two beams, caused by the slowing of the beam that had passed through the tissue, would produce a color. Unlike most applications of the light microscope, there would be no dye involved and no need to kill the cells beforehand. Huxley designed such an instrument and had the prisms and some of the other optical components manufactured for him; the rest he did himself.[22]

As he started on his new research project, Huxley took the precaution of reading all the old literature, most of it German, on muscle contraction. Interestingly, one of the English sources was the textbook that his grandfather, the illustrious T. H. Huxley, had written on the anatomy and physiology of the crayfish, published in 1880.[23] Andrew Huxley was all too aware that, during his research with Hodgkin on the nerve impulse, they had missed the paper of Overton's in 1902 that had described the importance of sodium in maintaining the excitability of muscle.[24] Had they been aware of it, they would have been bound to draw the right conclusions in their own 1945 paper on the squid axon. As he went through the old studies on muscle contraction, some of them going back 100 years or more, Huxley was impressed by what had been accomplished with the light microscope, often on the large muscle fibers in insect legs. To begin with, there were detailed descriptions of the light and dark bands that alternated along the muscle fiber. Further, the refractive properties of the dark bands were attributed, correctly, to the presence of the contractile protein myosin. When a muscle fiber contracted, the dark bands were seen to stay the same width, while the pale bands narrowed. One investigator even postulated, again correctly, that the myosin was in the form of "rodlets," stretching the full widths of the dark bands.[25] Of course, not everyone agreed on these points, and a favored hypothesis was that the muscle fiber shortened by the movement of fluid from the light to the dark bands. Though all the key observations and speculations needed for describing the true mechanism of muscle contraction had been made, much of the old work came to be forgotten or else ignored, and it was the faulty accounts that persisted. Scientists no longer used the microscope to study muscle, and the field had been taken over by the biochemists and physiologists. Although much new information had come to light, it had brought an explanation of muscle contraction no closer.

One of the physiologists working on muscle was A. V. Hill.[26] His viscoelastic hypothesis of muscle contraction, partly for which he had been awarded the Nobel Prize in 1922, had since been shown to be incorrect—the contrary findings being supplied by a young American, Wallace Fenn, working in Hill's own laboratory.[27] Notwithstanding the setbacks, and the later advances by succeeding generations of physiologists, Hill would continue to work on muscle until the very end of his long career. For the students and younger faculty at University College London, there was awe at the emergence from his laboratory of the tall, stern-faced, and very imposing figure. Yet if the white-haired old man could not tolerate fools, and was incapable of light conversation, he still had a sense of humor, as his later writings attested.

In contrast to the muscle physiologists, the chemists working on muscle had several solid achievements behind them, two of which had to do with the supply of energy needed for the contractile mechanism. One was the discovery of phosphocreatine in 1927,[28] and the other, two years later, was the finding of a still more powerful phosphorus compound, adenosine triphosphate (ATP).[29] Still later was the observation that muscle ATP could be split by the contractile protein myosin—the myosin, in this case, acting as an enzyme.[30] There was also the discovery of a second contractile protein, actin, which, at least in the test tube, had a high affinity for myosin.

Such, then, was the situation in 1953 when Andrew Huxley began to look at muscle with his new interference microscope. Rather than study a whole muscle, or even a bundle of fibers, Huxley had taught himself to dissect out a single fiber, and in good enough condition that it could still contract when stimulated. Working with Rolph Niedergerke, he saw that, as the fiber shortened, there was a change in its striations, the light bands becoming narrower and the dark ones remaining the same length. When the fiber was stretched, the dark bands again remained the same, but this time the light bands lengthened.[31] The patterns of response were those that had been described a hundred years earlier and had then disappeared from the literature. But what was the explanation for the changing patterns? For Andrew Huxley, the most likely one was that there were rods in the dark bands, and that these stayed at the same fixed length regardless of what was happening to the length of the fiber. It was another prediction that could have been found in the papers of the previous century. As it turned out, direct evidence that this interpretation was correct was already coming from another source.

The second source was the "other" Huxley—Hugh Esmor Huxley (Fig. 17–6)—who had been working in the Cavendish Laboratories at Cambridge as a Ph.D. student.[32] With a background in physics, he had attempted to learn about muscle

Figure 17–6 Hugh Esmor Huxley. (Courtesy of Professor Hugh Huxley)

structure by using X-ray diffraction. The method was to direct a beam of X-rays at a crystal and to work out the arrangement of the atoms in the crystal from the scattering of the X-rays. It was a technique pioneered by William and Lawrence Bragg, father and son, and had been applied first to inorganic crystals, and then, with very much greater difficulty, to biological ones. Two of Hugh Huxley's fellow students in the Cavendish Laboratory were a bright young American and a middle-aged British physicist, soon to become famous for their discovery of the DNA double helix—James Watson and Francis Crick.

In relation to muscle, Hugh Huxley used a low-angle X-ray beam and interpreted the diffraction pattern as indicative of two arrays of filaments running in the long axes of the myofibrils, the myofibrils being the bundles of contractile material that filled the insides of the muscle fibers.[33] To his eyes, it appeared as if the two sets of filaments extended without interruption from one end of a myofibril to the other. He realized, however, that the X-ray diffraction pictures were a very indirect way of examining the structure of the muscle fiber, and that a more accurate picture could probably be obtained with the electron microscope. With this idea in mind, he went to the United States in 1952 to gain the necessary experience. Working at the Massachusetts Institute of Technology, he made his first crucial observations in the same year. The dark bands were composed of thick filaments (rods), which he later showed to be the protein myosin, and the light bands were made up of thin filaments, actin. Further, when he examined transverse sections of the myofibrils, this in collaboration with Jean Hanson, it was apparent that the thin filaments surrounded the two ends of each thick filament in hexagonal arrays.[34] To both Huxleys, the mechanism of muscle contraction was now obvious: the fibers shortened as the thin filaments slid over the thick ones, increasing the amount of overlap as they did so.

The next question, obviously, concerned the nature of the mechanism that made the filaments slide over each other in this way. Once again, the answer came from Hugh Huxley's electron microscope. Working once more with Jean Hanson, this time at King's College, London, he saw, in extremely thin longitudinal sections of myofibrils, small projections from the sides of the thick filaments. These projections, the "cross-bridges," looked as if they might temporarily attach themselves to the thin filaments and move the latter towards the middle regions of the thick filaments (Fig. 17–7).[35]

Meanwhile, Andrew Huxley, aware of Hugh Huxley's work, realized that the "sliding filament hypothesis" could be tested by measuring the force developed by a muscle fiber at different lengths. Up to a certain point, the more the overlap between the filaments, the greater the number of cross-bridges that could engage and develop force. Such length–tension plots had been made before, even on single fibers, but there was a problem in that muscle fibers could not be stretched evenly: the middle regions of the fibers were stretched more than the ends. Andrew Huxley overcame this difficulty in a typically brilliant way. Not only did he dissect out a single, living muscle fiber, a very difficult task in itself, but he devised

Figure 17–7 Electron micrograph of part of a myofibril. In the center of the photograph, running from top to bottom like the rungs of a ladder, are 11 thick horizontal rods—the myosin filaments. Because of the angle that the section has been cut, the two ends of each rod are separated from those of its neighbor by two thinner structures—the actin filaments. Connecting the two types of filament are series of short, perpendicular projections—the myosin cross-bridges. Each myosin rod is approximately 1.7 μm long. (Courtesy of Professor Hugh Huxley)

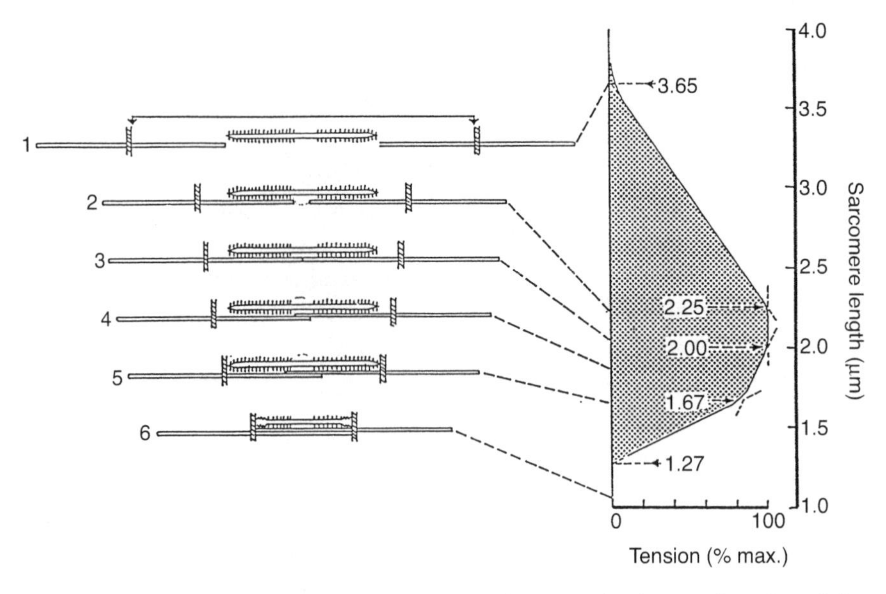

Figure 17–8 Relationship, in a single sarcomere, between the degree of overlap of the actin and myosin filaments and the force developed during a contraction. If the sarcomere is stretched too far, there is no overlap and hence no force generated (*top record*). Conversely, if the sarcomere is shortened too much, such that the actin filaments meet over the myosin filament, there is also no contractile force (*bottom record*). Otherwise the force depends on the number of myosin cross-bridges that can engage actin filaments (*intervening records*). (Adapted from Gordon AM, Huxley AF, Julian FJ. The variation in isometric tension with sarcomere length in vertebrate muscle fibres. *Journal of Physiology*; **184**: 170–192, 1966. Courtesy of John Wiley & Sons.)

an electronic feedback system to control the length of a very small part of the fiber. It involved sticking two minute pieces of gold leaf to the fiber, so as to interrupt a light beam, and attaching a servo-motor to one end of the fiber. The results were very clear, and fully consistent with the sliding filament hypothesis (Fig. 17–8).[36]

The new experiments were carried out at University College London, to which Andrew Huxley had been appointed Jodrell Professor of Physiology in 1960. Those working in the department at the time would retain vivid memories of the slim black-haired figure with the quick and decisive movements, walking briskly along the corridor or perhaps taking the stairs, two at a time, to the next floor. If they had caught a glimpse of his face, they would have seen the penetrating Huxley look, part of his grandfather's legacy and very evident in a photograph of the time (see Fig. 17–5). Had they attempted a conversation, they might have found it difficult going. Though Andrew Huxley was both attentive and unassuming, it was difficult to ignore the huge intelligence behind the dark eyes. The powerful brain was simply idling, effortlessly keeping pace, and ready to leap ahead whenever its owner wanted. It was the same story at the meetings of the Physiological Society, the organization his grandfather had helped to found in 1876. During the discussion following a presentation, the gray-suited man near the front of the auditorium would rise to his feet and, in his clear, well-modulated voice, ask a question. Invariably it would be something that no one else had thought of—a different interpretation of the results, perhaps, or a reference to a similar, but overlooked, study in the past. Nor did it matter what the subject was, for Andrew Huxley's knowledge seemed all-encompassing. Yet if he was held in admiration and awe, there was also the recognition of an innate kindness and modesty, and it was these last qualities that made him well-liked. Never was the affection more evident than on a bright fall day in October 1963 when, standing on the balcony overlooking the rear court of London's University College, Andrew Huxley acknowledged the cheers of the faculty and students for his share of the Nobel Prize for Medicine or Physiology. The award had been given for the work on the squid giant axon with Hodgkin.

So far as muscle was concerned, however, Hodgkin had demonstrated that the fibers had action potentials like those in nerve, Katz had shown how excitation was transmitted from the nerve endings to the muscle fibers, and the unrelated Huxleys, Andrew and Hugh, had worked out the mechanism of muscle contraction. There was still one part of the puzzle remaining—how did the action potential, propagating along the surface of the muscle fiber, signal to the myofibrils in the interior to contract? Once again, it was Andrew Huxley who provided the answer, and by a delicate but simple approach. He employed a microelectrode to stimulate the surface of a muscle fiber at different points, carefully noting the threshold at each. It soon became apparent that, in an amphibian fiber, the thresholds were always least in the centers of the light bands, and it was therefore probable that these were the sites at which the impulse was transmitted from the surface to the interior.[37] This possibility was made all the more likely by the report of very fine passages penetrating the interiors of the fibers at these locations—the transverse or T-tubules. Curiously, but in keeping with so much of the muscle research, the electron microscope had rediscovered what had been well described many years previously with the light microscope, but then forgotten.[38]

With the recognition of the transverse tubules, the last major feature of muscle excitation and contraction had been found, and it would be left to others to fill in the details. But as neurophysiology pushed forwards, so one of the great links with the past was broken. . . .

Eastbourne, March 4, 1952

The house, a large one, is not far from the promenade and the pebble beach. Inside, the Catholic sisters go about their business, ministering to the elderly and the infirm with the solicitude and the unstinting kindness of their order. In his room, the 94-year-old man sits by the fire. Even in his prime he was small, having to look up to those to whom he was talking. Now, in his old age, he seems to have shrunk further, filling only part of the large armchair. There is a small French volume on his lap, and though it is one he has read before, the eyes behind the spectacles continue to scan the lines with quiet satisfaction. He has always loved books, and no time was more enjoyable to him than that spent in the secondhand bookstores. Now all but a few of the books are gone, given away, some to the Colleges, some to his friends, many to the British Museum. When not reading, his mind has been returning increasingly to the past: the East Anglian boyhood, going up to Cambridge as a young man, the medical studies in London. And the people—his friends, his colleagues, his mother, yes, especially his mother. On the table beside him, written in his careful hand, lies the poem for her birthday. 1893, nearly 60 years ago. How tantalizing it was to search for, and how satisfying to find, that special phrase, that word, that captured the mood exactly.

A few minutes more and the book falls to the floor. There is no attempt to retrieve it. The hands, the same hands that did those countless dissections, are at rest.

Sir Charles Scott Sherrington, Grand Commander of the British Empire, holder of the Order of Merit, recipient of 22 honorary degrees, Nobel Laureate, the man who was recognized as the greatest physiologist of his day, has died. The flickering lights in the enchanted loom have been extinguished.

18

More Triumphs with Microelectrodes

John Carew (Jack) Eccles (Fig. 18–1) also read the paper by Hodgkin and Nastuk on muscle fibers, and immediately realized that intracellular recording might be the best way to advance his studies of synaptic transmission in the spinal cord. If the technique could be made to work on motoneurons, then it might be possible to carry his old studies with Sherrington to a satisfying conclusion. The problem was, would someone else get there before him?

Life had not been easy for Eccles since the days in Sydney, those days spent exploring the neuromuscular junction with Katz and Kuffler in the laboratories on the top floor of the Kanematsu Institute.[1] At first, everything had been so promising. The three men had enjoyed each other's company, often playing tennis together on Eccles' grass court at weekends. In the laboratory the work had gone well—at least in the neuromuscular junction, synaptic transmission was chemical, just as Sir Henry Dale had always insisted. With Australia's increasing involvement in the war, however, things had changed: Katz had left to join the armed forces and Eccles had discovered that the hospital administration, by building accommodation for residents above his laboratories, no longer supported him. It would not be the last time that promises to Eccles would be broken. There was something about Eccles, perhaps the combination of the compelling eyes behind the heavy glasses, the severe look, and the abrasive voice with its strong Australian accent, that turned people away. It was unfortunate, because Eccles, though a determined and driven man, was honest and fair in his own dealings. Moreover, there was a kindness and attentiveness that would reveal itself especially when younger scientists from overseas came to work with him. They were the same qualities that Sherrington had shown him in Oxford.

When the debacle at the Kanematsu Institute occurred, there were no openings in Australian universities for a physiologist of Eccles' stature. He received an offer from Liverpool University in Britain and may well have been tempted to accept. Despite the war, it would have brought him back to the country of which he had such pleasant memories. Moreover, he would have been following in the footsteps of his master, Sherrington, who had carried out his much of his finest work while head of physiology in that university. But in the end, Eccles declined. There was one other chance for him, this to take over the physiology department at the University of Otago in the little town of Dunedin, on the South Island

Figure 18–1 Sir John Eccles. (Courtesy of the John Curtin School of Medical Research, Australian National University)

of New Zealand. It was a small department, seriously understaffed, and it was as far from his beloved Oxford as the globe would permit.

New Zealand it was. His new position meant learning physiology all over again, and bringing himself up to date with the advances in the different body systems. As Erlanger had found, on moving to St. Louis as the new head of physiology, there were numerous lectures to prepare, in this case 75 for the second-year medical students alone, as well as practical classes to organize, and all the administration that a university asks of a departmental Chair. Inevitably, there was very little time for research, especially during the first few years. Yet if the university was small and remote, it had, for Eccles, some consolations. One of these was the presence of Karl Popper, a philosopher much interested in the way that scientists conducted their affairs. Popper, unlike many of his kind, was a lucid speaker and writer, and considered that scientific research was "deductive." By this he meant, in essence, that science proceeded by the testing of hypotheses, with the rejection of those that did not fit the experimental results.[2]

It was surprising that Eccles should have embraced a philosophy so much at variance with what actually happens in most biological laboratories. Physiologists, especially, are curious people. They like to observe, to look down a dissecting microscope, to prod and to poke, and to watch the changing patterns of electrical activity on an oscilloscope screen. True, they like to have some plan of attack, but they also know that once they become sufficiently familiar with their "preparation," the unexpected clues will start to appear; all that is required is the ability to recognize them. In this way had come Adrian's detection of the impulse code of the various sensory receptors, and Fatt and Katz's discovery of the miniature end-plate potentials. George Bishop, forever content to be in his laboratory in St. Louis, and always working on a wide range of animals and on different parts of the nervous system, was an especially good example of a curiosity-driven experimenter.

Eccles, himself, could also recognize the unexpected. Later in his career, he would carry out an experiment in which the nerves to two contrasting types of muscle in the cat hindleg had been switched over. The effect he had hoped to find in the spinal cord, a readjustment of the synapses on the motoneurons, had not taken place. It had been a long and arduous experiment and by now it was in the early hours of the next morning. Most experimenters would have switched off the electronic equipment, turned off the laboratory lights, and gone home. Instead, ever meticulous, Eccles insisted on examining the muscles themselves. Had the new nerve supply reached the muscles and made connections with the muscle fibers? And there—as he cut and reflected the skin—was the answer: not only had the regenerating nerve fibers got to the muscles, but they had completely transformed them. The previously pale gracilis muscle was now a rich red color, while the red crureus had become a pale muscle.[3] It was one of the most striking demonstrations of the sustaining, "trophic," properties of nerve cells, and the stimulus for countless experiments in laboratories around the world.

All of this came later, however. In 1943, and for the next few years, all Eccles could expect to do was teach the medical students the physiology he felt they needed to know, and to hope that better times, allowing more freedom for work in the laboratory, would eventually arrive.

And, gradually, the better times did arrive. There was also the presence of a talented electronics engineer, Jack Coombs, who designed and built the stimulators, timers, and amplifiers that Eccles needed for his experiments. Here Eccles differed from the physiologists of the Cambridge school, who took satisfaction in building their own apparatus and making do with the minimum required for their work—the "elegance in simplicity" principle. Eccles, in contrast, insisted on having the very best instrumentation. There were enough things that could go wrong in an experiment without the additional worry of the equipment breaking down. If the best could not be bought, it would have to be made by a professional engineer. As it was.

For the first of the new experiments on the spinal cord, Eccles had employed sharp steel needles, insulated to their tips, to record from the motoneurons. They had been good enough to pick up the discharges of the cells, when they had been excited by stimulating the sensory nerve fibers from the muscle spindles, but this was hardly a significant advance. However, if the motoneurons could be impaled with the same type of glass microelectrode that Hodgkin and Nastuk had used for their muscle studies, all kinds of new information might become available.

For Eccles, the most important issue in neurophysiology was still the nature of synaptic transmission between nerve cells. Was it brought about by the release of a chemical, as Otto Loewi had shown for the heart, and as Sir Henry Dale and his colleagues had demonstrated for the neuromuscular junction—a conclusion confirmed by Eccles himself, with Katz and Kuffler? Or was transmission in the brain and spinal cord electrical, due to currents flowing between one nerve cell and another? It was for electrical transmission at synapses that Eccles had argued so vehemently with Dale at the Physiological Society meeting in 1936. Nor was Eccles alone: Gasser and Erlanger also favored electrical transmission, and so did Lorente de Nó. Indeed, for Lorente, the validity of the electrical hypothesis would offer a chance at redemption after the mistakes with the frog sciatic nerve. If transmission proved to be electrical, then all the tedious experiments observing the effects of passing currents in and out of nerve fibers would be relevant after all.

It occurred to Eccles that it would be possible to apply the Popper philosophy for a crucial test. He would impale a motoneuron with a microelectrode and then stimulate an inhibitory nerve. He reasoned that, if synaptic transmission was electrical, the membrane potential should decrease, whereas, if transmission was chemical, it should increase. In the first instance, such were the electrical connections in his equipment that the trace on the cathode ray oscilloscope would go up, while in the second, it would go down. The decisive test was performed one day in August 1951.[4] A suitable motoneuron was penetrated, the inhibitory nerve was stimulated, and the oscilloscope trace went briefly . . . *down*. Transmission was chemical. Once again Dale's intuition had been correct.

It was now that Eccles showed his greatness by immediately accepting that he had been wrong all along. The ingenious arguments and electrical diagrams, so carefully developed and put forward over the years, were wasted. Eccles recognized the situation for what it was, that it was time to move on and, as quickly as he could, to exploit the opportunities opened up by the new technique of intracellular recording from nerve cells. The contrast with Lorente de Nó could not have been sharper. With the publication of the squid axon work by Hodgkin, first with Katz and then with Huxley, Lorente must have known that many of his conclusions in *A Study of Nerve Physiology* were wrong, or at least doubtful. Yet he persisted in arguing against the tides of experimental fact and scientific opinion. A great scientist had lost his objectivity.

Meanwhile, Eccles' fortunes were about to change for the better. In the year of the definitive experiment on synaptic transmission, 1951, he received an invitation from a most distinguished scientist. Howard Florey was also Australian, and, like Eccles, had worked at Oxford before the war. During the war, however, he had remained at the university and, despite all manner of obstacles, had been responsible for developing a method for manufacturing penicillin on a large scale—at one time hospital bedpans had been used as chambers for culturing the *Penicillium notatum* mold. It was work for which he would be knighted and would share the 1945 Nobel Prize in Physiology or Medicine with Alexander Fleming and Ernst Chain.[5]

Florey's proposition to Eccles was this: would he consider moving back to Australia and becoming the first Professor of Physiology at the new university under construction in Canberra? It was to be the national university, and to take postgraduate students only. At first, the physiology department would consist of no more than a temporary hut with three research rooms, to be built on a large tract of wooded grassland. There was, however, the promise of a handsome and well-equipped replacement. At this point in his life, Eccles was 48. The shock of hair, the flag of his youth, had turned prematurely gray and then largely disappeared. If he took the position, there would be only 17 years left before retirement, of which the first year or two would necessarily be spent mostly in organization and administration.

Without hesitation, Eccles accepted. With enthusiasm. The new university had strong government backing, including the provision of plentiful funds. After the years in the academic wilderness, Canberra must have seemed the promised land. And so began the most productive and successful phase of Eccles' research career. In the midst of the planning for the new department, however, he took time to return to Britain. It was his first visit since before the war and there was much to catch up on, including the exciting new work by Hodgkin, Huxley, and Keynes on the squid giant axon and by Katz and Fatt on the neuromuscular junction. There was also a meeting of the Physiological Society to attend, at which he formally announced his abandonment of the electrical hypothesis of synaptic transmission. While in temporary residence at Magdalen College, Oxford, he gave the Waynflete Lectures on *The Neurophysiological Basis of Mind: Principles of Neurophysiology*. The lectures, published as a monograph,[6] were highly praised, not least because they incorporated all the latest thinking on the ionic mechanisms responsible for the resting and action potentials. They showed, just as the renunciation of electrical transmission at synapses had done, how quickly Eccles could adapt and assimilate.

Eccles' time in Britain also allowed him to persuade Paul Fatt to leave University College London temporarily for Canberra. Fatt, the young American with the thick glasses and the unruly black hair, had carried out the definitive microelectrode study of the muscle fiber end-plate potential with Katz, and the two had discovered the small depolarizations produced by the spontaneous release of acetylcholine at the motor nerve endings.[7] Now Fatt would become Eccles' first postdoctoral student. In agreeing to leave the security of London for the unknown in Australia, Fatt was, though he may not have appreciated it at the time, following in the footsteps of Katz 13 years earlier. Eccles' time in Britain allowed him to visit someone else, someone especially dear to him, the person who had guided him so well through the first years of his research career. It was the same person who had also educated the young Australian, fresh from medical school in Melbourne, in the beauty of poetry, literature, and art, and who had given him a feeling for the history of medical science: Sherrington. Eccles went to the nursing home in Eastbourne just in time, for nine days later the great man was dead.

It is not known whether Eccles visited Cambridge during his stay in Britain. It is probable that he did not, for if he had, he might have discovered that someone else was attempting to impale motoneurons with microelectrodes and to study their synaptic potentials—Bryan Matthews.[8] Unfortunately no records of these experiments are to be found, but there is little doubt that, had he persisted, Matthews—with his formidable talent for devising new techniques, and with his experience in working on the spinal cord—would have provided serious competition in the new field.

A month after Sherrington's death, Eccles was on his way back to Australia, via the United States. There, at a symposium on the neuron held at Cold Spring Harbor on Long Island, he met the one other person who might sort out the details of the motoneuron before he, Eccles, could get started in Canberra. It was David Lloyd (Fig. 18–2). He, too, had also studied with Sherrington in Oxford and, after taking a position at the Rockefeller Institute in New York, had been able to continue his research without interruption during the war years. There had been a succession of excellent papers, mostly in the *Journal of Neurophysiology*, in one of which the time-courses of motoneuron excitation and inhibition had been described.[9] To obtain his results, Lloyd had stimulated the muscle nerves, or sometimes their sensory fibers just before their entry into the spinal cord. Like Lorente de Nó, in the latter's study of the eye-muscle nuclei in the brain stem, he had recorded the responses of the motoneurons with electrodes placed on their axons close to the cell bodies. Because the lengths of nerve fiber between his stimulating and recording electrodes were short, Lloyd had also been able to estimate the minimum times required to excite or inhibit the motoneurons quite accurately, and to measure the time-courses of excitation and inhibition (Figs. 18–3, 18–4 and 18–5).

Figure 18–2 David P. C. Lloyd. (Courtesy of the Rockefeller Archive Center)

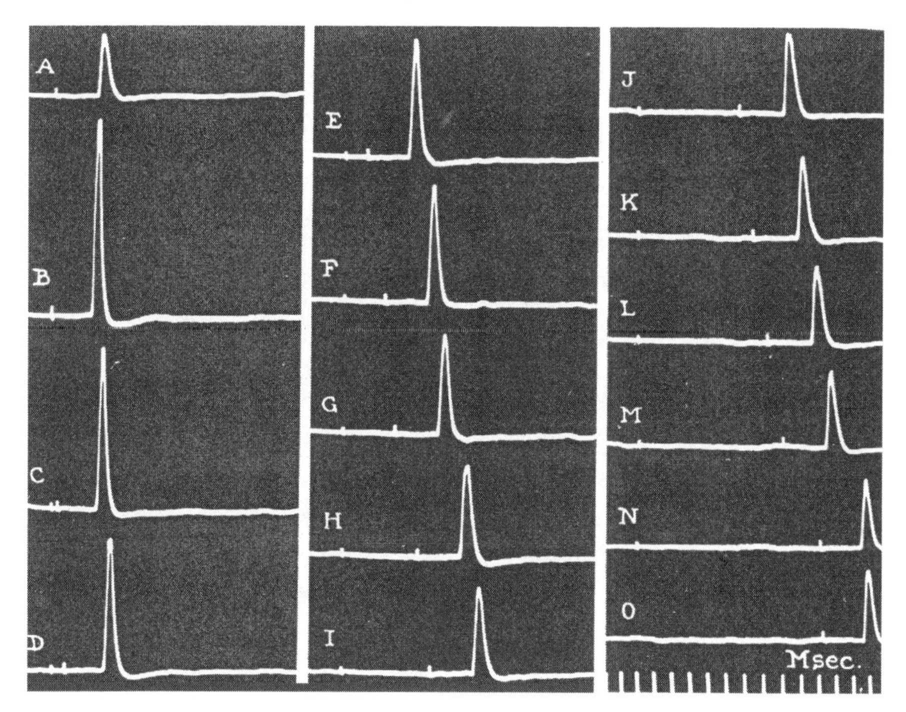

Figure 18–3 An example of temporal and spatial facilitation ("summation") from one of Lloyd's experiments, the recordings being made from a cat ventral root, and the stimuli being applied to two nerve branches to the same muscle; one stimulus was weak enough not to give a reflex response by itself, while the second gave the response in *A*. When the two stimuli were given simultaneously, the reflex response was much larger than when the stimuli were given separately (compare *B* with *A*); this is spatial facilitation. When the intervals between the stimuli were progressively increased so as to provide less temporal facilitation, the responses declined accordingly (records *C* to *O*). (Modified from Lloyd DPC. Facilitation and inhibition of spinal motoneurons. *Journal of Neurophysiology*; **9**: 421–438, 1946. Courtesy of the American Physiological Society.)

Important though Lloyd's work was, it suffered from its indirect nature. The results were only averages of motoneuron activity and the curves showing the decay of excitation or inhibition depended as much on the distributions of firing thresholds among the motoneurons as they did on the synaptic potentials. Clearly, as Eccles had realized, much more would be learned if one could record the changes in the membrane potentials of individual motoneurons. At the symposium on Long Island, there were two American neurophysiologists, Woodbury and Patton, who had, like Eccles in New Zealand, succeeded in recording intracellularly from motoneurons.[10] Fortunately for Eccles, they were interested more in the biophysics of the cell membrane than in the synaptic connections. Even more fortunately, Lloyd was unimpressed with their results: rather than the new technique contributing anything useful, he argued, it would only give information about dead and

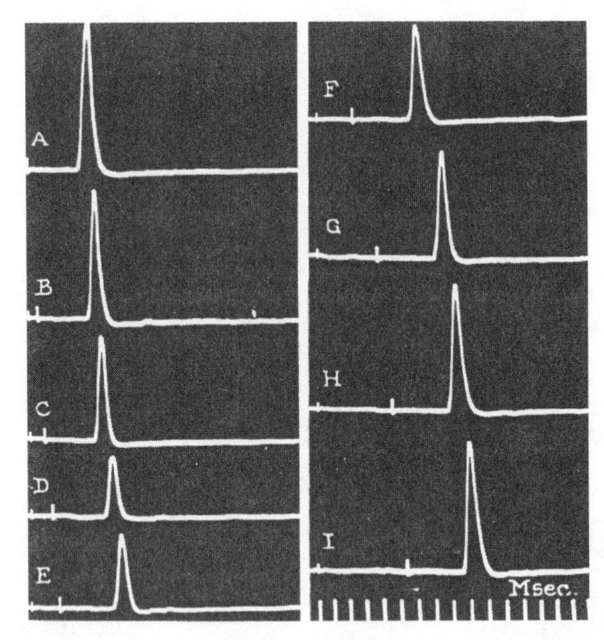

Figure 18–4 A demonstration of reflex inhibition, the recordings being made from a cat ventral root. The stimuli are applied to one muscle nerve alone, giving an excitatory reflex response (*A*), or else are combined with an earlier stimulus to an inhibitory nerve (increasing intervals from *B* to *I*). The inhibition is greatest in record *D*, in which the response is smallest. (Modified from Lloyd DPC. Facilitation and inhibition of spinal motoneurons. *Journal of Neurophysiology*; **9**: 421–438, 1946. Courtesy of the American Physiological Society.)

dying nerve cells. The negative attitude of his potential rival was more than Eccles could have hoped for. As quickly as he could, back in Canberra, he got his three laboratories in the temporary hut into working condition.

By March 1953 he was ready to begin the new experiments. To work with him there was Paul Fatt, newly arrived from Britain, and another postdoctoral fellow, Koketsu from Japan. There was also Jack Coombs, who had designed and built the superb stimulating and recording system for him in New Zealand. Another person who would come to play a vital role in the new enterprise, both by her own scientific endeavors and by instructing newcomers in the laboratory methods, was Eccles' daughter, Rose.[11] Within the next five years would come scientists from around the world—Sven Landgren and Anders Lundberg from Sweden, William Liley from New Zealand, Vernon Brooks from Canada, Robert Young and Benjamin Libet from the United States, Arthur Buller and Kris Krnjevic from Britain, and Ricardo Miledi from Mexico. They would mix together not only in the laboratory but in the lounge and dining room at University House, just as Eccles had mingled with the dons at the High Table in Magdalen College. It was as if Eccles had taken on the mantle of Sherrington, and Canberra had become the new

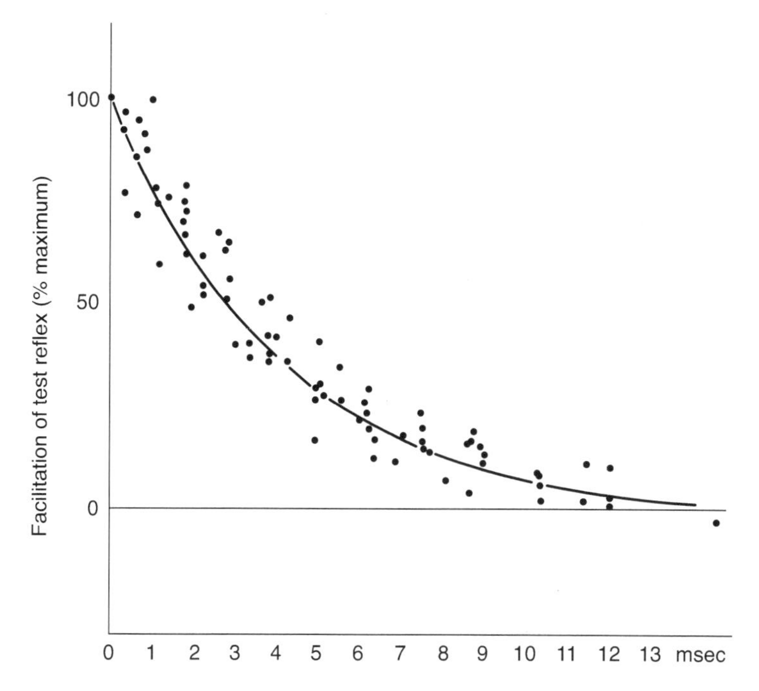

Figure 18–5 Time-course of facilitation, made from experiments such as that illustrated in Figure 18–3. (From Lloyd DPC. Facilitation and inhibition of spinal motoneurons. *Journal of Neurophysiology*; **9**: 421–438, 1946. Courtesy of the American Physiological Society.)

Oxford of neurophysiology. For younger neurophysiologists around the world, to visit Australia and work with Eccles became a rite of passage. The perspicacity of those who chose to come, and were accepted, would be amply rewarded.

The first major study, with Fatt and Koketsu, was the demonstration that motoneurons released the same chemical transmitter, acetylcholine, not only from the axons ending on the muscle fibers, but also from the recurrent branches that ended on some small inhibitory nerve cells in the anterior gray matter of the spinal cord.[12] These were the cells named after Renshaw, the Australian neuroscientist who had discovered them while working at the Rockefeller Institute in New York. The finding, that nerve cells secreted and released the same transmitter from widely separated axonal branches, was immediately accepted as a fundamental principle of the organization of the nervous system. Eccles had to tell Sir Henry Dale of the discovery, only for the eminent physiologist to write back, pointing out that he had anticipated Eccles' finding in a lecture he had given in 1934. Eccles, in return, only too pleased to have the great man's approbation, proposed that the sameness of the transmitter be known as "Dale's Principle."[13]

Eccles also recognized that the interposition of an inhibitory neuron, such as the Renshaw cell, was the mechanism employed by the nervous system to convert an excitatory effect into an inhibitory one. Other studies revealed the changes in the motoneuron membrane potential that accompanied excitation and inhibition, the very changes underlying Sherrington's old concepts of central excitatory and inhibitory states. The depolarization of the membrane responsible for the former looked remarkably similar to the end-plate potentials that Fatt and Katz had recorded in frog muscle fibers with their intracellular electrodes. Eccles termed the motoneuron depolarization the "excitatory postsynaptic potential" or EPSP (Fig. 18–6). Just as with the end-plate potential in the muscle fiber, once the depolarization reached a certain critical value in the motoneuron, it triggered an action potential. Further, if the initial EPSP was insufficient to fire a motoneuron, a second one—from the same or a different nerve—could induce a

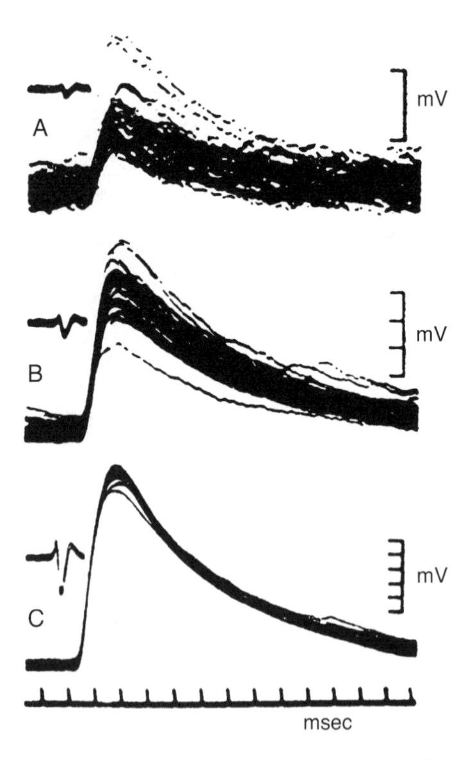

Figure 18–6 EPSPs (excitatory postsynaptic potentials) recorded from a single motoneuron with an intracellular electrode: many traces have been superimposed for each part of the figure. As the stimuli are increased, so the EPSPs become larger (records *A* to *C*). These slow waves are similar in appearance to muscle end-plate potentials and to sympathetic ganglion postsynaptic responses. (Modified from Coombs JS, Eccles JC, Fatt P. The specific ion conductances and the ionic movements across the motoneuronal membrane that produce the inhibitory postsynaptic potential. *Journal of Physiology*; **130**: 326–373, 1955. Courtesy of John Wiley & Sons.)

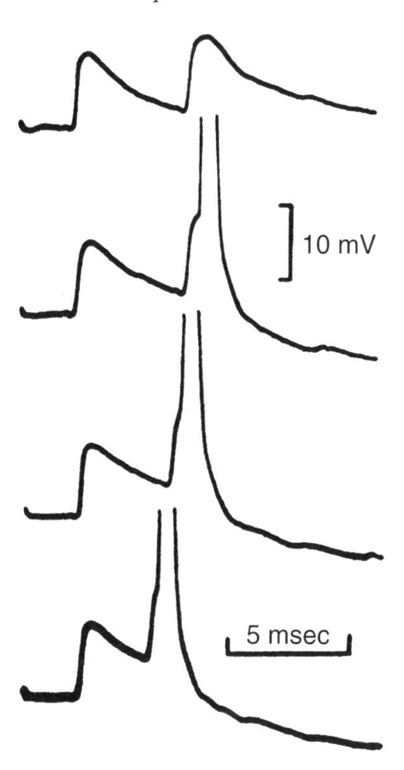

Figure 18–7 Temporal facilitation. EPSPs are evoked by two stimuli applied to the same muscle nerve. When the interval between the stimuli is too great, the summation of the two EPSPs is insufficient to discharge the motoneuron (*top trace*). When the interval is shortened, however, the motoneuron fires, and does so earlier as the interval is progressively reduced (*lower three traces*). (From Eccles, 1957)

discharge (Fig. 18–7). In contrast, the inhibitory process in the motoneuron generated not a depolarization, but an increase in membrane potential—the inhibitory postsynaptic potential or IPSP (Fig. 18–8).

Eccles now went a step further by passing current through the membrane and seeing how it affected the appearances of the postsynaptic potentials. To do this required the insertion of a double-barreled microelectrode into the motoneuron, one barrel injecting current and the other recording the change in membrane potential. The approach was similar to the one that Cole, and Hodgkin and Huxley, had used in their voltage clamp experiments on the squid giant axon. As in the squid experiments, it enabled the nature of the conductance changes in motoneurons to be deduced. And so, in volume 130 of the 1955 edition of the *Journal of Physiology*, four papers appeared from Canberra, each describing a different aspect of the motoneuron responses.[14] The thoroughness of the experiments, the beauty of the recordings, and the clarity of the writing were exemplary.

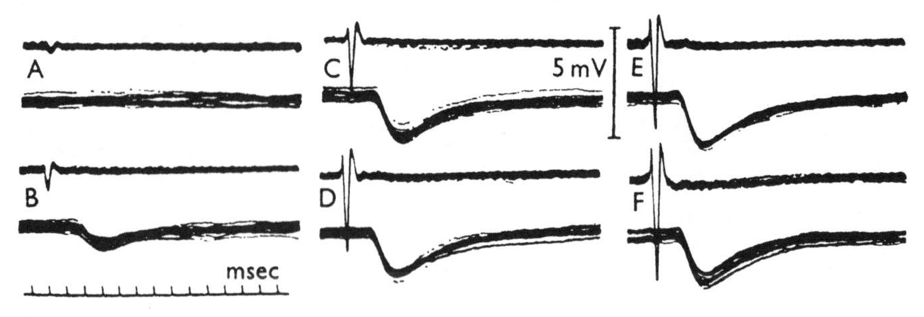

Figure 18–8 IPSPs (inhibitory postsynaptic potentials) recorded from a single motoneuron with an intracellular microelectrode. The lower of each pair of traces is the IPSP (many traces superimposed) and the upper is the impulse volley arriving at the spinal cord in the dorsal root. As the stimuli to the inhibitory nerve are made larger, both the arriving impulse volley and the reflex IPSP in the motoneuron increase (records *A* to *F*). (From Coombs JS, Eccles JC, Fatt P. The specific ion conductances and the ionic movements across the motoneuronal membrane that produce the inhibitory post-synaptic potential. *Journal of Physiology*; **130**: 326–373, 1955. Courtesy of John Wiley & Sons.)

The four papers were a foretaste of what was to come. There would be others, some to do with the effects on motoneurons of stimulating various classes of sensory nerve fiber,[15] others to do with a special type of inhibition in sensory fibers running up the dorsum of the spinal cord to the brain stem,[16] and others on the properties of neurons in that part of the brain known as the hippocampus.[17] Then a book would appear on the cerebellum, that large, enigmatic structure at the back of the brain.[18] In each case, the experiments and the conclusions were so thorough that nothing further could be done. First there would be an analysis of the currents flowing in a region of brain or spinal cord, which would indicate the locations of the active neurons, and then would come the intracellular recordings. Everything that could be measured was measured—the amount of depolarization or hyperpolarization and, above all, the timing of the responses. It was systematic and complete. To those watching the events in Canberra, it was as if an army had descended on a city and taken all the spoils. As if he were only too aware of the short time available to him, Eccles had succeeded in automating the exploration of the nervous system. It was a very different style to that of his former colleague in Sydney, Stephen Kuffler, who would produce supremely elegant, but technically demanding, techniques to answer questions that no one else had thought to ask. Neurophysiology needed both approaches, and was fortunate to have them.

In later life, Eccles would exult in the statistics of the Canberra era. By the end of the 14-year period, there would have been 74 visiting scientists from 20 countries, 411 scientific papers, and 4 books. The publications—he had actually weighed them, as if for *The Guinness Book of World Records*—amounted to 4.1 kg!

There was one more statistic: the award of a Nobel Prize in 1963. Sharing the platform in Stockholm would be Alan Hodgkin and Andrew Huxley.

The Photograph

It is 1964. In East Anglia, March is a dead month: it is cold, the skies are gray, and the buds on the trees have yet to swell and open into leaf. But in at least one small part of Cambridge, the paved area in front of the Physiology Building on Downing Street, there is life and excitement, and the buzz of many conversations. They have arrived from all parts of the world, invited for this special occasion. Never before, and probably never again, have so many distinguished neurophysiologists come together, among them eight who have won Nobel Prizes. Of the 76 scientists in attendance, only 4 are women, for this is an era when neurophysiology is still regarded as a man's pursuit. In front of the photographer the scientists slowly, and for the most part awkwardly, take their places, those in the front row seated on chairs, the second row standing, and the third row poised on low benches. Scattered among those in the third row, or else standing at the sides, are a dozen young men from the British universities, chosen because of their promise as research scientists.

The two chairs in the center of the first row have been reserved for the guest of honor and his wife, both of whom are wearing overcoats. Two places away, on the right of the guest of honor, and also dressed in an overcoat, is one of the guest's early collaborators. It is the Swede, Yngve Zotterman, who had performed that definitive experiment with him, when they had isolated a muscle spindle and recorded its impulse discharge as it was stretched. It was the first time that impulses had been recorded in a single sensory nerve fiber. The experiment had shown how the nervous system employed a frequency code to transmit information, each impulse the same as the next and the one before. At the far end of the row is another colleague from the old days, smartly dressed in a three-piece suit and his handsome face sporting a neatly trimmed beard. It had been his invention of an oscillograph that had enabled so many of the beautiful recordings to be obtained. And, among other achievements, he and the guest had demonstrated the human brain waves, the EEG, for the first time in front of an audience: Bryan Matthews.

There are two other faces from the past that the guest knows particularly well, those of Umrath and Von Muralt, who had also worked with him on single fiber preparations. But there is one face missing, that of the man who was perhaps his greatest friend as well as a gifted neurophysiologist. Alexander Forbes would have loved to be present, but, at the age of 82, he, who in his youth had sailed his yacht across the Atlantic, is now too frail to make the journey by a commercial airline.

The meeting to honor Lord Adrian, March 1964. In the first, seated row, from left to right, are: Sir Bryan Matthews, Stephen Kuffler, D. Whitteridge, G. Wiersma, P. C. Dell, M. G. F. Fuortes, R. Jung, Yngve Zotterman, Sybil Cooper-Creed, Lord Adrian*, Lady Adrian, H. Hoagland, H. Hartline*, H. H. Jasper, S. Gelfan, R. A. Granit*, Hallowell Davis, Denise Albe-Fessard, and W. Grey Walter.

In the second row, from left to right, are: Torsten Wiesel*, D. N. H. Hamilton, A. G. Brown, J. P. Griffin, H. G. Kuypers, W. K. Stewart, Y. Laporte, F. W. Campbell, G. Baumgartner, K. Umrath, G. Wald*, G. F. Poggio, A. Lundberg, Sir Bernard Katz*, O. Oscarsson, Vernon Mountcastle, C. G. Philips, A. Fessard, Sir John Eccles*, A. Iggo, V. E. Amassian, Ruth Hubbard, W. A. H. Rushton, Patrick Wall, G. Stella, C. Pfaffman, G. S. Brindley, G. M. Preston, R. Conroy, A. J. McComas, R. W. Gerard, and D. M. Lewis.

In the third row, from left to right, are: J. P. Wilson, D. Armstrong, G. W. Arbuthnott, Sir George Lindor Brown, Sir Alan Hodgkin*, B. Delisle Burns, D. H. Barron, A. von Muralt, P. O. Bishop, R. H. Nisbet, M. A. Armstrong-James, R. W. Wilmott, R. Hinde, D. H. Paul, F. W. Darwin, A. M. Nicholson, T. Angel, T. A. Sears, P. B. C. Matthews, C. R. Skoglund, J. A. B. Gray, G. Gordon, R. H. Adrian, R. J. Walker, and P. Hallett.

(*Present or future Nobel Laureate)

Immediately behind the guest of honor two of the foremost workers on the brain are engaged in conversation—Charles Phillips, of Oxford, who has been able to record from inside the cells of the motor cortex, and Vernon Mountcastle, of Johns Hopkins, who has carried out an extensive examination of the properties of cells in the somatosensory areas, those parts that receive information from the opposite side of the body surface. Next to Phillips stands Alfred Fessard, from the Collège de France, who, among other achievements, has worked on the electric ray. Then comes Eccles, who has come all the way from Canberra, and who finds it easier to smile these days, having, in the previous year, shared the Nobel Prize with Alan Hodgkin and Andrew Huxley. A few places further on, the American, Ruth Hubbard, can be seen talking to the former Cambridge physiologist, William Rushton, probably about visual pigments. There is another prize-winner in the front row, another Swede, Ragnar Granit, who, like Eccles, had been a student with Sherrington. Bernard Katz, the great authority on the synapse, is also there, and then, in the back, smiling as always, stands the remaining Nobel Laureate, Alan Hodgkin, justly famous for his work on the nature of the nerve impulse.

It seems that every part of the nervous system is represented by one or more of the scientists who study it. And in nearly every instance, it was the guest of honor who started the exploration and was able to describe the main working features of that part.

The photographer calls for attention and crouches behind his camera. There is a hush and, in his place at the center of the front row, the elderly man with the white hair and the spectacles smiles. The meeting, this great congregation of neurophysiologists, is in his honor, for this year Edgar Adrian, the first Baron Adrian of Cambridge, will celebrate his 75th birthday. For the past 40 years it has been his work, more than that of any other person, that has dominated neurophysiology.

19

The Single Ion Channel

Alan Hodgkin had liked the idea of a carrier. He had thought hard about it when it had become clear, in the early squid axon experiments, that the reversal of membrane polarity during the action potential was due to sodium. Hodgkin had envisaged molecules inside the nerve fiber membrane with negative charges to which the sodium ions would be attached. These molecules, the carriers, would be held by electrostatic forces at the outer face of the membrane in the resting condition. This situation would change when the membrane began to depolarize, either under the influence of a stimulating current or, in real life, by the approach of an impulse. The rising positivity in the interior of the fiber would allow the carrier molecules to move to the inner face of the membrane, where they would be "inactivated" by some constituent of the cytoplasm and would then release their sodium. When Hodgkin and Huxley had gotten their own voltage clamp system working, in the summer of 1949, they had looked for evidence of a small current caused by unattached carrier molecules crossing from one side of the membrane to the other, and separate from the larger current that would shortly follow and be carried by the sodium ions. They did not find it. There was also a quantitative problem. The number of sodium ions entering the fiber across a small patch of membrane was so great that there would have to be a correspondingly large population of carriers, and each of the carrier molecules would have to repeat its ferrying operation many times in the course of a single impulse. Since an impulse lasted only a millisecond, and since the current was not initially maximal, it all seemed improbable.

The alternative to a carrier was a channel in the membrane—a hole, or "pore," so specialized that it would allow only sodium ions to pass through. On the basis of the equation for membrane current during the action potential,[1] Hodgkin and Huxley had predicted that the channel contained charged particles that would "sense" the voltage across the membrane and move to new positions, thereby opening a "gate" through which sodium ions would flow. Though they did not use the term "gate" themselves, Hodgkin and Huxley proposed a similar mechanism to account for the slightly later flow of potassium ions outwards through the membrane. The sodium channel would differ from the potassium one, however, in having an "inactivating" mechanism that terminated the ionic current.

On this speculative note, the matter rested for the next 20 years. Although the voltage clamp technique was adapted to skeletal muscle fibers[2] and to the large cell bodies of mollusks, and although there was important work on the densities of the sodium and potassium channels in different types of membrane, and on the numbers of ions flowing though a single channel during an impulse,[3] further information on the structure and workings of an ion channel remained elusive.

One person who attempted to obtain more data was Kenneth Cole. Beaten by Hodgkin and Huxley in a race that was already half over by the time he became aware of the contest ("In a little over a year they had . . . caught up and run past me"[4]), he was burdened with administrative difficulties of one kind and another, first at the Naval Medical Research Institute in Bethesda and then at the National Institutes of Health. Nevertheless, with the assistance of younger workers, and of John Moore in particular, Cole returned to voltage clamping studies of the squid giant axon at Woods Hole. Though he made an interesting observation on the potassium current when the potential across the axon was greatly increased, the work did not advance the understanding of the channel itself.

And then, in the early 1970s, an advance *did* occur: the gating current was finally detected. As so often happens, there were two groups working on the same problem, and it happened that each included a researcher from the University of Chile. While that university had its own marine biological station, at Vin del Mar, shortcomings in the equipment led the two investigators to carry out research elsewhere. Thus, Francisco Bezanilla came to work with Clay Armstrong at Woods Hole, and Eduardo Rojas with Richard Keynes at Plymouth. The strategies employed in the two laboratories were very similar. The experiments were performed on the squid giant axon, and the current normally carried by sodium ions was eliminated by removing sodium from the bathing fluid and by applying the puffer fish poison, tetrodotoxin.[5] Further, the very small gating currents were enhanced by signal averaging, a technique pioneered 20 years earlier by George Dawson at London's National Hospital for Nervous Diseases. In the latter process the stimulus is repeated many times and, while the responses sum, the background noise, being entirely random, is self-cancelling.[6] Not only did the observed signals have the direction and time-course expected of gating currents (Fig. 19–1), but they also changed, as predicted, when the "stimulating" voltage across the nerve membrane was altered.[7,8]

The detection of the gating currents was the first confirmation of the Hodgkin–Huxley predictions. The next advance in the field came three years later from a new technique. It became possible to record the flow of ions through a single channel in an excitable membrane. For the neurophysiologists, the results were simply spectacular.

At the time Bert Sakmann (Fig. 19–2) and Erwin Neher (Fig. 19–3) were based in the Max Planck Institute for Biophysical Chemistry in Gottingen, in western Germany. As in Oxford and Cambridge, life in Gottingen centers around the university and the scientific institutes. Unlike the large German cities, the town was

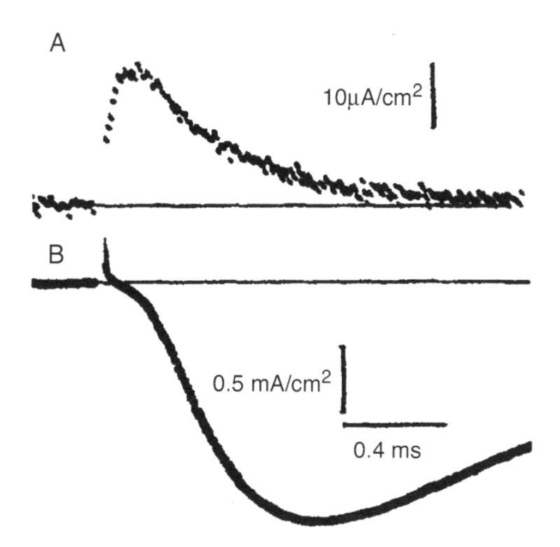

Figure 19–1 Gating current of sodium channel in squid giant axon (*top trace*) and the sodium current itself (*bottom trace*), following a depolarizing stimulus; 50 traces have been averaged. As expected, the gating current precedes the sodium current. Tetrodotoxin was applied to prevent the firing of an action potential. (From Armstrong CM, Bezanilla FM. Originally published in *Journal of General Physiology*; **63**: 675–689, 1974. Courtesy of the Rockefeller University Press.)

Figure 19–2 Bert Sakmann. (Courtesy of Professor Sakmann)

Figure 19–3 Erwin Neher. (Courtesy of Professor Neher)

spared in the 1939–45 war and has retained its history and its student traditions. Many of the old streets are still lined by half-timbered medieval buildings, and defensive ramparts still surround the original town. Though the university is considerably younger than those in Oxford and Cambridge, it has long held an enviable reputation in mathematics, both Gauss and Riemann having been among its professors.

Neher's training had been in physics, while Sakmann had studied neurophysiology with Otto Creutzfeldt in Munich and then with Bernard Katz at University College London. Although the transfer of excitation from nerve to muscle fiber had become an intensely competitive and fast-moving area since Katz had first recorded the miniature end-plate potentials with Paul Fatt in 1952, Katz's laboratory had remained pre-eminent throughout the 1960s and early 1970s. It was the place where other experts came to work for a year or two. It was not that the London laboratories were any better equipped than those elsewhere; rather, it was Katz's experience and, above all, his deep thinking about the subject that was the attraction. When Sakmannn had been with Katz in 1970–73, he had been exposed to the very latest ideas and techniques in the field.

In 1970 the University College London laboratory had reported another breakthrough. Katz, continuing his collaboration with the hugely talented Ricardo Miledi, showed it was possible to obtain information about the opening and closing of individual acetylcholine receptor channels at the neuromuscular junction. They noticed, as had others, that if they filled a micropipette with a weak acetylcholine solution and brought the tip close to the junction, there was a sudden increase in the "noise" that could be detected with a recording microelectrode inserted into the fiber. The effect was seen as a thickening of the trace on the oscilloscope. Katz and Miledi reasoned that the increased noise must be the result of acetylcholine diffusing out of the micropipette and combining with receptors on the surface of the muscle fiber. The London workers applied statistical methods to analyze the noise and were then able to estimate the average time that a receptor channel would be open, and the average amount of ionic current that would pass through it.[9,10] It was a clever approach and, at the time, it must have seemed that the information gained was the best that could be hoped for. It was an approach Neher and Sakmann came to use themselves, in a study of the acetylcholine receptors that spread along a muscle fiber a few days after its nerve supply is removed.[11]

In 1975, however, Neher and Sakmann realized it might be possible to go one step further if only they could isolate a channel physically and electrically from its neighbors. Nearly all the background noise would then disappear, and only the openings and closings of the one channel would be observed. To bring this about, they experimented with a recording microelectrode that, instead of being inserted through the membrane, rested on its surface. The trick, they found, was to make the seal between the membrane and the tip of the electrode as tight as possible so that it had a very high electrical resistance. They were able to do this by using an enzyme to remove the thin layer of connective tissue overlying the fiber, and by making the tip of the microelectrode as smooth as possible, with heat.[12] Even more effective, they later found, was applying gentle suction to the inside of the micropipette. With this combination of strategies, they were able to increase the electrical resistance between the wall of the micropipette and the membrane by more than a thousand-fold, to values in the giga-ohm range (giga = 10^9, 1 billion).[13] The results were impressive: the same single acetylcholine receptor channel could be seen to open and shut repeatedly, the switching time being extremely brief—less than 10 picoseconds (pico = 10^{-12}, one million millionth). As expected, the size of the current was small, a few picoamps, but its consistency indicated that the channels were either fully open or completely closed (Fig. 19–4). Apart from the significance of the open-times and currents, it was now possible to see, instantaneously, the effect of a single biological molecule, albeit a very complex one, changing its shape. It was an extraordinary advance.

The power of patch clamping having been demonstrated, it was not long before the technique was applied to those channels controlled by the voltage across the membrane rather than by combination with a chemical messenger such as acetylcholine. When single voltage-gated sodium channels were studied, they were found to have the rapid, brief openings that would have been predicted from the shape of the sodium current in the squid giant axon.[14] In contrast to the sodium channels, but again in keeping with the voltage clamp results of Hodgkin and Huxley, the potassium channels in the squid membrane opened rather later during

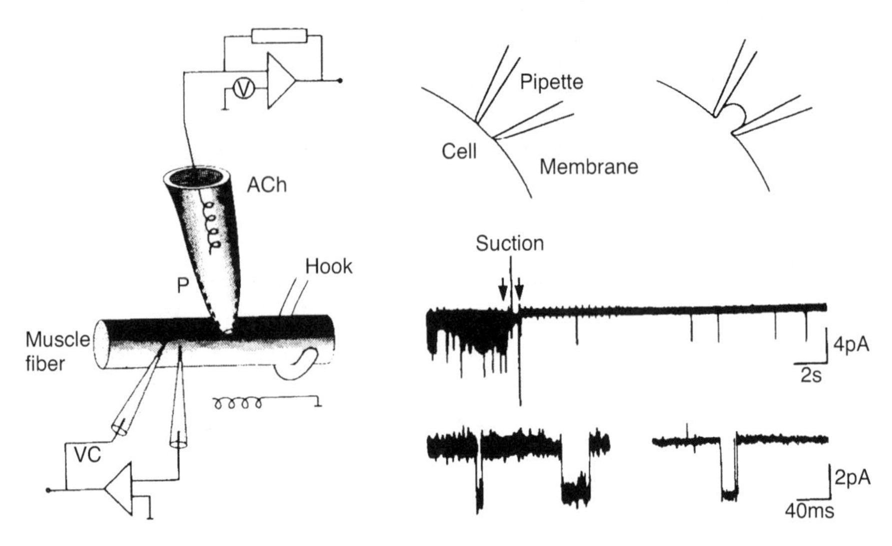

Figure 19–4 Patch clamp technique applied to the end-plate region of a muscle fiber. On the left is the experimental arrangement, with two intracellular microelectrodes for clamping the membrane voltage (*VC*, voltage clamp). The patch clamp micropipette (*P*) contains a low concentration of acetylcholine. When the micropipette is simply touching the fiber membrane, the openings of a single acetylcholine receptor are evident, but the thickness of the recording trace reflects the "noisy" background. Once suction is applied to the micropipette, however, the seal with the membrane becomes tighter and the noise level drops (*traces at right*). Note that the bottom traces are 50 times faster than the top ones. (Courtesy of Professor Bert Sakmann)

a depolarization, and they stayed open for longer. Once again, the predictions of the two young Cambridge men were being borne out.

As the work on ion channels continued, the original patch clamping technique of Neher and Sakmann was sometimes modified. The tip of the micropipette, instead of resting on the surface of the cell, was used to remove a small piece of membrane, which could then be studied. In the case of small nerve cells, the tip could be made to punch a hole in the membrane and to seal round the edge, enabling the investigator to record all the cell currents from channels of one type or another. At the same time the technique was extended to many different kinds of cell, and not only those with excitable membranes such as neurons and muscle fibers. The results confirmed and extended findings that had been made either by voltage clamping or by applying drugs and toxins to the membranes. From voltage clamp studies, it was already apparent that, even for the same species of ion, there were different types of channel. Some channels opened to depolarizations, others to increases in membrane potential. In some the opening was rapid, in others slow, and the same was true for the inactivation process. Some channels passed more current than others, and there were differences in the way that they were affected by pharmacological agents. The variety in channels was especially true for potassium and calcium, less so for sodium. Intriguingly, ion channels were found in bacteria, algae, and yeasts, a finding that led to speculation on the evolutionary processes responsible for the diversity found in more complex organisms. Bertil Hille, a noted authority on channels,[15] suggested that the potassium and calcium channels were ultimately derived from "transporters," molecules in the membrane that supply the cell with sugars, amino acids, and other nutrients, consuming the energy compound, ATP, as they do so. Sodium channels evolved later, probably through the mutation of a calcium channel gene.[16] Once they had appeared, the sodium channels enabled more effective excitation of the cell membranes to occur.

After their breakthrough with the acetylcholine receptor, Neher and Sakmann continued at the forefront of patch clamping, initially directing their attention to the detailed structure of the same receptor. Subsequently Sakmann, working with other colleagues, applied the patch clamping technique to those channels in brain cells that respond to chemical transmitters such as glutamate, GABA, and glycine.[17] In 1992 the two investigators shared the Nobel Prize.

While Neher and Sakmann were beginning their patch clamp studies in the mid-1970s, the protein chemists were also busy. Attempts were being made to purify the acetylcholine receptor and sodium channel and to learn more about their chemical structures. The task was formidable, for, as would emerge, there were around 2,000 amino acids in a single receptor or channel. Nevertheless, the protein chemists had a powerful tool to help them in the purification process. They could identify the fractions containing the receptor or channel by treating them with toxins that were known to combine with the proteins and that could be made radioactive. Alpha-bungarotoxin, found in the venom of a snake, the banded krait, was employed for the acetylcholine receptor, while saxitonin and tetrodotoxin were used for the sodium channel. The chemists were also helped by having very rich sources of material, since both acetylcholine receptors and sodium channels are found in high concentrations in the electric organs of the electric ray (*Torpedo*) and the electric eel (*Electrophorus electricus*). As described earlier, in both species the electric organ is made up of stacks of muscle fibers, each fiber being depolarized on only one side by a nerve impulse.[18] As the purification of the proteins proceeded, the different components were separated from each other by passing a weak current through a gel (polyacrylamide gel electrophoresis).

In the case of the acetylcholine receptor, the protein was found to consist of five subunits, two of which, the alpha subunits, were the same, with each containing a site for binding acetylcholine.[19] The sodium channel, in contrast, consisted of a long single chain of amino acids. But to work out the sequence of 2,000 amino acids by conventional techniques would be virtually impossible. How, then, to proceed? It was a challenge taken up in several laboratories around the world, but in the end a group of Japanese workers were the first to achieve success.

In the minds of most people, the name "Kyoto" is associated with the international agreement to reduce greenhouse emissions into the atmosphere. There is much more to Kyoto than this, however. For more than a thousand years Kyoto, surrounded on three sides by mountains, was the capital of Japan and the residence of the Emperor. Today, despite its large population and its flourishing industries, Kyoto remains a beautiful city. To balance the factories are the many shrines, the traditional houses with their wooden pillars and tiled roofs, the parks and pavilions, the museums and No theaters, and above all, the Kyoto Imperial Palace and the Nijo Castle. With its history, its buildings, and its culture, Kyoto, more than any other city, is the soul of Japan.

Not surprisingly, Kyoto has an excellent university, and it was there that Shosaku Numa in the Department of Medical Chemistry became the first to deduce the amino acid sequence of the acetylcholine receptor.[20] It was a considerable feat, and it could not have been achieved alone. Helping him was a team of more than a dozen investigators, including Michael Raftery, who had been one of the first to purify the acetylcholine receptor and who had worked out a small part of the amino acid sequence. Knowing the partial sequence, it was possible to construct a matching DNA probe. This probe was then tested against a DNA "library" consisting of hundreds of thousands of "clones," each constructed by the enzyme, reverse transcriptase, working on messenger RNA from the electric organ. Eventually a clone in the library was found that combined (hybridized) with the DNA probe. Since the clone was larger than the probe, it was possible to deduce more of the amino acid sequence and then to create a new, expanded, DNA probe. Once again the DNA library would be searched and the whole process repeated over and over again until the clone for a whole receptor subunit was found. Once the DNA pattern of this clone was known, the entire amino acid sequence of the subunit was also known. The ultimate proof that these clones coded for the acetylcholine receptor subunits was obtained by injecting toad oocytes (eggs) with matching RNA, and showing, electrophysiologically, that the cells produced receptors. The idea of injecting an oocyte with RNA for such a purpose was Ricardo Miledi's, and had been used by him in Katz's department in London.[21] It should be added that a DNA clone, produced in this way, corresponds, in the intact nucleus of a cell, to a gene that has had its introns (non-coding sequences) removed by cutting and splicing.

The amino acid sequence of a protein is referred to as the primary structure. The way that the amino acids are arranged in helices and sheets is the secondary structure. The ultimate goal, of course, is to go one step further and determine the three-dimensional shape of the protein, the tertiary structure, for only then can one truly know how the protein achieves its function in the body. At the time that Shosaku Numa and his colleagues cloned the genes for the acetylcholine receptor, it was already known that the five subunits were arranged around a central pore. In 1988 Nigel Unwin and Chikashi Toyoshima, in the Medical Research Council's

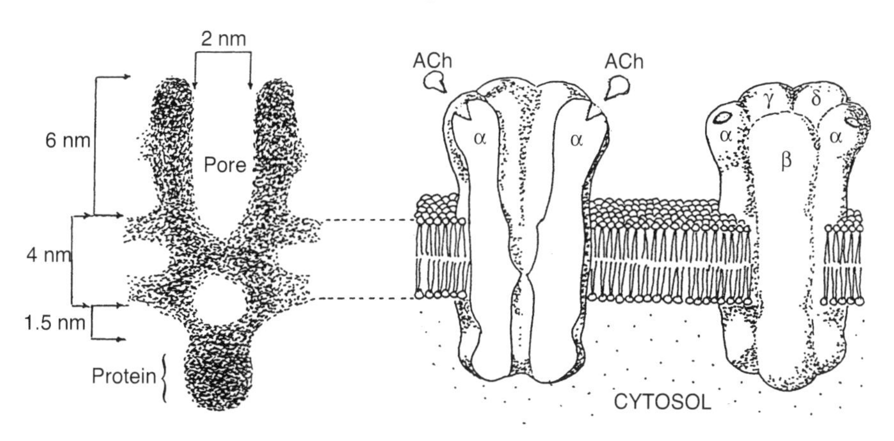

Figure 19–5 *Left*: The acetylcholine receptor as seen by Toyoshima and Unwin, retouched slightly and with dimensions added. (Reprinted by permission from Macmillan Publishers Ltd. (Nature). Toyoshima C, Unwin N. Ion channel of acetylcholine receptor reconstructed from images of post synaptic membranes. *Nature*; **336**: 247–250, 1988.) *Right*: Three-dimensional structure of the receptor, with all five subunits, embedded in the fiber membrane. *Middle*: Beta subunit removed to show central pore, through which Na^+ and K^+ ions pass after two molecules of acetylcholine have combined with the two alpha subunits. (Reprinted, with permission, from McIntosh BR, Gardiner PF, McComas AJ. *Skeletal Muscle: Form and Function*, 2nd ed. Champaign, IL: Human Kinetics, 2006.)

Laboratory of Molecular Biology at Cambridge, obtained the pictures that people had been waiting for.[22] They had been able to crystallize receptors in ice and examine them by X-ray diffraction and electron microscopy before reconstructing the images. The acetylcholine receptor, seen in profile, was about 25 Å (1 Å = 10^{-7} mm) across and 110 Å long, much of it protruding well above the membrane. The cavity, initially wide, narrowed within the membrane and then opened out again, giving the appearance of an hourglass (Fig. 19–5).

Even if the details were not understood, it was not difficult to imagine two acetylcholine molecules combining with the two alpha subunits of the acetylcholine receptor. A small change in the shape of the subunit proteins would then open up the central pore for both sodium and potassium ions to run through. A millisecond or so later, the acetylcholine molecules would be split by the enzyme acetylcholinesterase and the alpha subunits would regain their former shape, narrowing the central pore and stopping the flow of ions. The sodium channel, however, promised to be more complicated. If Hodgkin and Huxley were right, there should be a voltage sensor, composed of several parts, that would be able to move within the membrane and open a gate for sodium ions to enter the channel. There would be another gate, normally open, that would then close, inactivating the channel and stopping the flow of ions. Further, the central pore, at some point, must be able to distinguish between sodium and potassium ions, accepting the former and excluding the latter.

The results of Numa and his colleagues did not disappoint. Using the strategy that had worked so well for the acetylcholine receptor, they were able to clone the DNA for the sodium channel as well.[23] They found, as Hodgkin and Huxley had inferred, that the channel had several parts, though four rather than the predicted three. Each of the parts, or domains, had amino acid chains that completely spanned the membrane. One of these chains, the S4 segment, differed from the others in having a number of the positively charged amino acids, arginine and lysine. The Numa group suggested that this segment, because of its highly charged nature, might act as a voltage sensor. Once the amino acid sequence had been established, it was possible to learn more about the structure and function of the sodium channel by exposing it to an enzyme, or, more frequently, to a drug or biological toxin. There was no shortage of toxins to work with. The long list included not only tetrodotoxin and saxitonin, but also scorpion venom, batrachotoxin (a powerful poison secreted by a species of South American frog), aconitine from buttercups, and pyrethrins from chrysanthemums. The drugs included local anesthetics and strychnine. The function of the channel was also interfered with,

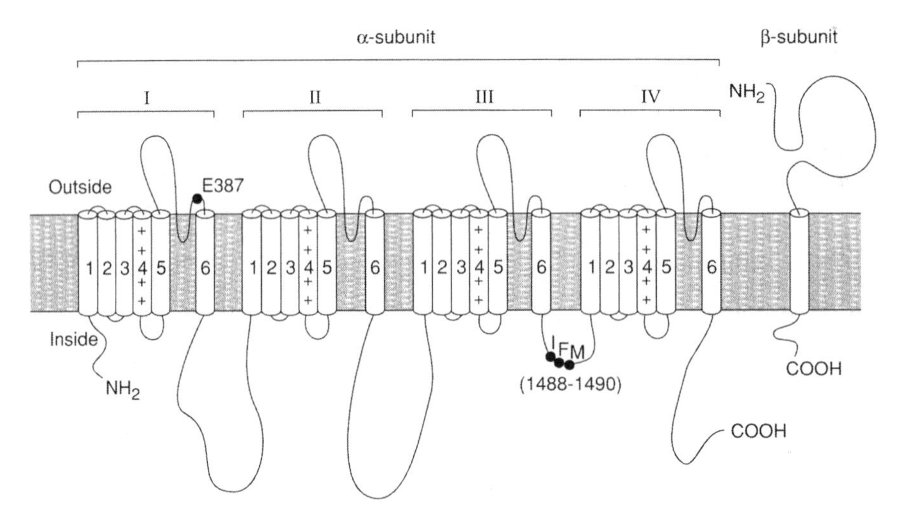

Figure 19–6 Secondary structure of the voltage-gated sodium channel, based largely on the work of Numa and his associates (Noda et al., 1984). There are four repeats of the basic unit, each of which has six membrane-spanning segments. The heavily charged segment 4 in each repeat was thought to be part of the voltage sensor-gating mechanism, while the cytoplasmic link between the third and fourth repeats was considered to be the inactivating gate. The links between segments 5 and 6 in each repeat were assigned to the central pore. (From Ashcroft FM. *Ion Channels and Disease.* Academic Press, 2000. Courtesy of Elsevier.)

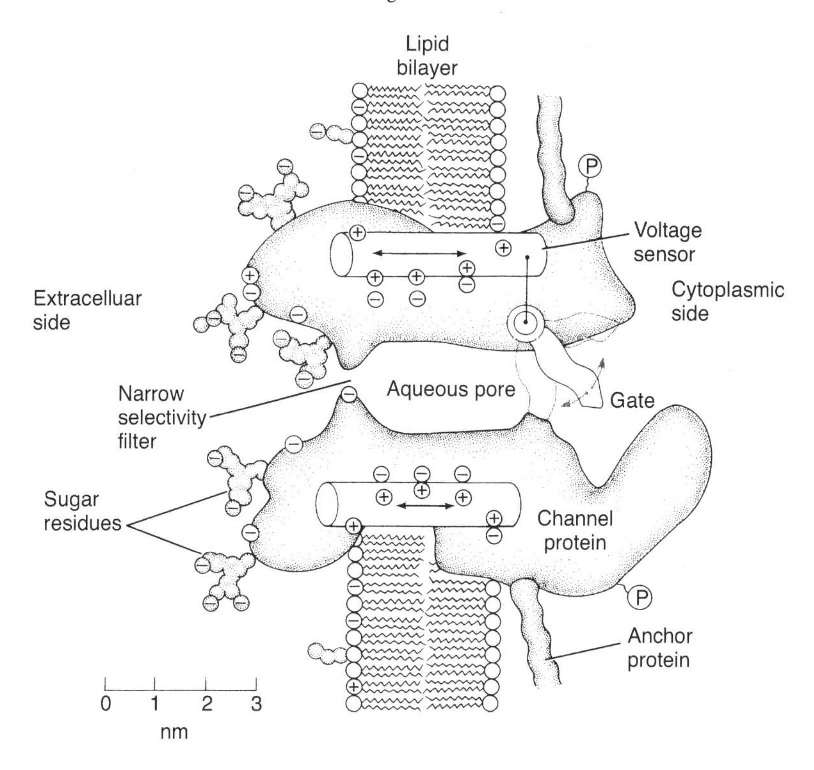

Figure 19–7 Hypothetical structure of a channel prior to the definitive studies of MacKinnon and his colleagues. (From Hille B. *Ion Channels of Excitable Membranes*, 3rd ed., 1992. Reproduced with permission of Sinauer Ltd.)

by blocking parts of it with antibody. Especially informative were experiments in which the normal amino acid sequence in the channel was altered by modifying the channel DNA (site-directed mutagenesis).[24]

As a result of these interventions, it became apparent that the pore of the sodium channel was formed by the amino acids linking segments 5 and 6 (Fig. 19–6). As to the inactivating gate, Clay Armstrong and Francisco Bezanilla had previously shown that this must correspond to a cytoplasmic part of the channel, since they could prolong the sodium current by digesting the proteins inside the squid axon with an enzyme.[25] The same workers also suggested that the inactivating mechanism might operate like a ball on a chain, the ball moving to plug the pore of the channel. After the amino acid sequence of the channel had been worked out by Numa's group, it was thought likely that the ball and chain were formed by a loop of amino acids dipping into the cytoplasm between the second and third of the four channel domains. But what about the amino acid identity of Hodgkin and Huxley's voltage sensor? It was agreed that this was formed by the S4 segment, and there was no shortage of suggestions as to how this might work. In his book *Ionic Channels of Excitable Membranes*, Bertil Hille gave 11 possible mechanisms, as well as a diagram of a hypothetical channel (Fig. 19–7).[26]

Which one was right? And might it be possible to "see" a voltage-gated channel in much the same way that Nigel Unwin and Chikashi Toyoshima had been able to visualize the acetylcholine receptor? Before the breakthrough came, something else very interesting would be discovered about ion channels.

20

Myotonic Goats and Migraines

While Hodgkin and Huxley were bringing their analysis of the voltage clamp work to its triumphant conclusion, they had wondered what implications, if any, their findings might have for clinical medicine: a better understanding of pain and anesthesia, they thought, but otherwise not very much.[1] That there might be naturally occurring disorders of ion channels was a possibility that no one was considering. At the same time, however, it was known that neurologists would occasionally be confronted by patients with bizarre neuromuscular disorders that defied satisfactory scientific explanation. It was the same story in the veterinary world, in which some of the strangest animals were "nervous" or "myotonic" goats. The origin of these animals was shrouded in mystery, though all those in the United States were descended from a small flock in Tennessee.[2]

At some point, as the goats began to spread throughout the state, they had been cross-bred with Angora and other strains, and now had brownish markings. Their extraordinary behavior, however, was unchanged. As with the original flock, a sudden noise or a threatening sight would cause all four legs to stiffen so that, instead of running off, the animal would remain rooted to the spot or would topple over (Fig. 20–1). Only after several seconds would the limbs regain their mobility and the goat be able to trot away.[3] Until the late 1930s research studies had achieved little beyond providing clinical descriptions of the peculiar disorder, showing that the blood chemistry was normal, and noting that quinine improved the muscle stiffness. One of the investigators of these animals at Johns Hopkins University was the neurologist Lawrence Kolb.[4] In the same department as Kolb was a young physician, A. McGhee Harvey, who had already taken part in some studies on human patients with muscle stiffness or weakness.[5] Becoming interested in Kolb's goats, Harvey realized that electrophysiological studies were needed to define the nature and mechanism of the muscle stiffness. Familiar with the work on neuromuscular transmission by Dale's group in London, partly for which Dale had received the Nobel Prize in 1936, Harvey decided that the goats must also be examined in England.

Harvey, accompanied by a number of the goats, had arrived in 1937. By then Dale had given up laboratory work, but the person who could best help Harvey, by providing the necessary electrophysiological expertise, was still there. George Lindor Brown had performed the close arterial injections and the recordings of

Figure 20–1 A myotonic ("nervous") goat that has evidently been startled and is lying helpless with all four limbs rigidly extended. (Courtesy of Mr. Jim Knapp, Jr.) Several videos of these fascinating animals are available on the Internet and can be easily accessed through Google.

electrical and mechanical activity that had provided the last, but much needed, evidence that acetylcholine was the chemical transmitter at the neuromuscular junction. It was a logical decision to apply the same techniques to the two myotonic goats that Harvey and Brown had selected for their experiments.

They were big experiments.[6] Apart from the sizes of the animals, extensive dissections and fixations were needed to prepare the muscles and tendons for recordings, and the arteries for injection. As with his earlier work, with Feldberg and Dale, Brown used the four-stage amplifier he had built himself. For displaying the mechanical and electrical activity of the muscles, however, Brown favored a cathode ray oscilloscope rather than the Matthews instrument.[7] Unlike the early models that Gasser, Erlanger, and Bishop had exploited so well, the cathode ray oscilloscopes now had sufficiently dense electron beams for single traces to be photographed satisfactorily. Further, the instruments were easier to use than the Matthews machine and had better frequency responses, enabling them to record extremely rapid variations in potential. It was a change in instrumentation that would be made in the other British laboratories. For fine recordings from the muscles, Brown and Harvey used concentric needle electrodes, the kind pioneered by Adrian and Bronk in their studies of muscle contraction in human subjects.[8]

The first important finding in the myotonic goats was that the unusual muscle stiffness was associated with long runs of action potentials in the muscle fibers. But what produced those action potentials? They could still be provoked after the nerves to the muscles had been cut, so the abnormal discharges could not have originated in the brain or spinal cord. To Brown and Harvey's surprise,

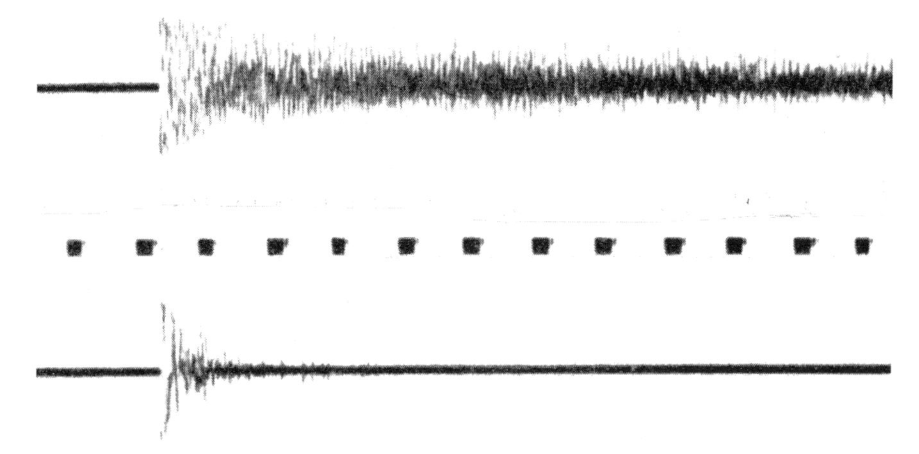

Figure 20–2 Recordings from a denervated leg muscle in a myotonic goat. *Top trace*: Although the muscle has been curarized, the muscle fibers still fire action potentials in response to a tap on the muscle belly, indicating that it is the muscle fibers themselves that are hyperexcitable. *Bottom trace*: The hyperexcitability is greatly reduced following an injection of quinine, a drug known to benefit myotonic patients. *Middle trace*: 0.2-second time intervals. (From Brown & Harvey, 1939, courtesy of Oxford University Press)

the phenomenon, this time elicited by tapping the muscle, persisted after an injection of curare (Fig. 20–2). The only possible conclusion was that the muscle fibers themselves were hyperexcitable.

And there the matter was left. Harvey returned to Johns Hopkins and, after the United States entered the 1939–45 war, served in Australia, meeting Stephen Kuffler in Sydney and even carrying out some wartime experiments with him. Brown was also involved in medical research during the war, either in the laboratory or through committee work. Unlike the wartime experiences of Hodgkin, Huxley, and Katz, it had nothing to do with radar or direction finding. Instead, Brown worked on oxygen toxicity and carbon dioxide narcosis in underwater divers exposed to high pressures.[9] After the war, neither Brown nor Harvey did anything further on the myotonic goats. The next step was taken by a young professor in the Department of Pharmacology at the University of Cincinnati.

Shirley Bryant had read about the goats in a veterinary journal. By then the animals had reached Texas. Intrigued, he arranged for some of the goats to be shipped to Cincinnati for study. Rather than carry out massive, heroic experiments on whole animals, in the manner of Brown and Harvey, Bryant removed small samples of external intercostal muscles, the muscles that connect the each rib to its neighbor. Since the muscle fibers were short, running the few millimeters separating one rib from the next, it was possible to excise them without damaging them. Then, under the dissecting microscope, he was able to insert glass microelectrodes for stimulating and recording, just as Nastuk and Hodgkin had done in the frog, and as Katz was then doing in his study of the neuromuscular junction.

One of the first things Bryant found was that the membranes of the muscle fibers had high electrical resistances—that is, ions were not able to move through the membranes as freely as they would in normal fibers.[10] Since, in the resting state, normal muscle fibers had appreciable permeability only for potassium and chloride ions, the channels for one or both of these ions must have been defective. Further experiments indicated to Bryant and his colleagues that it was the chloride channels that were affected. One important piece of evidence was that myotonic discharges could be induced in normal muscle fibers if their chloride permeabilities were abolished by bathing the fibers in solutions containing an aromatic acid.[11]

But how could a reduction in chloride permeability make the muscle fiber membranes unstable? Just as Harvey had done a decade and a half earlier, Bryant decided to take the goats to Britain to find out. This time Harvey's colleague was unavailable for help, for Sir Lindor Brown, having been appointed after the war first to the Jodrell Professorship of Physiology at University College London, and then to the Waynflete Chair at Oxford, had recently died. It did not matter in any case, for Bryant had already chosen to work with Adrian in Cambridge—Richard Adrian, who would succeed to his famous father's baronetcy in 1977 (Fig. 20–3). Once again, single intercostal muscle fibers were impaled with glass microelectrodes and the responses of the fibers to stimulation were observed.[12]

Figure 20–3 Richard Adrian, the second Baron Adrian of Cambridge (1927–1995). (Courtesy of Professor Ian Fleming)

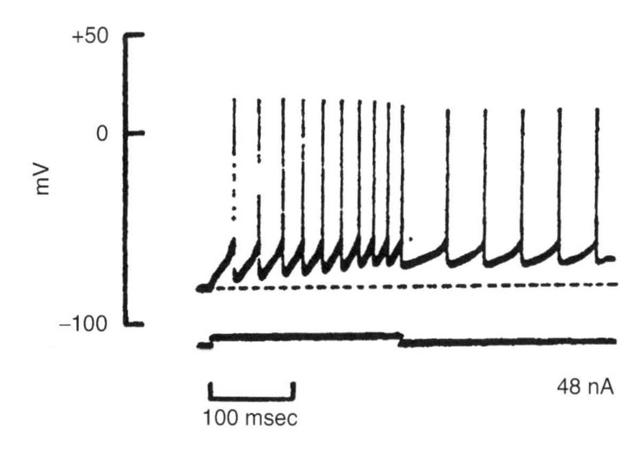

Figure 20–4 Intracellular recording from an intercostal muscle fiber of a myotonic goat. Unlike a normal muscle fiber, which might respond once to a small stimulating current, the myotonic fiber fires a series of action potentials, and continues to discharge after the stimulus is terminated. In addition, a negative after-potential develops, indicated by the elevation of the trace above the starting potential (*interrupted line*). (From Adrian RH, Bryant SH. On the repetitive discharge in myotonic muscle fibres. *Journal of Physiology*; **240**: 505–515, 1974. Courtesy of John Wiley & Sons.)

It was noted that, as before, the myotonic fibers had lower than normal thresholds and tended to fire repetitively (Fig. 20–4); in addition, there was a depolarizing after-potential that built up but could be abolished by treating the fibers with glycerol. Glycerol was used because it was known to break the transverse tubules, the same tubules that Andrew Huxley had shown to conduct the impulses from the surfaces of the fibers into the interiors.[13] On the basis of their observations, Adrian and Bryant suggested that, as the abnormal fibers started to discharge, potassium ions collected in the narrow transverse tubules. Without the normal stabilizing effect of the chloride permeability, the myotonic fibers were depolarized by the build-up of potassium in the tubules and the fall in membrane potential triggered further action potentials. The explanation was simple and ingenious, and it had the ring of truth to it. But was there any relevance of all this work on the goat to humans?

Shortly before the goats had mysteriously appeared in Tennessee in the late 1800s, a paper had been published from Denmark by a certain Dr. Thomsen.[14] A physician, Thomsen had suffered muscle cramps whenever he attempted sudden movements. It was a problem that had, to his knowledge, affected at least 20 other family members in four generations. Following Thomsen's description, other cases of muscle spasms during voluntary movement were soon recognized and reported by some of the leading physicians of the day, especially in Germany, where academic medicine was at its strongest. The human condition became known as myotonic congenita or Thomsen's disease. With the passage of the years, it became apparent that, while in families such as Dr. Thomsen's the disorder was inherited through a dominant gene, tending to affect half of the children, in other families a recessive inheritance must have been involved.

The individual stories of sufferers from myotonia were as fascinating then as they are now. There was, for example, the lady who was unwilling to give firm handshakes during job interviews for fear that she would be unable to let go of her prospective employer's hand. There was also the woman who, when traveling in a streetcar, would have to get up from her seat while the car was moving and still some distance from her destination. Had she waited until the car had stopped, the sudden attempt to stand would "bind" her to her seat and, embarrassingly, would draw the attention of other passengers to her predicament. Another lady, having stepped off the bus, had been unable to let go of the rail at the side of the exit and had been dragged by the bus. Then there were tales about the myotonic goats, one of them very amusing:

> "I've heard a story about a new hired man who, a few days before a barbecue, was given a scatter gun and told to go into the back pasture and kill one of the goats. He was advised that the goats were very shy and that he should be very careful not to apprise them of his presence until ready to shoot. After crawling up cautiously, he picked out a nice fat kid, took careful aim, and fired. Goats dropped in every direction, some thirty animals collapsing simultaneously. Aghast, and without waiting for the resurrection, he ran back to the house. 'I don't know how it happened,' he panted. 'I only fired once but I killed every damn one of them goats.'"[15]

Though there was persuasive evidence, largely on the basis of Shirley Bryant's work, that human and goat myotonias were due to abnormal chloride channels, the final proof would have to come from molecular biology, in which mutations could be shown in the DNA coding for the chloride channel.[16] At about the time that this happened, in 1994, another demonstration of a channelopathy, at the level of the gene, was being made. Once again it involved a muscle disorder, and once again it seemed to run in families, but this time there was little or no muscle stiffness. Further, the preceding electrophysiological studies had been even more impressive than those in myotonia.

It is possible that an instance of slowly developing paralysis, with complete recovery, in an otherwise normal individual, had been seen and recorded by William Musgrave in 1687.[17] In the late 19th century, when academic medicine was at its zenith in Germany, more definitive reports appeared, while the first case in the English literature was that described by Singer and Goodbody in 1901.[18] As the papers accumulated, a picture began to emerge. Thus, the condition was rare and, when it did occur, there were usually other family members, male and female, similarly affected. The paralysis, which often began in the legs, would take several hours to become fully established but would usually spare the respiratory muscles. There were precipitating factors, too, especially heavy exercise, emotion, or a large carbohydrate meal. That there was something wrong with the muscles themselves, rather than with the brain or spinal cord, was shown by the failure of the muscles to contract when they were stimulated directly, and this was recognized by some of the early German neurologists.[19] Already the nerve impulse, or in this case the muscle impulse, was beginning to play a role in clinical diagnosis. Equally important was the later finding of a change in the potassium level in the blood during the attacks, and this enabled two types of the periodic paralysis to be defined, those in whom the serum potassium rose and those in whom it fell.[20]

Following the reports of Hodgkin and Nastuk, and of Ling and Gerard, on the use of glass microelectrodes to record the membrane potentials of frog muscle fibers, it was natural to apply the same technique to the muscles of patients with periodic paralysis. It might have been predicted, on the basis of the Goldman equation,[21] that in those patients in whom the serum potassium rose, the membrane potential would fall, and this was what was found.[22] Indeed, even between attacks there was some reduction in potential. These observations led to the suggestion that the sodium permeability of the membranes was abnormally increased.[23] That this was so was subsequently shown by Frank Lehmann-Horn and his colleagues, Reinhardt Rudel and Ken Ricker, in Ulm, Germany. In most cases, the problem appeared to be a failure of the sodium channels to inactivate,[24] and this was confirmed by patch clamping (Fig. 20–5).

In the same year that the sodium channel gene had been identified and cloned, it was shown that the majority of mutations in this type of paralysis did, indeed, involve the inactivating gate.[25] Thus, well before the molecular biologists were able to supply the definitive proof, the electrophysiologists, with their microelectrode recordings, had identified the nature of the membrane problems in both myotonia congenita and the high-potassium (hyperkalemic) form of periodic paralysis. There was, however, a surprise. In those patients in whom paralysis was associated with a *fall* in the serum potassium concentration (hypokalemic periodic paralysis), the genetic abnormality did not affect either the sodium or the potassium channels directly; instead, it was the calcium channels in the muscle fiber membranes that were abnormal.[26] Even now, it is uncertain how such a defect can lead to paralysis.

By the end of 1994, then, three neurological disorders, and a further two that were closely related, had been identified as channelopathies, and all of them involved skeletal muscle. In addition, there was myasthenia gravis, which was also a channelopathy, though not one caused by a genetic mutation. It was a chance finding. At the Salk Institute Jim Patrick and Jon Lindstrom had been trying to raise antibodies against the acetylcholine receptor. They had used the electric organ of the ray (*Torpedo*) as the source of receptor, since the latter are especially dense in the specialized muscle fibers. When Patrick and Lindstrom injected the purified receptor into rabbits, however, in the hope that the animals would produce antibody, the rabbits invariably became weak and floppy and died. Though her name

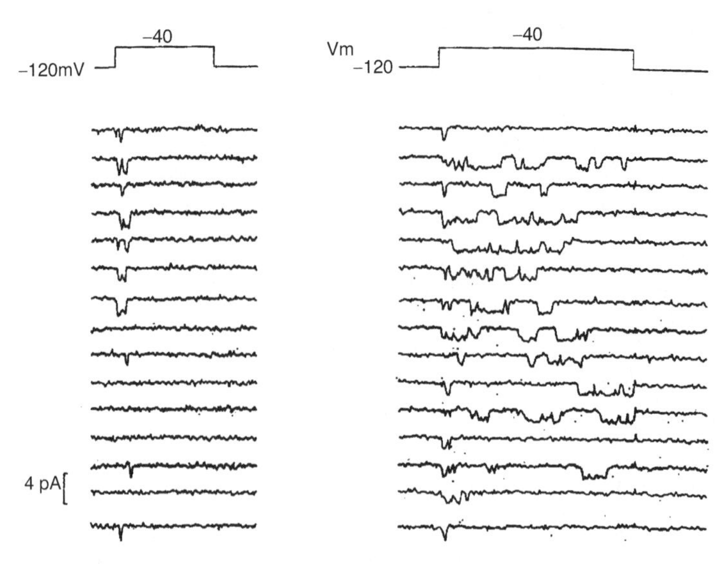

Figure 20–5 Patch clamp recordings of sodium channel openings in a normal muscle fiber (*left*) and in a fiber from a patient with familial hyperkalemic periodic paralysis (*right*); in both fibers a depolarizing current has been applied. Whereas the channels usually open once, and only briefly, in the normal fiber, in the patient's fiber there are repeated and prolonged openings due to a failure of the inactivating mechanism. (Modified by Ashcroft (2000) from: Cannon DP. A sodium channel defect in hyperkalemic periodic paralysis: potassium-induced failure of inactivation. *Neuron*; **6**: 619–626, 1991. Courtesy of Elsevier.)

did not appear on the subsequent publication, it was a visiting neuroimmunologist from Australia who immediately realized what was happening.[27] The rabbits were developing myasthenia gravis, a condition that, in humans, is characterized by fluctuating weakness and that was known to involve the neuromuscular junction. Following the recognition of the condition in the rabbits, and the discovery of its cause, it was shown that, in humans, too, myasthenia was an autoimmune disorder and that the acetylcholine receptor was the antigen involved.

More surprises were to come, for it was next found that some diseases of the brain were also caused by abnormal ion channels. These included an unusual form of migraine, one that ran in families and produced transient hemiplegia in association with the headache. There were, too, conditions in which patients had attacks of unsteadiness, or else became permanently unsteady because of progressive degeneration of nerve cells in the cerebellum, the large structure behind and below the cerebral cortex. All three disorders were discovered to be due to abnormal calcium channels.[28,29] At about the same time, it was found that defective potassium channels were responsible for a benign form of epilepsy in infants. Nor were the channel diseases restricted to the muscles and brain. It transpired that both sodium and potassium channel abnormalities could cause an electrical disorder of the heart, the long QT syndrome, a condition that can cause sudden death in young adults from ventricular fibrillation. As the list of channelopathies became longer, it was evident that the abnormal mutations could affect different parts of the channels. In some it was the activating gate, in others the inactivating mechanism, and in yet others it was the central pore that was defective. And then, to make the list of channel diseases even longer, there were, in addition to myasthenia gravis, other disorders of the ligand-operated channels, those channels that normally open not because of a change in membrane potential, but after combination with a chemical messenger.

And the list of channelopathies continues to grow. The conditions include diseases of skin and kidney, of the eye and inner ear, and, in the case of cystic fibrosis, of multiple tissues. It may well be that there is an even longer list of conditions in which cells either die or function improperly because channels, otherwise normal, are affected secondary to some alteration in the chemistry of the cell. For example, cells may die after taking up abnormal amounts of calcium, or because an excessive intake of water causes them to burst. Indeed, interest in the ion channels, far from being restricted to neurophysiologists concerned with the nerve impulse and its effect, is now shared by clinicians and researchers in nearly every branch of medicine.[30] Further, with a complete understanding of the channel abnormalities responsible for the various diseases are coming, and will continue to come, specific drug treatments.

For the physiologist, and for scientists in general, none of this is surprising for, as has been shown over and over again, curiosity-driven research sometimes has the greatest consequences. What *is* remarkable is that, in the case of the ion channels, the fundamental work was accomplished by a relatively small number of individuals, so quickly and with so little financial and technical help.

And, especially in the case of Alan Hodgkin and Andrew Huxley, with such elegance and depth of understanding.

21

The Swinging Gate

If ion channels had human characteristics, the potassium channels would have appeared rather dowdy and unexciting compared to their more flamboyant sodium cousins. Thus, in the squid giant axon, the potassium channel was found to open only after the sodium channel, and then to allow a smaller current without any inactivating gate to stop it. Further, whereas the sodium current generated the action potential, the potassium current merely brought the membrane back to the resting state.

And yet, as research on ion channels developed, the potassium channels were found to have interesting features as well. For one thing, there were so many different types, far more than the varieties of sodium channel, and this reflected the different functions that the channels were used for in the cell. Not only were there channels that were affected by the potential across the cell membrane, but there were also channels that responded to a rising calcium ion concentration inside the cell, and others that opened when there was a reduced availability of ATP, the energy compound. The potassium channels also differed a hundredfold in the amounts of current they could pass, and while it was true that the potassium channel in the squid axon did not have an inactivating gate, other potassium channels did. So far as the action potential was concerned, the potassium channels not only served to restore the resting membrane potential, but in addition they determined the duration of the action potential and the maximum frequency that the cell could fire impulses.

The potassium channels also differed from each other, and from the sodium and calcium channels, in the agents that could block them. Not surprisingly, there were ions, drugs, and venoms, including that of the honeybee, and there was also a particularly effective and useful compound, TEA (tetraethylammonium). Ironically, in view of the disappointing reception of his mammoth study on peripheral nerve, it was Lorente de Nó who had first synthesized and used TEA.[1] The compound became one of the few legacies of all those years of painstaking work on the frog sciatic nerve.

The DNA cloning of the potassium channel proved even more difficult than that for the sodium channel and acetylcholine receptor, since there was no rich source of potassium channels. The purification of the channel and the construction of a DNA probe for part of the amino acid sequence was not possible. Instead it was necessary to identify the probable site of the channel gene by studying a mutation, and then to "walk" along the chromosome with a series of probing clones, the first of which was developed for a neighboring region. Eventually, it was possible to make a probe sufficiently large to cover the part of the chromosome containing the mutation, and hence the gene for the channel.[2,3] The mutation that provided the clue as to the general whereabouts of the potassium channel was one responsible for leg shaking in the fruit fly, *Drosophila melanogaster*, when the flies were exposed to ether.

The results of the cloning were rather surprising. Whereas the sodium and calcium channels had separate genes for each of their four distinct domains, there was only a single gene for the potassium channel. Since the potassium channel had almost certainly been the first to appear in evolution, its gene could be regarded as the primordial one. The various types of potassium channel were shown to result from the different ways the gene was spliced when messenger RNA was to be made from it. It was to the potassium channels in the bacterium *Streptomyces lividans* that Roderick MacKinnon (Fig. 21–1) first directed his attention.

At the time MacKinnon was a postdoctoral fellow, having just completed medical studies at Tufts University in Boston. Like so many other newly qualified doctors before him—Cajal, Sherrington, Adrian, Gasser, Erlanger, Eccles, Katz, and Kuffler among them—he had felt a strong attraction for basic science, and he had returned to the biochemistry laboratory of Chris Miller at Brandeis University. It was 1988, and the gene for the *Shaker* potassium gene had been cloned in the preceding year. Initially with Miller and then with other colleagues, MacKinnon was able to work out the sequence of amino acids that lined the pore of the

Figure 21–1 Roderick MacKinnon.

channel and that were responsible for its selectivity for potassium. To do this, they made mutations of the channel gene and determined which of these affected the normal ability of scorpion toxin to block potassium ion conductance.[4,5] The research also showed that a single channel consisted of four identical subunits of channel protein fitting together and forming a central pore. Hodgkin and Huxley's prediction of four particles in the squid axon potassium pathway, represented in their current equation by the term m^4, was confirmed.

Impressed, as others had been, by the remarkable efficiency of an ion channel in transferring ions of a particular species across a cell membrane, MacKinnon then set out to visualize, in three dimensions, the critical features in its structure that made this possible. There was only one way of obtaining the very high spatial resolution needed for such an analysis, and this was X-ray crystallography. The decision to enter a new field, one in which he had no experience, was a bold one. Scientists tend to persist with techniques in which they have become expert, applying them to this problem and that throughout their careers. MacKinnon had one problem of supreme interest and he would learn the methodology that might solve it.[6]

Might solve it. X-ray crystallography was, and is, a notoriously difficult field. It had started in Germany just before the 1914–18 war, less than 20 years after the discovery of this form of radiation. A photographic plate had been placed behind a crystal of copper sulfate exposed to a source of X-rays, and, when developed, had shown a pattern of spots. The young Lawrence Bragg, a student in his father's laboratory at Cambridge, had had the inspiration to deduce what had actually happened to produce the spots. Rather than having to postulate different wavelengths of X-ray, Bragg suggested that the atoms in the crystal effectively formed "lattices," each lattice acting like a minute mirror capable of reflecting X-rays in a particular direction. Which lattices would take part in the overall reflection from the crystal would depend on the plane of the lattice in the three dimensions of the crystal, and on the angle presented to the X-rays by the crystal. As the X-rays penetrated further into the crystal, however, the reflections from the various lattices would interact, sometimes adding and sometimes subtracting—depending on the distances between the lattices. When the reflections added, the spot on the photographic plate or film would be correspondingly dark. The addition was maximal when successive lattices were parallel to each other, and therefore reflecting X-rays at the same angle. The problem was that the lattices existed in different planes, producing reflections that interacted with each other in different ways and giving spots of varying intensities. Nevertheless, a principle existed. From the positions of the spots on the photographic plate, it should be possible to work backwards, as it were, and deduce the orderly and repetitive arrangement of the atoms in the crystal.

It was one thing to work out the structure of an inorganic crystal, such as sodium chloride or copper sulfate, as Bragg had been able to do very quickly, but quite another to decipher the infinitely more complex structure of a protein.

Most would have thought it impossible. In 1934, however, a publication appeared in *Nature* by J. D. Bernal and Dorothy Crowfoot Hodgkin,[7] who, working together at Cambridge, reported that an X-ray diffraction pattern had been obtained from crystals of the digestive enzyme pepsin.[8] Later, beginning immediately after the 1939–45 war, Dorothy Hodgkin used X-ray diffraction to determine the structure of penicillin.[9] At the time it was the largest molecule to have been worked out in this way, but it took her four years to solve a structure that had a molecular weight of "only" 300. Max Perutz, resuming his research in Bragg's Cavendish Laboratory at Cambridge, would have to spend 30 years on the much larger molecule, hemoglobin, before finally establishing its structure in the 1970s.[10] In contrast, the revelation, in 1953, that DNA was a double helix had come easily: it had needed only a glance at a single photograph for James Watson to realize immediately that some kind of helical structure was present.[11]

Since the early days there had been significant advances in X-ray crystallography. For one thing, the apparatus providing the source of the rays was now a particle accelerator of one type or another. Being more powerful and precise than the high-voltage gas tubes originally used to produce X-rays, the accelerator made it no longer necessary to expose crystals for hours or even days to obtain a satisfactory print. Further, as successive protein structures were worked out, it became possible to correlate certain characteristic features of a diffraction pattern with parts of a protein. Again, instead of having to refer to books of tables, the very formidable mathematics involved in the solution, much of it based on Fourier analysis, could now be undertaken by computer. Yet despite the technical advances, there was still a considerable art involved, especially in the growth and mounting of the crystals. It had been Rosalind Franklin's skill in these that had resulted in the superb DNA photograph seen, perhaps surreptitiously, by Watson in 1953.[12] To make proteins form into crystals was very difficult, and in some cases impossible. There was also an art in guessing the sort of structure that might be present and upon which the analysis might be based.

All of these points would have been considered by MacKinnon, but it was not in his nature to give in easily: "From my perspective I had little choice because I wanted to understand K+ selectivity and I knew that the atomic structure provided the only path to understanding. I would rather fail trying than never try at all."[13]

In his favor was the fact that, having worked on potassium channels for several years, he was already familiar with their functional properties and with the amino acid sequences involved in their different parts. Moreover, there was a psychological advantage in that, having taken up mathematics again as a medical intern, he was comfortable with the equations employed in X-ray crystallography. Lastly, even if he were to fail in his ultimate goal, he knew, or at least strongly suspected, that interesting findings would emerge along the way. Such is the nature of discovery, and how different to the Popperian philosophy of scientific investigation discussed earlier![14]

To undertake this new line of work, and not be distracted by further pursuit of the familiar, MacKinnon resigned his assistant professorship at Harvard, an extraordinary action in itself, and moved to the Rockefeller University in New York. The research team could hardly have been smaller, consisting of one post-doctoral scientist, MacKinnon himself, and—to provide additional help as well as company—MacKinnon's wife, Alice. It was not easy, but enthusiasm and a strong sense of purpose ultimately prevailed: "Working with membrane proteins was very difficult as expected. We had our periods of despair, but every time we felt left without options something good happened and despair gave way to excitement."

The first publication on the new work appeared in the April 3, 1998, issue of *Science.*[15] MacKinnon had used a bacterium, *Streptomyces lividans*, as his source of potassium channels, introducing the channel gene into the common intestinal bacterium, *E. coli* (*Escherichia coli*). The *E. coli* was allowed to multiply, producing potassium channels as it did so. The channels were then labeled and separated from the other cell constituents before being placed in a potassium-rich solution in which they formed crystals. The analysis of the X-ray reflections was facilitated by a number of strategies, including the selection of the best data, averaging the results from several experiments, and site-directed mutagenesis of the channel. Like all the other potassium channels studied until that time, those in *Streptomyces lividans* were found to consist of four identical subunits surrounding a central pore. Each subunit was formed by an inner and an outer alpha-helix linked by a sequence of 30 amino acids. It was this sequence that lined the pore of the channel and created its remarkable selectivity, permitting 10,000 times as much potassium to traverse the channel as sodium. MacKinnon's admiration and fondness for the potassium channel was reflected in his description of it. The four inner helices were tilted, causing the subunit to "open like the petals of a flower facing the outside of the cell." Another analogy was to "an inverted tepee" (Fig. 21–2). The use of such analogies was helpful to the specialist and non-specialist. Indeed, unlike the technical jargon that clogs so many contemporary papers, especially in molecular biology, the contents of MacKinnon's paper, and of those that were to follow from his laboratory, were easily understood.

The initial study revealed more. There was a central cavity in the channel that was large enough to allow a potassium ion to keep its surrounding shell of water, and thereby to overcome the electrostatic barrier imposed by the lipid membrane. The critical diffusion distance, rather than being the full thickness of the membrane, 45 Å, was reduced to the short length of the selectivity filter, 12 Å. When a potassium ion entered the narrow selectivity filter, its shell of water was removed and the ion was contacted instead by the negatively charged oxygen atoms of the amino acids lining the filter. The tight fit between the oxygen atoms and the potassium ions was crucial, since it maximized the attractive forces between the two. The gap that would exist between the wall of the channel and a relatively small sodium ion evidently diminished the electrostatic attraction of the filter and

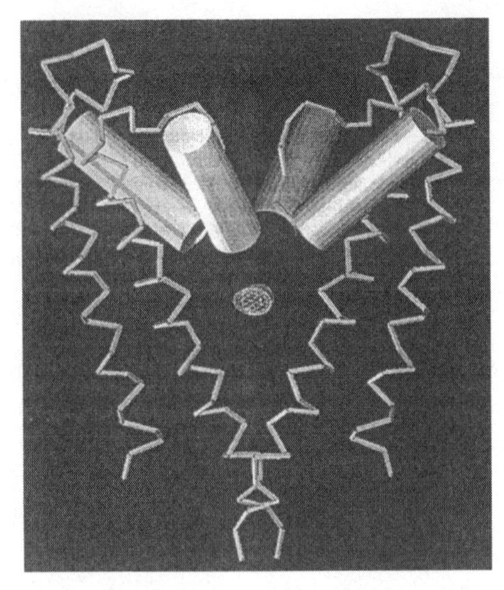

Figure 21–2 Model of part of a potassium channel. The main-chain traces of two of the channel subunits (repeats) are shown, together with the pore helices (*white cylinders*) of all four subunits. The helices are in the form of an inverted tepee, with their negatively charged carboxyl ends pointed towards a potassium ion in the water-filled cavity. (From Roux D, MacKinnon R. The cavity and pore helices in the Kcs K+ channel: electrostatic stabilization of monovalent cations. *Science*; **285**: 100–102, 1999. Reprinted with permission of the American Association for the Advancement of Science.)

conferred its astonishing selectivity. At any given instant there were two potassium ions in the selectivity filter, which repelled each other.

The importance of the new work from the Rockefeller laboratory was recognized immediately by protein chemists and neurophysiologists alike, and scientists and journal editors would subsequently refer to the initial paper as a landmark study. Fittingly, the author of one of the later review articles would be Chris Miller, MacKinnon's former supervisor during his undergraduate days at Brandeis University.[16] The publications from MacKinnon's laboratory were a remarkable triumph for a small team of investigators that had been obliged to start from scratch.

In the following year, 1999, there was another paper from the same laboratory, in which Poisson analysis was used to confirm the importance of the central cavity in overcoming the dielectric barrier of the lipid membrane.[17] How important mathematics was for this type of work!

In 2001 a paper appeared on a different aspect of an ion channel, the mechanism enabling the inactivating gate to close the channel.[18] Through experiments employing mutations and the application of toxins, it had been known for some years that the inactivating gate was at the cytoplasmic, intracellular, end of the channel. Further, there had been a proposal that the gate might operate rather like a ball and chain, the ball swinging across and plugging the opening into the internal pore. Although this proposal had been widely accepted by those in the field, MacKinnon and his colleagues now showed that, on the basis of new experiments on a mammalian voltage-sensitive potassium channel with an inactivating gate, there was a more probable alternative. The experiments involved a number of techniques, including induced mutations of the channel, heavy ion penetration followed by crystallization and X-ray diffraction analysis, and patch clamping. The results pointed to a more complex, but very effective, mechanism—a three-step process in which the gate binds initially to the intracellular surface of the channel, as previously supposed, but then enters the internal pore and central cavity as an extended peptide. As MacKinnon pointed out, the new proposal gave more satisfactory explanations for various features of inactivation.

A paper published in the same year from the MacKinnon laboratory returned to the issue of the selectivity filter. Once again X-ray diffraction was employed, but at a still higher resolution. To make this possible, the potassium channel crystals were grown in the presence of a channel antibody, the antibody fragments providing a skeleton for the crystal. The resolution achieved in the analysis of the diffraction pattern, 2 Å, was far greater than the thickness of the membrane (45 Å) and not much more than the diameter of a potassium ion (1.33 Å). The 50 water molecules surrounding a potassium ion in the central cavity of the channel were found to create a space in the shape of a prism and, as the potassium ion entered the selectivity filter, the antiprism was replaced by one constructed of the oxygen atoms lining the walls of the filter. The same study also showed that the selectivity filter narrowed if it was exposed to a low concentration of potassium ions, as would happen when the gate near the intracellular mouth of the channel closed. Another paper in the same issue of *Nature* showed that the selectivity filter actually had four binding sites for potassium ions.[20] At any given moment, however, only two sites would be occupied by potassium ions, the intervening ones being filled by two water molecules. As MacKinnon's group had shown earlier, when a third potassium ion entered the filter, the ion at the far end was displaced into the extracellular environment (Fig. 21–3). As with the preceding publications, the two 2001 papers had been written clearly and, as much as possible, in language simple enough to be understood by the non-expert.

Ever-mindful of the pioneering work of the British neurophysiologists 40 years earlier, MacKinnon wrote to Hodgkin, pointing out that the crystallographic findings confirmed the prediction that Hodgkin had made with Keynes. It was the prediction that had come from the simple experiment with the steel balls in the two connecting chambers—that, at any given instant, an open channel would have several ions within it. The letter came at a time when Hodgkin—Sir Alan Hodgkin—was seriously ill, but if the hands were barely capable of holding the letter, the quick mind could appreciate its contents. A gracious and appreciative reply came from Lady Hodgkin. It was one of those curious twists of fortune that, 60 years earlier, Lady Hodgkin's father, the Nobel Laureate Peyton Rous, had been

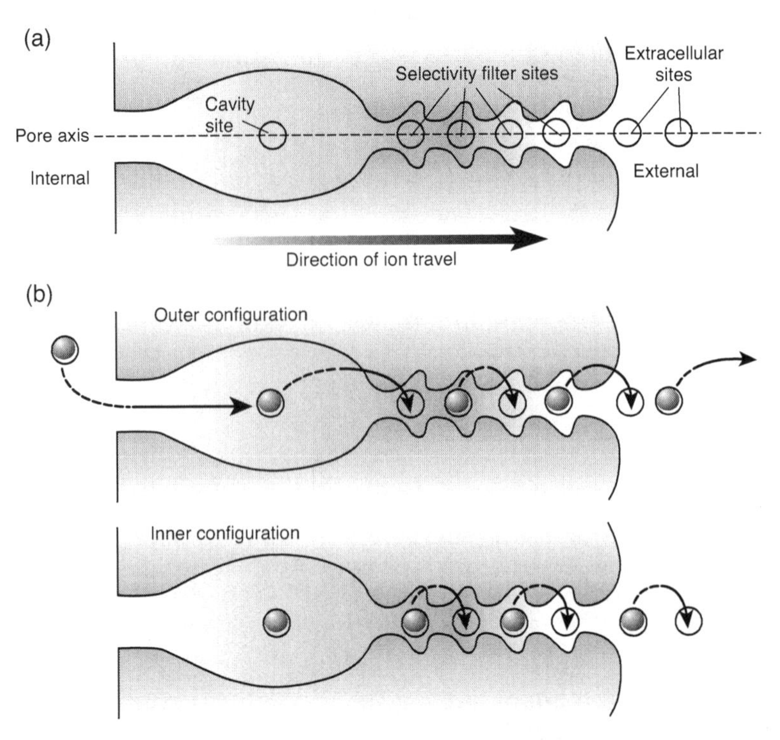

Figure 21–3 Movement of potassium ions through a potassium channel. *Top*: There are seven sites available, of which four are in the selectivity filter, one is in the water-filled cavity, and two are just beyond the filter. Though the cavity site is always occupied by a potassium ion, at any given instant only three of the remaining sites are taken by potassium, the others being occupied by water molecules. *Bottom*: As the potassium ion moves from the cavity to the filter, it repels the nearest potassium ion and the furthest ion is, in turn, pushed out in a "bucket brigade" manner. (Reprinted by permission from Macmillan Publishers Ltd: Nature. Miller C. See potassium run. *Nature*; **414**: 23–24, 2001.)

working at the Rockefeller within a few minutes' walk of where MacKinnon now had his laboratory.

There was still the matter of the sensing mechanism that would open up a voltage-gated channel, and this was perhaps—in view of its long squid axon history—the most intriguing question of all. The answers all came in 2003. Hitherto, much of the research had been conducted on potassium channels that opened in response to a change in hydrogen ion concentration. Now it switched to bacterial channels that, on the basis of their amino acid sequences, appeared to be voltage-sensitive. Unfortunately, repeated attempts to make satisfactory crystals from the bacterial potassium channels were largely unsuccessful. For this reason MacKinnon turned to one of the oldest forms of life on the planet, *Aeropyrum pernix,* a bacterium found in hot sea water vents. If there were voltage-sensitive channels, it was likely that, because of their functioning at high temperatures, they would be exceptionally stable. First, however, MacKinnon had to show that *Aeropyrum pernix* had potassium channels that, like those in nerve fibers of animals, were voltage-dependent. Not only did the bacterial channels contain the same sequence of amino acids lining the selectivity filter, but they were also blocked by the same toxins, including those of the scorpion and the Chilean rose tarantula, that were so effective in mammalian nerves. The remarkable nature of the last finding was not lost on MacKinnon: "There is no evolutionary pressure for a tarantula native to the dry scrublands of Chile or a scorpion from the deserts of the Middle East to recognize receptor sites on a K_v channel from an archaebacterium that lives near an oceanic thermal vent off the coast of Japan."[21]

As MacKinnon knew, the effects of tarantula or scorpion toxins on bacteria were a side show. There *was* an evolutionary pressure at work, and it had come about because the channels, developed in the earliest forms of life, and capable of operating in a voltage-gated mode, had become so useful to the various species that had developed over the next billion or so years. The success of a species had depended, in part, on conservation of a primordial channel DNA.

In the second of the three papers in *Nature*, MacKinnon's group described the general features of the bacterial potassium channel, confirming that its selectivity filter was the same as that in the previously investigated *Streptomyces lividans* channel.[22] It was in the bacterial voltage sensor that the main interest lay. The diffraction pattern, obtained from crystals of the isolated sensor, indicated that, in each of the four subunits, the transmembrane segment S4—with part of S3—formed the key part of the sensor, as had been anticipated. However, instead of lining the pore, as expected, the four voltage sensors (one for each subunit) were on the outside of the channel and lay perpendicular to its axis within the membrane. The four sensors strongly resembled "paddles" that could bend at their attachments to the channel. If the paddles could move within the membrane, then their attachments to the channel could open and close the pore. Was this, in fact, the secret of Hodgkin and Huxley's predicted gating mechanism?

In the third of the *Nature* papers published in 2003, MacKinnon and his colleagues showed that it was. This time the channel, inserted in a lipid membrane, was made to open or close by adjusting the voltage across the membrane. At the same time the position of a voltage-sensing paddle was determined by binding the paddle with tarantula spider venom or with the compound avidin. It was found that when a negative potential was maintained across the membrane, mimicking the resting potential of a nerve or muscle fiber, the potassium channel was closed with the four paddles lying within the membrane against its cytoplasmic face (Fig. 21–4). When the membrane was depolarized, however, as it would be in the approach of an action potential, the opening of the channel was associated with a change in the position of the paddles. They had swung up so that their tips were now against the extracellular face of the membrane, and it was evident that the positive charges on the paddles, provided by the four arginine amino acids, were

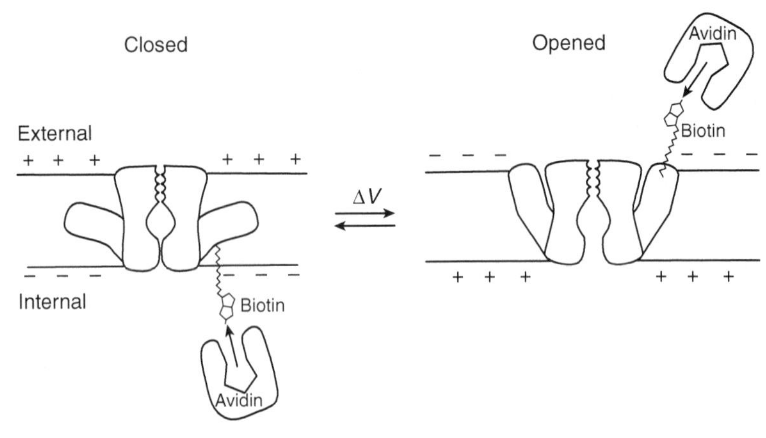

Figure 21–4 The last major part of the puzzle—the gating mechanism. Each of the four subunits of this voltage-gated potassium channel has a positively charged gating "paddle." *Left*: With a cell in its resting state, the paddles are nearer the negatively charged cell interior and the channel is closed. *Right*: When the membrane is depolarized, the four paddles move towards the negatively charged cell exterior, opening up the channel. In this study, biotin and avidin were used to locate the positions of the paddles. (Reprinted by permission from Macmillan Publishers Ltd: Nature. Jiang Y, Ruta V, Chen J, Lee A, MacKinnon R. The principle of gating charge movement in a voltage-dependent K+ channel. *Nature*; **423**: 42–48, 2003.)

responsible for the movements. Thus the positive charges, carrying the paddles with them, would always be attracted to the more negative side of the membrane.

Without any previous experience in a notoriously difficult field, X-ray crystallography, MacKinnon had achieved the seemingly impossible. At the outset he had been told that his aspirations were "altogether unrealistic." It had taken him eight years, but by the end he had elucidated the structure of a voltage-gated ion channel and had shown how each part—the central cavity, the selectivity filter, and the voltage sensor—was exquisitely suited to carry out its role in the operation of the channel. There was an elegance in the structure of the channel, just as there had been an elegance in the structure of DNA when it had first been revealed in 1953 by Watson and Crick.[24] There was an elegance, too, in the way the results had been obtained and described—again, just as those for DNA had been reported. But it must be added that, compared with DNA, MacKinnon's task had been much the greater. As a reviewing editor of *Nature* would remark: "the structure of [the] voltage sensor, so simple and, with hindsight, so obvious, is a wonderful end to a 50-year-old mystery."[25] In solving the mystery, MacKinnon had confirmed the bold predictions of Alan Hodgkin and Andrew Huxley in 1952 and, to all intents and purposes, had completed the line of research that had begun two centuries earlier in the laboratory of Luigi Galvani. From frog leg to squid axon to archaebacterium, it had been an extraordinary journey. In 2003 Roderick MacKinnon, the most successful of the later travelers, was awarded the Nobel Prize in Chemistry.[26]

22

Departures

Electrophysiologists never die, they only depolarize.
Silvio Weidmann

Most of them, those who had studied the nerve impulse and what it might do at the end of the nerve fiber, lived to a good age (Fig. 22–1). Sherrington had been 94, dying quietly in his nursing home. The death of the person who had been regarded as the foremost physiologist of his day was widely reported in the newspapers and in the scientific journals. Yet, curiously, there seemed little interest in preserving anything of his work in Oxford. At St. Thomas' Hospital in London, a few yards from the south bank of the Thames, where he had been appointed lecturer in 1887, there was the new Sherrington School of Physiology. In Liverpool, where he had become the Holt Professor of Physiology in 1895, the university had honored him, on the occasion of his 90th birthday, by instituting a biennial lectureship. Adrian, with whom he had shared the Nobel Prize in 1932, was the first recipient. In the years to come, there would be three biographies of Sherrington. But in Oxford, where Sherrington had occupied the Waynflete chair for almost a quarter of a century, and established his great school of reflexology, there was nothing.[1] It was puzzling, especially since his successor as Waynflete Professor, Liddell, would write a thoughtful and generous obituary. But there would be no Sherrington Room, no preservation of the great man's equipment, not even the optical myograph, which he had invented and of which he had been so proud. Instead, through the intercession of his last student at Oxford, William Gibson, the Sherrington memorabilia—the academic gowns, the medals, the notices of the honorary degrees, and even the laboratory notebooks—would end up 7,000 miles away in the University of British Columbia. Had anyone visited Oxford for some memory of Sherrington, the best they could do was to sit down at the desk in his old study—that massive, double-sided, desk with its numerous drawers—and think of the many paper records from the laboratory that he would have laid out on top for measurement, the drawings he would have prepared, and the manuscripts he would have written. It was a strange oversight on the part of the university.

Adrian, whom Queen Elizabeth II would honor by an appointment as the first Baron Adrian of Cambridge, was still active in the laboratory at the time of

Figure 22–1 Adrian (*left*), Sherrington (*center*), and Alexander Forbes (*right*). The meeting took place in the garden of Sherrington's retirement home in Ipswich in 1938, when Sherrington was 80. It had been six years since Adrian and Sherrington had shared the Nobel Prize. (Courtesy of the Harvard Medical Library in the Francis A. Countway Library of Medicine)

Sherrington's death and continued to experiment for several more years. To the end he remained a reclusive figure in the Cambridge department, experimenting entirely alone. An anesthetized animal, usually a rabbit, would be left in a basket outside the door of Adrian's laboratory in the morning, together with charged lead-acid accumulators for running the amplifiers. At the end of the working day, the dead animal would be outside the door, awaiting collection. Inside the laboratory, if one had somehow managed to get inside the room, one would have noticed dust everywhere—and one macabre relic. For there, inside the door and facing the intruder, "was this Keith Lucas gas mask from the First World War and the two cylindrical eyes would be staring at you through stalactites of matted cobwebs and it was incredible."[2]

Incredible, indeed. In the end, the long career was brought to its conclusion by accident. A tap was left on in the room above Adrian's, and when the rubber tubing to which the tap was connected burst during the night, the water went through the ceiling of the great neurophysiologists's laboratory. Almost everything was ruined—all the electronic equipment and even the circuit diagram of his first amplifier, still hanging on the wall after almost 40 years. He was too old to start again: "As it would have meant the rebuilding of all my apparatus and learning how to use new radio instruments I had not the necessary energy to start work again in earnest."[3] Adrian survived Sherrington by 25 years, dying in 1977 at the age of 88. In his later years, the man who, throughout his long career, was often tense with others and was happiest working by himself seemed more relaxed. There was continuing correspondence of a very informal kind with former colleagues, notably Yngve Zotterman and Detlev Bronk. As with Sherrington, there were degrees awarded by other universities, and honorary memberships and fellowships of learned societies, but it is doubtful if any award meant more to him than his becoming Master of Trinity College and, a few years before that, President of the Royal Society. In 1964, in anticipation of his 75th birthday, the leading neurophysiologists came from around the world and gathered in Cambridge to celebrate the scientific life of their most eminent member.[4] Keith Lucas's pupil had long been the undisputed master.

Bryan Matthews, who, in his turn, had been pupil to Adrian, also prospered. Had he committed himself more fully to the laboratory, in the manner of Adrian, he, too, would have achieved the highest honors. It was not that his scientific output was meager, or that his research had ever been other than of the highest quality. There had, after all, been the first studies of single muscle receptors, the detection of the dorsal root potentials, the confirmation of Berger's EEG waves, the invention of the iron-core oscillograph, and the innovation of the differential amplifier. The problem was the inevitable comparison with Adrian, and, especially in regard to sailing, the business of experiencing life to its fullest. But if Matthews' career had faltered in the university, it had gained strength again in the 1939–45 war, during which he had carried out important and daring research on high-altitude physiology for the Royal Air Force; a knighthood followed. When Adrian retired as head of the physiology department at Cambridge, Matthews was his choice to succeed him, despite the outstanding success of the younger Hodgkin in the same department. It was a position that Sir Bryan, now very much the elder statesman, gave every appearance of enjoying. Nor was he detached from the students, as a person of his eminence could so easily have been. Instead, there would be appreciative memories of the distinguished-looking gentleman, always immaculately dressed, with his gray hair smoothed back above the ruddy complexion, and who, drawing on his vast experience, made the physiology of the nervous system so fascinating (Fig. 22–2). Nor had his inventiveness—and playfulness—deserted him. Indeed, those attending one of the annual Cambridge meetings of the Physiological Society would be amused and intrigued to see the first single frog-powered motor, a gastrocnemius muscle turning a flywheel as it contracted and so enabling the next stimulus to be delivered to the sciatic nerve.[5] Besides the inventions and the publications there was a legacy of a different kind. Just as Adrian's son, Richard, would become a neurophysiologist, specializing in the properties of muscle fiber membranes, so would Matthews' son, Peter, establish himself as a leading authority on muscle receptors and their effects on posture and movement.

The one at Cambridge, the other at Oxford.[6] Professor Sir Bryan Harold Cabot Matthews died in 1986 at the age of 80.

Figure 22–2 Sir Bryan Matthews late in his career, at the time that he was head of the Cambridge Physiological Laboratory. He is holding one of the robust and reliable oscillographs that he had designed and built many years earlier, and that gave such superb records. (Courtesy of Professor Bill Harris, of the Department of Physiology, Development and Neuroscience, University of Cambridge)

What of the three Americans who had pioneered the use of the cathode ray oscilloscope for studying the nerve impulse in the 1920s? Gasser (Fig. 22–3) left St. Louis and, despite his natural reticence—and the high-pitched voice—became a very successful Director of the Rockefeller Institute. The award of the Nobel Prize in 1944 was a surprise to him, an embarrassment even, and came at a time when, due to his administrative responsibilities and the disruption caused by the 1939–45 war, his own research had stopped. Following his retirement as Director of the Institute in 1953, there was the opportunity to start again. In keeping with Gasser's nature, it would be modest in its aims, and there would be no attempt to compete with the new school of single-fiber neurophysiologists. Nor would he have a colleague: he would do the work himself, and complete what he saw as unfinished business from the old days in St. Louis. The first matter was the structure of the very fine, unmyelinated, fibers responsible for the C wave in the compound action potential, the wave that Peter Heinbecker had been the first to see and that had almost destroyed George Bishop's career.[7] And so Gasser became an electron microscopist. In this he was aided by Keith Porter and George Palade, who had used the first such microscope at the Institute to examine the fine structure of different types of body cell, devising their own methods for cutting and embedding extremely thin sections as they went along. Gasser saw, in his own

Figure 22–3 Gasser at his desk in the Rockefeller Institute. (Courtesy of the Rockefeller Archive Center)

cross-sections of nerve, that the C fibers had a unique relationship to the Schwann cells. Whereas a large A fiber consisted of an axis cylinder surrounded by a succession of Schwann cells, each of which produced a segment of the myelin sheath, a C fiber was quite different. Gasser observed that large numbers of these thin fibers could be engulfed in the cytoplasm of a single Schwann cell, though the persistence of a narrow cleft enabled the fiber to communicate with the extracellular fluid (Fig. 22–4). The paper describing these features, in both skin and olfactory nerves, was beautifully written and illustrated, and it well fulfilled the sentiment of the opening sentence:

> "Morphology reaches the acme of its usefulness when it is brought into relationship with the function of the structures."[8]

Gasser's last paper, published in 1960, reported that some of the deflections in the compound action potentials, the waves of which Erlanger had been so enamored, were, in fact, recording artifacts.[9] Notwithstanding the irony of his last finding, for Gasser it was a productive retirement, and there must have been satisfaction in still being able to carry out original, and technically demanding, research in his early seventies. Perhaps there would have been more. Two years later, not having fully recovered from a stroke, Gasser died from pneumonia in the New York Hospital, aged 74. When the time came to clear out the belongings from his rooms at the Institute, there was a complete record of a scientific life lived to

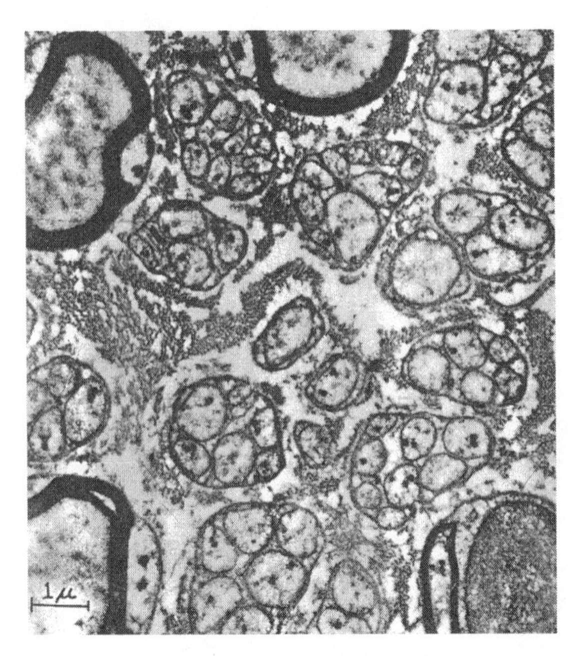

Figure 22–4 Electron micrograph of unmyelinated (C) fibers. Groups of four to eight fibers are enclosed by Schwann cells. Also shown are parts of three myelinated nerve fibers (*left* and *top*), much larger than the C fibers. (From Gasser HS. Comparison of the structure, as revealed with the electron microscope, and the physiology of the unmedullated fibres in the skin nerves and in the olfactory nerves. *Experimental Cell Research*; Suppl 3: 3–17. Reproduced by courtesy of Elsevier.)

the full, from the notebooks he kept as a student in Wisconsin to the hundreds of slides of nerve cross-sections and oscilloscope photographs of his later years. On the wall was a pencil portrait of Lord Adrian and a drawing of a nerve by Cajal. With Gasser's death, on May 11, 1963, the first of the St. Louis pioneers had gone.

Erlanger outlived his former associate by two years. As he approached retirement, he may well have felt that he had achieved all that was within his range. He had been the founding professor in two physiology departments, first at the University of Wisconsin in Madison and then at Washington University in St. Louis. In St. Louis, where he had remained head of physiology for 35 years, he had not only organized a strong department, propelling it into international prominence, but had quickly established himself as a major force in Washington University's reorganized School of Medicine. In the laboratory he had carried out original and important research in two fields—the cardiovascular and nervous systems—and he had served as President of the American Physiological Society for three years. There were also honorary degrees and invited lectureships. And then, to round everything off, had come the Nobel Prize in 1944. By then Erlanger was 70, and his research activity had already dwindled, the only experiments being those with Gordon Schoepfle, a junior member of the department. His last scientific paper, published in 1951, was on the phenomenon the very existence of which he had previously denied in his dispute with the young Alan Hodgkin in 1937—the local response in a nerve fiber.[10] After his retirement, Erlanger continued to live in St. Louis, attending symphony concerts with his wife, reading in the library, and walking in the country.[11] His mind remained sharp until the end. Erlanger died on December 5, 1965, aged 91.

George Bishop, the man who should have shared the Nobel Prize with Gasser and Erlanger, also lived long. In his 80th year, he spoke at the 1969 Erlanger Memorial Service at Washington University. Neither on that occasion, nor on any other, was there any rancor against the man who had deliberately minimized his, Bishop's, research contributions and so deprived him of the supreme accolade. As for Erlanger's terrible anger over the C fiber recordings, Bishop later dismissed it as having arisen from a misunderstanding. It was a charitable remark and one reflecting Bishop's attitude toward research, that it was the work that mattered, and that the honors and rewards were secondary. In the event, Bishop's banishment to the Department of Ophthalmology proved advantageous. Not only did it remove him from Erlanger's influence, but it also prompted him to study the electrical responses of the brain, and to make the first detailed recordings of the visual evoked potentials. Throughout his early years as a neurophysiologist, it was his strong engineering background that gave him his greatest advantage and enabled him to improve on the original Gasser and Erlanger experiments. After transferring his attention to the cortex, not only was he able to work out the directions in which the currents were flowing in the brain, and hence the locations of the active cells, but he was also the first to realize that the slow waves in the brain were not summated action potentials in large bundles of axons, but prolonged potentials in the cell bodies and dendrites—the excitatory and inhibitory postsynaptic potentials that Eccles later demonstrated in motoneurons.[12]

Even the advent of the 1939–45 war, and the disappearance of his laboratory staff, did not hold Bishop back from his research. By then in his early 50s, he decided to study pain, using himself as the subject and taking innumerable skin biopsies from different parts of his body for microscopic study. At the end of the war, as his laboratory staff increased, he maintained his interest in pain and in its fiber pathways in the brain and spinal cord. But, as always, there were so many other projects, so many interesting questions and observations. In this Bishop always differed sharply from Erlanger, who regarded himself as a pure "axonologist," confining himself to analysis of the potentials recorded from stimulated peripheral nerves. Bishop also differed from Erlanger in enjoying the collaboration with others, the junior colleagues, postdoctoral fellows, and students who always seemed to be around him. After Peter Heinbecker, the young Canadian who had unwittingly and temporarily misappropriated the C fiber response, came James O'Leary, Howard Bartley, Margaret Clare, and William Landau. And others. Indeed, until his early 80s, Bishop remained active in his laboratory, publishing one or two papers a year.[13] He died on October 11, 1973, aged 84, leaving a host of discoveries, ideas, and trained colleagues. For Bishop, it would have been a better legacy than a share in the 1944 Nobel Prize.

Helen Tredway Graham, the person perhaps responsible for drawing the attention of the Nobel Committee to the claims of the St. Louis investigators, also enjoyed a full and successful research career of her own.[14] Recruited by Gasser, she worked with him on drug effects, on after-potentials, and on the slow responses generated in the spinal cord after peripheral nerve stimulation—all this while bringing up a family and coping with the social responsibilities of being wife to the professor of surgery. After Gasser left St. Louis, she corresponded with him over her work and visited him in New York, evidently continuing to regard him as her scientific mentor. With Gasser's departure, however, the fascination of peripheral nerve dimmed and she switched her attention to histamine, devising a method for measuring it and demonstrating its presence in white blood cells and platelets, and in the tissue mast cells. In 1954, after 29 years of first-rate research, Helen Graham was finally made a full professor. Still active in research, and funded by the National Institutes of Health, she died in 1971 at the age of 79.

Other than Gasser, Helen Graham had also worked with Lorente de Nó, during his first years in the United States, and had been responsible for the Spaniard's initiation into neurophysiology. However, it was in New York, at the Rockefeller Institute, that Lorente spent the remainder of his working life. Much of that time, a full 10 years, had been occupied with the ill-fated study of frog sciatic nerve, during which he had erroneously concluded that neither sodium was responsible for the action potential, nor potassium for the resting potential. The failure of that long and arduous study, and his prolonged and quixotic defense of his conclusions, detracted from his many remarkable achievements. Among these were the studies on the cerebral cortex and the first realization that the cortical neurons were organized in columns. There were anatomical explorations of the acoustic and vestibular nuclei in the brain stem, and the definitive division of the hippocampus into various regions. Also in the past had been the investigations of synaptic function in the nuclei for the eye muscles, and the conclusion that the dendrites were the part of the nerve cell where incoming messages could be added and subtracted. And much more. Of all the neuroscientists working at that time, he had been the one with the most extensive expertise in both structure and function, and, as he continued his work, he made that expertise freely and graciously available to the younger scientists at the Rockefeller Institute. Fortunately, with the passage of time, the failure of the gargantuan *Study of Nerve Physiology* was forgotten, and it was the beautiful anatomical drawings of the different regions of the brain, and the conclusions reached from these, for which Lorente would be remembered. Lorente, too, lived a long life, dying in 1990, a week before his 88th birthday.[15]

Those who had worked to show that impulses released chemicals at the endings of the nerve fibers also lived to respectable ages. The first to go was Otto Loewi at 88. As a Jew he was imprisoned in Austria following the Nazi occupation in 1938, but he managed to leave before the outbreak of the war, first staying with his old friend and fellow laureate Sir Henry Dale, and then accepting an academic position in New York. He continued to lecture and write and enjoyed spending summers with fellow scientists at Woods Hole. As for Dale himself, there was also writing and lecturing in his retirement years, as well as the occasional visit to a meeting of the Physiological Society. From the front windows of his London home he would have seen, almost directly opposite, the house in which Burdon Sanderson, a century earlier, had invited the British physiologists to come and discuss the formation of a Society.[16]

Unlike Dale and Loewi, Wilhelm Feldberg continued to work in the laboratory long after the normal retirement age. He had returned to the U.K. from Australia shortly before the outbreak of the 1939–45 war and had become head of the Physiology and Pharmacology Division at the National Institute of Medical Research in London—the same Institute (though in a different location) where he had spent those few wonderfully productive years with Dale and Brown, and with Gaddum, showing that acetylcholine was the transmitter at the neuromuscular junction and in the sympathetic ganglia. Though his resources were much greater than before, with a whole division at his disposal, and though he worked hard on such topics as temperature and blood sugar regulation by the brain, the results never had the same clarity, nor the same impact, as the acetylcholine work. Undeterred, he might have continued to experiment even longer had not the anti-vivisectionists intervened. Feldberg died within a few weeks of his 93rd birthday.

It would be nice to report that Kenneth Cole, the person who, with Howard Curtis, had first shown the great increase in membrane permeability during the passage of the nerve impulse, and who had first employed the voltage clamp technique, had also ended his scientific career contentedly. The evidence, however, indicates otherwise. Despite the completeness and brilliance of the Hodgkin–Huxley 1952 papers describing the British work on the squid giant axon, the preparation to which he had drawn Hodgkin's attention in 1938, Cole continued his own work on nerve. At the Naval Medical Research Center in Bethesda he gathered a team of young scientists, but, due to all manner of problems, including the political intrusion of Senator McCarthy, the promise was never fulfilled. Next would come a move literally across the road, to the National Institutes of Health, and it was there, and in the summers at Woods Hole, that Cole spent the remainder of his working life. With John Moore as a colleague, the voltage clamp work on the squid was resumed. In addition, the Hodgkin–Huxley equations, the climax of the Cambridge work, were put to the test on a computer. Even here, Cole was unlucky. A potential flaw in the Hodgkin–Huxley equations was later traced to a programming mistake by the Americans. Andrew Huxley, who had performed his thousands of arithmetic steps on a hand-operated calculating machine, was correct all along. Nevertheless, there were some small gains. More was learnt about the development of the potassium current that terminated the action potential.

A greater contribution was the publication of *Membranes, Ions and Impulses* in 1968, a book that summarized Cole's own work in the field, as well as that of others.

Respected and admired as a pioneer in the study of the nerve impulse, the recipient of medals and honorary degrees, Kenneth Cole was not content. This kind and unassuming man continued to resent the fact that *his* preparation and *his* voltage clamp had been used by Hodgkin without due acknowledgement. The bitterness was all too evident in Los Angeles on the evening of October 19, 1981, when Cole rose to give the first lecture in the Presidential Symposium of the Society for Neuroscience.[17] For some in the audience there was disbelief, for others surprise, and for others embarrassment. Many wondered how such a thing could have mattered after 30 years. Others might attribute Cole's anger to his exclusion from the 1963 Nobel Prize. However, no one listening to the 81-year-old man doubted his sincerity or ignored his passion. In less than three years he was dead, and with his death the controversy ended. It was fitting that the fair-minded Andrew Huxley was the one to write Cole's obituary for the National Academy of Sciences.[18]

Alan Hodgkin had also continued to work on the squid giant axon after the publication of the 1952 papers. In collaboration with Richard Keynes, he had used radioactive tracers to show, as the Hodgkin–Huxley equations predicted, that with each impulse a small amount of sodium did actually enter the nerve fiber, and a similarly small amount of potassium left it.[19] The two investigators had also obtained indirect evidence for a pumping mechanism in the membrane that continually removed the sodium ions that leaked into the fiber during the resting state and during the action potential. After that there were the investigations with Baker, Caldwell, and Shaw on squid axons that were emptied of their contents. Those simple but compelling experiments showed that nothing more was required for the resting and action potentials than a membrane with the appropriate concentrations of ions on either side. There were also the pioneering microelectrode studies of skeletal muscle fibers. In 1970, however, had come a change in direction.

Like his Trinity College predecessor Adrian 20 years before him, Hodgkin was elected President of the Royal Society and there was no longer time for summers at the Marine Biological Laboratory in Plymouth. Hodgkin left the nerve fiber and took up vision research instead. It was a route that peripheral nerve physiologists had followed before: Adrian had done the same, with his experiments on the optic nerve of the conger eel, Bishop had been obliged to study vision after his expulsion from Erlanger's department, and Rushton, Hodgkin's colleague at Cambridge, had taken up color vision. Hodgkin chose to work on the light receptors in the retina, the rods and cones. In this he was helped initially by Michelangelo Fuortes, an Italo-American neurophysiologist who had been able to insert especially fine microelectrodes into the receptors to study their electric responses. It was a very competitive field and Hodgkin was a late entrant, but once again imagination and technical ingenuity enabled him to make important contributions. Unlike Cole, perpetually bedeviled by problems of one kind or another, success always seemed to follow Hodgkin.

During his Presidency of the Royal Society, Hodgkin was knighted. Then, almost half a century after entering Trinity College as an undergraduate, Sir Alan Hodgkin became its Master, once more following in the footsteps of Adrian. In 1988 Hodgkin wrote his last paper on vision and started on his autobiography, *Chance and Design*. A little before this time he had begun to have serious medical problems, one of which would leave him without position sense in his legs and unable to walk without support. The nerve impulses, whose ionic nature he had so convincingly demonstrated during his life, were no longer able to reach their destinations. Alan Hodgkin, Knight Commander of the Order of the British Empire, Nobel Laureate, and holder of 15 honorary degrees, died on December 20, 1998, aged 84. What a life it had been!

And the three men who, in the early 1940s, would meet at weekends to play tennis in Sydney, what of them? Of the three, it was Bernard Katz whose later life was the least eventful. Having won the Nobel Prize in 1970, he simply carried on with his experiments on the neuromuscular junction at University College London. For many years he continued to collaborate with Ricardo Miledi, the diminutive Mexican who came to him after working with Eccles in Canberra. Miledi, like Stephen Kuffler at Harvard, had imagination and could get the most difficult preparations to work. Among many other achievements, he was the first to think of injecting a toad's egg with RNA to see the protein that would be manufactured. After Miledi there were other postdoctoral fellows and visitors to Katz's department, including Bert Sakmann, who shared the Nobel Prize with Ernst Neher for the introduction of the patch clamp, and, with it, the ability to study the opening and closing of individual channels in the cell membrane. Knighted in 1969, Katz continued to collect awards and honors and to be invited to speak all over the world. And the lectures were remembered, his deep voice having made the intricacies of the nerve fiber membrane and the synapse so easily understood by his audience. Katz, too, lived long, dying at the age of 93. University College London would name a building after him.

During the time of his great triumphs in Canberra, Eccles resolved to continue working intensively well after the normal retirement age. And why not? Despite his age, he was still at the top of his form, and there were few left with such a rich background in research. Had he not, 40 years earlier, studied reflexes with Sherrington, and jousted with Henry Dale over synaptic transmission? The prospects for an emeritus professor in Canberra were not promising—a small salary, a single room, and little support. There had, however, been generous offers from the University of British Columbia and from the newly established Institute of Biomedical Research in Chicago. Eccles chose Chicago. At first everything went well and then, just as had happened in Sydney, a promise was broken. The retirement age was lowered, and Eccles would have to leave at 68. There was one more option for him, this time in Buffalo. Eccles took it, working there for seven years and building a happy and successful team around him, one that concentrated its research on the cerebellum. It was not Canberra, but it was an enjoyable and productive end to a long career. At 75 Eccles retired, with his second wife, to a villa in Switzerland. There, surrounded by his books and journals, and outside by mountains and valleys, Eccles returned to thinking about the relationship of the brain to the mind. It was this puzzle that had taken him, as a newly qualified 22-year-old doctor, from Melbourne to Oxford 50 years previously. But for once Sherrington was not able to help, dying with the belief that the problem would remain forever insoluble. Eccles, though, was determined to make the attempt while he still could, enlisting the aid of his former colleague from his days in New Zealand—Sir Hans Popper. And so *The Self and its Brain* appeared, with its nebulous "perceptrons."[20] It was a brave attempt, and admired as such, and did little to damage the reputation that Eccles had so arduously established as an experimental neuroscientist. Indeed, for several more years Eccles would remain in demand for international symposia, emerging from his villa to fly around the world and to give the weight of his presence to the proceedings. Eccles lived to the same age as Sherrington, dying on May 2, 1997, at the age of 94.

That left Stephen Kuffler (Fig. 22–5). At the end of the 1939–45 war he had left Australia to work in Ralph Gerard's Physiology Department in Chicago, and then, after little more than a year, had taken a position at Johns Hopkins Hospital in Baltimore. It was not in the Physiology Department but in the Wilmer Institute of Ophthalmology. It was there, in the basement of the building, that he began

Figure 22–5 Stephen Kuffler in his office at Harvard. (Courtesy of Dr. U. J. McMahan)

studies of ganglion cells in the mammalian retina, and of the fine structure and function of the muscle spindles, the sensory endings that had led Adrian to his discovery of a frequency code of impulses in sensory nerve fibers. There were also the large and complex stretch receptors in the tail muscles of lobsters and crayfish to explore, and for that study Kuffler would work at the Marine Biological Laboratory in Woods Hole during the summer, buying a house near the beach at Buzzard's Bay. In the receptor he was able to obtain impressive recordings of generator potentials, the depolarizations of the nerve membrane produced by stretching that were responsible for initiating the trains of impulses in the sensory nerve fiber.[21] It was as Adrian had envisaged 30 years previously but had never been able to show.

After 12 years at Hopkins, the rooms in the basement had become overcrowded with students, postdoctoral fellows, and visiting scientists eager to work with Kuffler, and he accepted a position in pharmacology at Harvard, taking many of the young people with him. A few years later, his success was recognized by the establishment of his own department at Harvard, the first department of neurobiology in the world. The title reflected the scope of Kuffler's work—not only electrophysiology, but also microscopy, biochemistry, and pharmacology, and later molecular biology as well. During the Boston years, Kuffler kept his summers for working at Woods Hole. It was there that his work on the crustacean stretch receptor had already led to the discovery of a special type of inhibition, one in which the nerve fiber was prevented from releasing the full amount of its excitatory chemical transmitter.[22] He then showed that this effect was brought about by GABA (gamma-aminobutyric acid), later found by others to be the most potent and plentiful inhibitory transmitter in the brain. There was much other research, equally elegant and original, and most often involving a new type of preparation. But just as important was the training and opportunity that Kuffler gave to the younger people around him, many of whom would come to occupy important and influential positions themselves. In their turn, they adored Kuffler.

It could not last. In his mid-forties Kuffler developed diabetes, and from this arose complications—glaucoma, cataracts, and hypoglycemic episodes with loss of consciousness. And heart disease. After one of his long swims in Buzzard's Bay Stephen Kuffler collapsed and died. He had only recently turned 67, and he would, in the summer of the same year, have received an honorary degree from Oxford University and an invited lectureship in Munich. He died at the pinnacle of his career and was buried where he always wanted to be, in Woods Hole—the final resting place of Otto Loewi. It is difficult to believe that, had Kuffler lived one year more, he would not have shared the Nobel Prize with two junior colleagues from his own department, whom he had introduced to the neurophysiology of vision while at Hopkins—David Hubel and Torsten Wiesel.[23] Stephen Kuffler, the small man with the inexhaustible fund of jokes, who had so disliked physiology as a medical student in Vienna, had, long before his death, become one of its most famous practitioners.

September 7, 2006

Situated on the east side of Manhattan, between a busy city street and a riverside expressway, Rockefeller University presents a sharp contrast to its surroundings. To appreciate this, however, it is necessary to do something that few of its Manhattan neighbors ever do, and step inside the main gate on York Avenue. Walking up the slight incline to Founders Hall, the noise of the traffic quickly falls away. On either side are lawns, with the shade of tall trees, mostly planes, providing welcome relief from the heat of a summer day. The first buildings to be seen are old, ivy-covered, and majestic in their proportions. Their history is given in the carved inscription that runs across the top of Founders Hall: The Rockefeller Institute of Medical Research.

On this summer day there are few people to be seen. The unbroken quiet, the venerable buildings, and the beauty of their immediate surroundings could be those of a seminary. And, indeed, within the buildings there is much contemplation concerning the nature of life, though it has more to do with the intimate structure of living things than with the place of humankind in the cosmos. So powerful has been this thinking, and so brilliant the experiments to which it has given rise, that no fewer than 23 Nobel Laureates have been associated with the Rockefeller.

Whereas the early work, carried out when the Rockefeller was an institute and not a university, was done in the older buildings in the southern part of its small campus, the present research is nearly all conducted in the two towers at the north end. One of these, the Rockefeller Research Building, juts out over the riverside expressway, and it is here, on the second floor, that the MacKinnon Laboratory is to be found. At the center of the modestly sized complex is its nucleus. Dr. MacKinnon's office, facing out over the East River, is bright with the morning's sun. With its profusion of journals and papers on the desk and table, it is the office of a busy man.

The mood inside the surrounding rooms, for there are several work areas, is one of calm efficiency, though now and then there is laughter among the young postdoctoral fellows and graduate students. Yes, there are more people now than at the beginning, when there were just the three of them—MacKinnon himself, a solitary graduate student, and Alice, MacKinnon's wife, who had volunteered to help out. Alice is still here, keeping the laboratory running smoothly, but now there are 12 postdoctoral fellows and 6 graduate students as well. It is as many, perhaps rather more, than MacKinnon is comfortable with.

Inside the office, the head of the laboratory is dressed informally and sensibly—it is a warm day—in navy shorts and a T-shirt. Looking at the wiry frame, it is easy to believe that he had been a good gymnast, though his exercise now comes mainly from his daily walk to the university from his apartment. He is not a morning person, arriving usually between 9:30 and 10 a.m., but only after a full 12 hours or so will he leave.

Despite having won the Nobel Prize, Roderick MacKinnon feels that his work is only at the beginning of a story that is much larger than the structure of the ion channels. It is not enough for him to have worked out the architectural features that give the channels their remarkable properties—their selectivity, their opening and closing, and their astonishing efficiency. No, his quest embraces ion pumps as well, those specialized proteins that move ions against their electrical and concentration gradients, and that are as essential to life as the ion channels. Dr. MacKinnon's work is also beginning to throw light on the structural features that enable ion channels to activate second messengers, those molecules that can then move into the interior or the cell and modify the function of its enzyme systems and its DNA and RNA. Though it is to the future that he looks, he is mindful of the past and, in particular, of the electrophysiological studies, and their brilliant analysis, by Hodgkin and Huxley 50 years previously. Indeed, in the bookcase beside his desk is a copy of Hodgkin's Chance and Design, a book with which he is thoroughly familiar.

View across the East River from the terrace at Rockefeller University. The Rockefeller Research Building, which includes the MacKinnon Laboratory, is close to the terrace.

Within the world of the very small, a world beyond the range of the most powerful microscope, the vista that Roderick MacKinnon contemplates is large, and he knows there will be enough work for many years. All this he looks forward to with excitement, anticipating that, just as with the ion channels, there will be surprises along the way.

Courteous and unassuming, unspoiled by success, Roderick MacKinnon is an excellent role model for those around him.

23

Postscript

What happened after the neurophysiological giants passed on? The answer is that there were other great physiologists but, because there were more of them and because the most fundamental problems had already been tackled, they could never attain the stature of those who had come before. But the next generation accomplished much. Indeed, those who worked during that period came to regard the 1950s and 1960s as the golden age of neurophysiology. It had started with the brilliance of the Hodgkin–Huxley voltage clamp experiments, and the extraordinary calculations and deductions as to the nature of the ion channels. The two Cambridge men had shown how a nerve impulse was generated. Then Katz and, a little later, Eccles, had revealed what happened when the impulse reached the ends of the fiber. With its foundations so securely established, and with the new microelectrode recording technique waiting to be used, neurophysiology was set to expand. George Bishop's analogy with the exploration of a new continent became even more apt. The frontiersmen had passed through, and now it was time for the settlers to move in and begin their work.

And they did. The new neurophysiology went on around the world, but it was the scientists in the United States and, to a lesser extent, in Sweden, Britain, and Japan, who did the bulk of it. Favoring the American contribution were that country's numerous universities and the generous funding from the National Institutes of Health and other public granting agencies. There was also the enviable attitude of everything being possible. More than any others, the Americans made the new neuroscientific era exciting.

The whole of the nervous system was open for exploration, just as Gasser had foreseen in preparing his report to the Board of Scientific Directors at the Rockefeller Institute in 1936. In the spinal cord, there were further studies on motoneurons. Robert Burke, at the National Institutes for Health, showed, by stimulating motoneurons with intracellular electrodes, how it was possible to relate the electrical properties of the cell bodies to the muscle fibers that they innervated. The largest motoneurons supplied the greatest numbers of muscle fibers, and these were the fibers that had the fastest contractions and fatigued most readily.[1] Others showed that these motoneurons were especially active in quick, explosive movements such as running and jumping.[2] The smallest neurons, in contrast, supplied the least numbers of fibers, and these fibers were resistant to fatigue and contracted slowly. These motoneurons were the ones that were active in maintaining posture. Elwood Henneman, at Harvard University, by dissecting out single motor nerve fibers and then stimulating them, reached similar conclusions about the sizes and properties of the muscle fiber colonies. In addition, he showed that there was a relationship between the sizes of the motoneurons and the order in which they were recruited, with the smallest neurons initiating slowly developing or sustained contractions.[3] Ragnar Granit, in Stockholm, who had been a young postdoctoral fellow with Eccles in Sherrington's laboratory, also adopted intracellular stimulation of motoneurons and showed further differences in electrical properties between cells.[4] For his work on the spinal cord, and his more extensive studies on the neurons in the retina, Granit became another neurophysiologist to win a Nobel Prize.

Working on the spinal cord were other Swedes—Sven Landgren, Olaf Oscarsson, and Anders Lundberg. This time the interest was in the neuronal circuitry, including that of the tracts of nerve fibers ascending to, or coming down from, the brain. There was special interest in the small internuncial neurons in the spinal cord that elaborated the messages to the motoneurons by grading the amounts of excitation and inhibition. This type of work, extraordinarily difficult to carry out and analyze, was continued by Elizabeta Jankowska.[5] At the other end of the nervous system, Charles Phillips, in Oxford, was obtaining the first intracellular records from neurons in the cerebral cortex. Because of his interest in movement, and because the cells in its deeper layers were some of the largest in the brain, he chose the motor cortex for his studies.[6]

Glass microcapillaries, first exploited in the study of muscle by Ling and Gerard in Chicago, and by Hodgkin and Nastuk in Cambridge, were not only used to measure the potentials across the membranes of nerve cells. They also became micropipettes, their insides filled with solutions of substances under consideration as transmitters. Further, it became possible to join several micropipettes together, each containing a different substance, so that the same nerve cell could undergo multiple tests. Acetylcholine had already been identified as one transmitter in the spinal cord, by Eccles and colleagues. Now, with the pioneering work of Krnjevic and Phillis, a number of other candidates were examined, small amounts being driven out of the micropipettes, by electric pulses, onto the surfaces of the neurons in the brain and spinal cord.[7] In this way, the excitatory actions of glutamate and aspartate were discovered, and the inhibitory effects of gamma-aminobutyric acid and glycine. Later, from Hillarp, Fuxe, and Carlsson in Sweden came new histochemical techniques that showed fine fibers ascending to the cerebral cortex from nuclei in the brain stem, fibers that released other transmitters—noradrenaline, serotonin, and dopamine—in addition to acetylcholine.[8]

Glass microelectrodes also became, for many years, the electrodes of choice for recording the impulse discharges in many areas of the brain, and that could be accomplished with the tips of the electrodes resting outside the bodies of the cells—a much simpler proposition than inserting the tips through the membranes into the interiors. In this way there were important studies of the sensory receiving areas of the cerebral cortex.

As regards the body surface, Adrian, towards the end of his career, had already mapped out the regions of cortex responding to touch on the opposite side of the body and face. He had gone on to show that the amount of cortex devoted to the different parts of the body depended on the importance of that part to the animal and varied from one species to another. Animals that depend very much on scenting odors, for example, had relatively large areas given to the snout. Similarly, dextrous animals such as squirrels and monkeys had strong representations for the forepaws. Adrian also found that there was a second sensory area of cortex adjacent to the primary one. While he was carrying out these investigations in the 1940s, in Cambridge, another group of neurophysiologists were doing much the same thing at Johns Hopkins University. These scientists, notably Philip Bard and Thomas Woolsey, were content to study the extracellular slow waves generated by hundreds, if not thousands, of nerve cells, just as Adrian had done.[9] With the advent of the glass microelectrode, with its ease of manufacture and its fine tip, Vernon Mountcastle and his colleagues, part of the new generation of neuroscientists at Johns Hopkins, were able to extend the earlier studies into an examination of the receptive properties of individual neurons. Situated mostly behind the motor cortex, these cells were found to respond to light touch on the skin, others to stronger pressure, and yet others to movements of the joints or stretch of muscles. Some of the cells would discharge only when the stimulus involved extremely small areas of skin, while others were less selective.

Of great significance was the discovery that, as Mountcastle recorded from cells lying deeper and deeper in the cortex, the part of the body surface to which they responded remained the same. It appeared as if the neurons in the somatosensory part of the cortex were organized in columns perpendicular to the surface of the brain.[10] Given the enormous complexity of the brain, with its billions of neurons, and trillions of synaptic connections, it was an important advance to have discovered one of its organizational principles. Mountcastle, steeped in the history of his science, did not claim all the credit for the discovery. There had been someone else, a young man who had studied the fine anatomy of the cortex with the gold and silver staining methods of the time, and had reached the same conclusion—Lorente de Nó. He had prepared a chapter on the structure of the cerebral cortex for John Fulton's *Physiology of the Nervous System*, published in 1938. Looking at Lorente's beautiful drawings, Mountcastle would have noted Lorente's prefatory statement: "One of the reasons why writing it has been so laborious is that I have verified in my collection of brain sections the truthfulness of every statement in the text and of every line in the drawings."[11] It is a chapter still widely cited today. If only Lorente had continued down that path!

It was not from the somatosensory receiving part of the brain, however, that the greatest advances in cortical physiology would come, though Johns Hopkins was also—for a time, at least—involved. On this occasion it was through Stephen Kuffler. As already noted, Kuffler had left Australia towards the end of the 1939–45 war, moving first to Chicago and then to Baltimore.[12] It was there, in a small room in the basement of the Wilmer Eye Institute, that he started his studies on vision. Using glass microelectrodes to record from single cells in the cat retina, he mapped out, on a screen in front of the animal, the region of the visual field in which a spot of light would cause the cell to discharge. The fields, he found, were invariably circular and in many the central excitatory areas were surrounded by zones producing inhibition. Other fields were just the opposite—inhibition in the centers and excitation around them.[13] Kuffler found cells with similar properties, with circular fields of light or dark, at the next stage of the visual pathway to the brain, the lateral geniculate nucleus. In 1955 Kuffler took on Torsten Wiesel, a recently qualified Swedish physician, and then, three years later, David Hubel (Fig. 23–1).

Born in Canada, Hubel had grown up in Montreal and had studied medicine at McGill University, where he had had the good fortune to come under the influence of one of the great pioneers of neurophysiology, Herbert Jasper. Perhaps more than anyone else, Jasper had developed EEG from the interesting phenomenon demonstrated by Adrian and Matthews to the Physiological Society in 1934 into an extremely useful clinical discipline. Indeed, Jasper's ability to locate brain tumors with the EEG so impressed Wilder Penfield, the neurosurgical founder of the Montreal Neurological Institute, that Jasper was invited to move from the United States to become the Institute's chief of clinical neurophysiology. There, in addition to his work in EEG (and EMG), Jasper recorded from the exposed brains

Figure 23–1 David Hubel (*right*) with his former mentor, Herbert Jasper, then aged 90.

of patients in the neurosurgical operating theater and embarked on his own studies of single nerve cells in the cortex and deep nuclei of animals.[14] Encouraged and enlightened by Jasper, Hubel decided to specialize in neurology, moving to the United States for residency training and ending up doing research with Kuffler at Johns Hopkins. Hardly had Hubel settled in than the entire laboratory moved to Boston, Kuffler having been appointed to the Department of Pharmacology at Harvard University. Five years later, Kuffler had his own kingdom, the world's first Department of Neurobiology.

Prior to the move, Wiesel and Hubel had already begun to extend Kuffler's visual studies (Fig. 23–2). As before, spots of light were shone onto a screen in front of the animal, but this time it was the neurons in the cortical receiving area at the back of the brain that were explored with a fine metal electrode. The electrode was of a type developed by Hubel and consisted of a sharpened tungsten wire with a platinum tip. It was rugged but also discriminating, and ideal for making prolonged observations on the same cell. The behavior of the cells, though, was puzzling. The responses to spots of light were not very striking, certainly nothing like those observed in the eye and in the lateral geniculate nucleus. Then, just as it had with Adrian and with Katz and Fatt, chance intervened. One of the slides projected onto the retina elicited a strong burst of impulses and, upon further examination, it was found that, rather than the black dot in the center of the slide being the effective stimulus, it was the edge of the slide itself.[15] Rather than respond to circles of light and shade, the cells in the cortex preferred bars and edges.[16] The nerve fibers coming into the cortex had evidently been reorganized, and then reorganized again, as the message was passed on from one area of cortex to another. Just as in the experiments of Mountcastle, where the effects of touching the body surface were examined, the cells in the visual cortex were organized

Figure 23–2 David Hubel (*left*) studying the cat visual cortex with Torsten Wiesel. (Courtesy of Professor David Hubel)

in columns, each column sensitive to a particular orientation of the bar or edge. Additional experiments showed that the synaptic connections in the brain, to be effective, depended on the eye being used in the first few weeks of life.[17] The visual studies of Hubel and Wiesel provided the first detailed insights into the way that the cortex worked, and led to the award of the Nobel Prize in 1981. Had Stephen Kuffler, their mentor and friend, lived another year, he would surely have shared it with them.

Despite the increasing success of the neurophysiologists in their exploration of the brain, there were problems. One of these, due to movement of the tissue from pulsation of the small arteries and from respiration, was the difficulty in keeping the tip of the microelectrode inside the nerve cell, or, when extracellular recordings were being made, close to the same cell. One approach was to mount the microelectrode in a small chamber attached to the skull, and then to increase the pressure inside the chamber until the pulsations had been dampened sufficiently. Among those who used this technique were Mountcastle, at Johns Hopkins, and Edward Evarts, at the National Institutes of Health. Like Charles Phillips in Oxford, Evarts was especially interested in the activity of cells in the motor cortex, and, with the pressure chamber, was able to record their discharges as the animal carried out various movements with its arms.[18]

For those neurophysiologists concerned not with the long-range effects of nerve activity, but only with the properties of membranes, ion channels, and synapses, a solution to the brain instability problem appeared from an unexpected source. Harold McIlwain had come into his own as a biochemistry graduate student at the University of Newcastle upon Tyne in Britain. Wishing to measure the brain's utilization of energy compounds, McIlwain decided to cut slices of brain tissue, place them in a bath containing a nourishing solution, and stimulate them with wire electrodes; the experiment would conclude with the biochemical analysis.[19] The new approach proved very successful. Many years later, as an established professor at Imperial College in London, he and Harold Hillman took the further step of examining the viability of the nerve cells in the brain slices. For this, Hillman used glass microelectrodes to record resting membrane potentials of the neurons, and found that the values were remarkably normal. The findings were not immediately accepted, and there was disbelief when they were presented at a meeting of the Physiological Society.[20] But Hillman and McIlwain were right, and before long the brain slice preparation was being widely used.[21]

The advent of the brain slice brought an answer, a partial one, to a much greater problem than brain instability. To varying extents, many physiologists had a diffidence about experimenting on animals, well exemplified by Bernard Katz. On the one hand, they could reassure themselves by reflecting that large animals, such as cows, sheep, and pigs, were slaughtered in colossal numbers every day for food. Further, physiology was one of the medical sciences and, as such, provided information of potential benefit to the human species. In most countries, too, there were strict regulations about who could conduct research on animals, and the type of experiment permitted. In Britain, for example, investigators required a license and were subject to visits from Home Office inspectors.

All well and good. And yet the hesitancy persisted, especially in those recording impulse activity in the brain of an animal. There, on the oscilloscope screen, was the discharge of a neuron, a series of brief spikes on a green trace and, over the loudspeaker, an accompaniment of audible clicks. It was exciting, wondrously so, but there was also a feeling that one was trespassing on the inner life of another living creature. This, after all, was a brain that had seen, and heard, and smelled. A brain that had successfully explored the animal's world, and had learned different patterns of behavior in the process. Above all, a brain that might—who knows—have been capable of thinking. To add urgency to the debate were the activities of the antivivisectionists, some of whom, in Britain, thought nothing of invading and destroying laboratories or, even more alarming, of sending bombs through the mail to the scientists and their families.

Other than use brain slices or cultured nerve cells, or limit experiments to the lowly rat, there was one more option for the neurophysiologist. This was to perform the experiments, wherever possible, on human volunteers. There was a strong tradition for such an approach, for physiologists had been stimulating and recording from the nerves, and even the brains, of themselves and their subjects for many years. As long ago as 1918, Paul Hoffmann, in Würzburg, using a galvanometer to record calf muscle potentials, had made a detailed study of a spinal reflex[22]—accomplishing in humans what Sherrington had been doing at the same time in cats. As another example, the recording of the nerve compound action potential, work for which Gasser and Erlanger had shared the Nobel Prize in 1944, had become routine in diagnostic EMG laboratories throughout the world, the testing of the patient usually being undertaken by a technician. At a somewhat higher level, needles or wires or even sticky patches on the skin could be used to pick up the discharges of individual muscle fiber colonies, and the impulses in single sensory or motor nerve fibers could be recorded in human subjects with tungsten microelectrodes.[23]

In the case of the brain, Hans Berger's EEG recordings, carried out in Jena and dutifully described in his laboratory notebook, have already been mentioned. For those responses evoked in the brain by sensory stimulation, however, something more was needed. Not only were the potentials, recorded through the scalp, extremely small, but they were also mixed in with all the ongoing EEG activity, which acted as unwanted "noise." The solution to this particular problem was provided by George Dawson (Fig. 23–3) at the National Hospital for Nervous Diseases in London. Dawson's first approach was to superimpose faint oscilloscope traces on photographic film. The EEG noise, being random, remained faint, while the small signals from the cortex, being locked to the stimulus, became increasingly definite.[24] Dawson, a very practical physician who ascribed his engineering prowess to a misspent youth, then found an even better way to extract small biological potentials from noise. He designed an averaging machine in which a motor rotated

Figure 23–3 George Dawson, whose invention of signal averaging did so much to advance neurophysiology. (Courtesy of Professor Susan Schwartz-Giblin)

a disc of capacitors, each of which stored part of the signal at a certain moment after the stimulus had been delivered. After a number of repetitions, the charge on each capacitor was measured, giving an average potential for a corresponding instant following the stimulus.[25] Dawson's machine, which worked perfectly well, can now be seen in the National Science Museum in London. It was the only mechanical averager ever built, being superseded, almost immediately, by digital devices that worked on the same principle—repetition of the stimuli and summation of the responses.

While these developments were going on in London, American neurophysiologists, led by Vernon Mountcastle at Johns Hopkins, were beginning to employ full-size commercial computers. It was an ambitious undertaking: the software programs had to be written by the researchers, and the computers themselves were far too large to be accommodated in the laboratories. Even so, it was the computer that would solve the third problem of neurophysiology, how to analyze the enormous amounts of data collected from the discharges of nerve cells, particularly in the brain.

The fourth problem in neurophysiology was, and is, an intellectual one, a problem that may yet be solved, but probably not in a way that we can presently envisage. It is the problem of the neural mechanisms underlying consciousness, the same problem that occupied Eccles towards the end of his life, and the problem that Sherrington thought would remain beyond reach. It is the challenge to which Francis Crick, co-discoverer of the DNA double helix, devoted the last decades of his life, collaborating with Christof Koch at the Salk Institute in California.

While it is true that the days of the pipe-smoking neurophysiologist in his tweed jacket, with his fondness for mountain climbing and long-distance sailing, are long past, some would say that the golden age of neurophysiology is still with us. Certainly, more money is being spent on nervous system research than ever before, membership of the Society for Neuroscience continues to increase, and the 1990s were officially designated the decade of the brain. Yet those who lived and worked in the years before could, with justification, claim otherwise. They would point out that neuroscience, in its great expansion, had become fragmented. Until the early 1970s, for example, it was possible for a neurophysiologist to retain an intellectual mastery of the whole field, and to lecture with confidence on any aspect of the nervous system. Now, however, an entire career could be devoted, say, to one particular class of ion channel, or to one small part of the brain. Much of neuroscience had become molecular biology and cellular neurochemistry, important topics in themselves, but there were fewer interested in the ways that whole systems of neurons worked and were prepared to carry out the necessary experiments. There were no longer any Adrians and Sherringtons, or Eccleses and Jaspers, to unify and make sense of the masses of laboratory findings. Even worse, as university librarians continued to replace shelves of earlier science journals with computer terminals, the time was approaching when it would be impossible to know what the great neurophysiologists thought in their own time.

Of equal concern was the changed attitude to physiology, including neurophysiology, in the medical schools: there had been a realization that a student could become a competent physician without having to study physiology. This was undoubtedly true, but it was at least debatable whether, without knowledge of the way the body worked, the student had received an adequate education. It was a concern that would have been shared by Abraham Flexner, had he still been alive, for it was on the basis of his thoughtful recommendations that medical education in the United States had been reorganized and had flourished for so many years.[26] As a consequence of the new thinking, however, physiology, rather than continuing to form a vital part of the university structure, had become an irrelevance, and many of its practitioners had been obliged to make their careers in industry, mainly with pharmaceutical companies. Unfortunately, there are few companies sufficiently enlightened to allow their physiological employees the time and opportunity to indulge their curiosity. Organizations such as the Wellcome Trust remain anomalies.

Yet another factor militating against neurophysiology, and some of the other disciplines, was the changing perception of what constituted good research. As an example, far more attention was given to the dubious science of functional brain imaging than to the electrical recording of brain activity. There was also the influence of granting agencies and the editors of the more general medical journals. While fundamental, curiosity-driven, research was not actually discouraged, it began to receive much less attention, and certainly far less money, than large multicenter drug trials or the sort of soft science that reported on health risks associated with obesity, or on the behavioral consequences of watching violence on television. There was, in addition, the assumption of a correlation between the amount of money spent, which for a clinical trial could be tens of millions of dollars, and the scientific worth of a study. If one was looking for merit, surely it should have been in the inverse correlation.

One would like to think that the situation will change. Perhaps one day it will, when a new Hodgkin or a Huxley emerges, having done something so brilliant on the nervous system that the world has to take notice.

The neural basis of consciousness, perhaps.

November 9, 2004

The slight figure in the gray raincoat and the brown fedora makes his way swiftly through King's Cross Underground station, darting first one way and then another, easily finding the gaps in the rush of people making for the escalators on the Piccadilly Line. Sir Andrew Huxley, a former President, is on his way to a meeting of the Royal Society. The agility is remarkable in someone who, next week, will turn 87.

Tonight is the night of the annual Ferrier Lecture, given in honor of the Scotsman who mapped out some of the functions of the cerebral cortex more than a century ago. In the circles in which Sir Andrew moves, including tonight's meeting of the Royal Society, he will be instantly recognized and deferred to, as he enters into first one conversation and then another, his head turned slightly to the side and the dark brown eyes, the intense Huxley eyes, gleaming under the bushy white eyebrows. While the fierce intellect is undimmed, with the years the face has taken on a greater kindness, a benignity, and this is reflected in his manner. For those close enough to hear, the voice has a strong melodic line and, as they always have been, the thoughts are expressed in perfectly formed sentences.

He is still extremely busy. Earlier in the day he had driven in to Trinity from his Grantchester home, and lunched in the Great Hall, where his portrait now hangs beside that of his former colleague and fellow Master, Sir Alan Hodgkin. He had then gone through some of the mail and papers among the several piles on the table in his office. In that room, just as in the large, well-furnished living room at home, there is an original Collier portrait of his grandfather above the mantelpiece. In one corner of his office, standing on top of a filing cabinet and hidden beneath a plastic cover, is the Brunsviga calculating machine on which he worked out those remarkable equations half a century ago. The scientific part of his mind now, however, is much less engaged with the nature of the nerve impulse than with the structure and movement of those molecules in muscle fibers that serve as motors, propelling the slender actin filaments along the myosin rods, and causing the fibers to shorten. Though it is some time since he last worked on this problem in the laboratory, the molecular motors continue to fascinate him and he has serious reservations about some of the recent conclusions of others.

Tomorrow morning, having returned by the last train of the night to Cambridge, he will go into the Physiological Laboratory, this time to hear a lecture on another of his interests, the detailed function of the inner ear. It is an interest he has carried for many years and, indeed, had discussed at one of the meetings of the Cambridge Natural History Club when he and Alan Hodgkin were both students.

Almost 70 years ago.

NOTES

Preface

1. Huxley, A (2002). From overshoot to voltage clamp. *Trends in Neurosciences*; **25**: 553–558.
2. Piccolini, M (1998). Animal electricity and the birth of electrophysiology. The legacy of Luigi Galvani. *Brain Research Bulletin*; **46**: 381–407. Only after completion of the manuscript did I discover that Dr. Piccolini was, in fact, writing a book, in conjunction with Dr. Marco Bresadola (*Frogs, Torpedoes and Sparks: Galvani, Volta and Animal Electricity*). As the title implies, this book deals very extensively with the Galvani–Volta controversy.
3. Finger, S (2000). *Minds Behind the Brain: A History of the Pioneers and their Discoveries*. Oxford: Oxford University Press.
4. Lorente de Nó (1947). A study of nerve physiology. *Studies from the Rockefeller Institute for Medical Research*; **131–2**: 496 & 548 pp. New York: Rockefeller Institute.
5. Eccles, JC (1977). My scientific odyssey. *Annual Review of Physiology*; **39**: 1–18.
6. Helmholtz, HLF (1878). The facts of perception. In *Helmholtz on Perception: Its Physiology and Development*, edited by Warren, RM, & Warren, RP. New York: John Wiley & Sons, 1968, pp. 207–231.
7. Dr. C. Barber Mueller, now an Emeritus Professor at McMaster University, began his medical studies at Washington University in 1938. Other than seeing Dr. Erlanger in the physiology practical classes, he worked for several months in Dr. George Bishop's laboratory, analyzing oscilloscope recordings. Dr. Mueller, a medical historian himself, recently wrote *Evarts A. Graham: The Life, Lives and Times of the Surgical Spirit of St. Louis*. Hamilton: BC Decker, 2002.
8. See Figure 22–1, page 302.

Chapter 1: Introduction

1. Sherrington, CS (1940). *Man on his Nature. The Gifford Lectures, Edinburgh, 1937–1938*. New York: MacMillan.
2. Hartline, HK. (1934). Intensity and duration in the excitation of single photoreceptor units. *Journal of Cellular and Comparative Physiology*; **5**: 229—247.
3. See *The Voltage Clamp*, page 214.
4. Watson, JD, & Crick, FHC (1953). A structure for deoxyribonucleic acid. *Nature*; **171**: 737–738.
5. Avery, OT, MacLeod, CM, & McCarty, M (1944). Studies on the chemical nature of the substance inducing transformation on pneumococcal type I: Induction of transformation by a desoxyribonucleic acid fraction isolated from pneumococcus type III. *Journal of Experimental Medicine*; **79**: 137–158. Many biochemists and molecular biologists consider that Avery, as senior investigator, deserved at least a share of a Nobel Prize for this crucial study.
6. Watson, JD (1968). *The Double Helix. A Personal Account of the Discovery of the Structure of DNA*. New York: Athenaeum.

7. MacKinnon, R (2004). Nobel Lecture. Potassium channels and the atomic basis of selective ion conduction. *Bioscience Reports*; **24**: 75–100.

Chapter 2: The Spark

1. Galen (129–99) has been much written about, and a recent account of his life and work is included in Finger, S (2000). *Minds Behind the Brain. A History of the Pioneers and their Discoveries*, pp. 39–55. Oxford: Oxford University Press.

2,3. Clarke, E, & O'Malley, CD (1968). *The Human Brain and the Spinal Cord. A Historical Study Illustrated by Writings from Antiquity to the Twentieth Century*. Berkeley: University of California Press.

4. The word *electrica* was introduced by Sir Thomas Gilbert, Physician to Queen Elizabeth I of England, in the 17th century. Gilbert used it to describe objects that, when rubbed, would attract other, lighter, ones to them—the phenomenon of static electricity.
5. Scribonius Largus (1529). *De Compositionibus Medicamentorum. Liber unus*. Ed, J. Ruel. Paris: Wechel. Cited in Finger (2000).
6. Electric stimulation of the brain is now being used, with considerable success, as an experimental treatment for epilepsy and migraine, following suggestions by the author's colleague Dr. Adrian Upton of McMaster University. The author has himself treated many episodes of familial "hemiplegic" migraine, using transcranial magnetic stimulation to provide the electric pulses; in every instance the migraine symptoms were abolished instantaneously. See McComas, AJ (2007). *The Artful Chameleon. An Exploration of Migraine and Medicine*. 2nd ed. West Flamborough, Canada: Alkat Neuroscience.
7. Not surprisingly, much about Galvani is available on the Internet, in addition to numerous books and journal articles. Among the latter, special mention should be made of Piccolini, M (1998). Animal electricity and the birth of electrophysiology: The legacy of Luigi Galvani. *Brain Research Bulletin*; **46**: 381–407.
8. The original lecture theater, situated in the old university building, the Archiginassio, was completely destroyed in the Second World War. Its replacement has been rebuilt as close to the original design as possible, including extensive wood paneling, wooden statues, and a cedarwood ceiling.
9. *"Dans les champs de l'observation, le hasard ne favorise que les éspirits prépares."* ("In the fields of observation, chance favours only the prepared mind.") Cited by Cannon, WB (1945). *The Way of an Investigator*. New York: WW Norton, p. 205.
10. It is now thought that the house in which the Galvanis performed their experiments is number 2, Via Alfredo Testoni. Although the house, part of a terrace, still stands, there is as yet no historical plaque honouring Galvani. However, the house in the Via Marconi in which Galvani was born does have such a plaque; the house is now a bank.
11. Galvani, L (1791). De viribus electricitatis in motu musculari. Commentarius. *De Bononiensi Scientarum et Artium Instituto atque Academia Commentarii*; **7**: 363–418. (Translated paragraph given in Piccolini, 1998.)
12. As with Galvani, there is much information on Volta on the Internet, but the author is also indebted to Dr. Marco Piccolini for his analysis of the controversy with Galvani (see Note 7).
13. The statue, by Adalberto Cencetti, was completed in 1879 and offers a more vigorous and appealing depiction of Galvani than does his portrait.

14–16. See Moruzzi, G (1996). The electrophysiological work of Carlo Matteucci. *Brain Research Bulletin*; **40**: 69–91. This is by far the most comprehensive account of Matteucci's life and work. The discovery of Moruzzi's article, originally published elsewhere in 1964, and its translation from Italian to English, is due to Dr. Marco Piccolini.

17. A recent account of du Bois-Reymond's life and background is given in Finkelstein, G (2006). Emil du Bois-Reymond vs Ludimar Hermann. *Comptes Rendus Biologies*; **329**: 340–347.
18. *Encyclopaedia Britannica*. 15th ed., vol. 2, p. 1035, 1979. Much information on this and other 19th-century equipment used in nerve and muscle physiology has been provided by Dr. Timothy Lenoir: Lenoir, T (1986). Models and instruments in the development of electrophysiology, 1845–1912. *Historical Studies in the Physical and Biological Sciences*; **17**: 1–54.

19,20. Cited in Moruzzi (1996).

21. If the explanation given were true, that the action currents in the muscle fibers had excited the overlying nerve fibers, one might expect the same phenomenon to be demonstrable in the intact preparation, in which bundles of nerve fibers run through the belly of a muscle to the innervation zone. However, even when the entire muscle is stimulated, only a very small proportion of the motor nerve fibers are excited ("backfiring"), and even this response is probably due to acetylcholine depolarizing the nerve terminals. See Katz, B (1969). *The Release of Neural Transmitter Substances.* Liverpool: Liverpool University Press, pp. 45–46.
22. Marey carried out his work in a very handsome building near the Bois de Boulogne in Paris and is buried in the grounds. After the 1939–45 war, the same building, now the Institute Marey, became an important neurophysiological research center due to the skilled but divergent activities of Madame Denise Albe-Fessard and her husband, Dr. Alfred Fessard. One of the latter's interests was the electric ray, *Torpedo.*
23. The electric ray contains a large, flattened, kidney-shaped electric organ on either side of the head. Each organ is made up of several hundred columns, each column consisting of stacks of electric plates, rather like a pile of coins. The plates are themselves derived from muscle fibers that, during embryonic development, lose their contractile proteins and become specialized current generators instead. Only one side of a plate receives a nerve supply and can develop an impulse (action potential). This property and the arrangement of the plates on top of each other ensure that the voltages developed during the impulses add together (like batteries connected in series). Depending on the species of ray, the voltages can be as high as 400 v, though the current flowing from the ray will depend on the resistance of the surrounding water. The rays, like other "strongly" electric fish, use their electric organs to deliver brief bursts of powerful, rapidly repeating, shocks—either to stun their prey or to ward off predators.
24. Lorenzini, S (1678). *Observazioni interne alle Torpedini.* Florence: Per l'Onofri. Cited by Finger, S (2000).
25. Until 1871 "Germany" consisted of a loose confederation of 39 states, of which Prussia was the most powerful. Both Königsberg and Berlin were in Prussia, the latter having long replaced the former as the capital.
26. For a succinct account of Helmholtz's experiments on muscle contraction and impulse conduction in nerve, see Bennett, MR (1999). The early history of the synapse: From Plato to Sherrington. *Brain Research Bulletin*; **50**: 95–118. For information on Helmholtz's life and his many other scientific achievements, the Internet may be consulted and there is also the excellent biography by David Cahan (see Note 27).
27. Olesko, KM, & Holmes, FL (1993). Experiment, quantification and discovery. Helmholtz's early psychological researches, 1843–1850. In Cahan, D (ed). *Hermann von Helmholtz and the Foundation of Nineteenth Century Science.* Berkeley: University of California Press, pp. 50–108.
28. Müller, J (1838). Handbuch der Physiologie des Menschen für Vorlesungen. Coblenz: J Hölscher.
29. Cited by Buchthal, F, & Rosenfalck, A (1966). Evoked action potentials and conduction velocity in human sensory nerves. *Brain Research*; **3**: 1–122. Von Haller arrived at the correct answer, 50 m/s, prompting Du Bois-Reymond to comment that the value "was not a little remarkable, in its wonderful coincidence with Helmholtz's measurements of conduction velocity, considering that every single step in Haller's reasoning was erroneous. In this case the *Aeneid* really has proved a book of oracles."
30. Helmholtz, H (1850). Messungen uber den zeitlichen Verlauf der Zuckung animalischer Muskeln und die Fortpflanzungsgeschwindigkeit der Reizung in den Nerven. *Archiv fur Anatomie, Physiologie une Wissenschaftliche Medicin*; pp. 276–367.
31. Formerly a professor of physiology at the universities of Zurich and Vienna, Ludwig became the founding director and professor at the Physiological Institute in Leipzig, where he remained for the rest of his career. An innovative and brilliant experimenter, Ludwig exerted an enormous influence on the development of physiology as a science.
32. See Grundfest, H (1963). Julius Bernstein, Ludimar Hermann and the discovery of the overshoot of the axon spike. *Archives of Italian Biology*; **103**: 483–90. A more detailed account of

Bernstein's life and scientific achievements has been recently provided by Seyfarth, E-A (2006). Julius Bernstein (1839–1917): pioneer neurobiologist and biophysicist. *Biological Cybernetics*; **94**: 2–8.

33. Bernstein, J (1868). Ueber den zeitlichen Verlauf der negativen Schwankung der Nervenstroms. *Pflüger's Archiv fur die Gesamte Physiologie des Menschen und der Tiere*; **1**: 173–207.
34. Hermann, L (1879). Allgemeine Nervenphysiologie. In *Handbuch der Physiologie*; **2**: 1–196. Leipzig: Vogel.
35. Bernstein, J (1902). Untersuchungen zur thermodynamik der bioelektrischen ströme. *Pflüger's Archiv fur die Gesammte Physiologie des Menschen und der Tiere*; **92**: 521–562.
36. Höber, R (1905). Über den Einfluss der Salze auf den Ruhestrom des Froschmuskels. *Pflügers Archiv fur die Gesamte Physiologie des Menschen und der Tiere*; **106**: 599–635.
37. Bernstein, J (1912). *Electrobiologie.* Braunschweig: Fr. Vieweg. This was not Bernstein's only book, for in 1876 he brought out *The Five Senses of Man,* at a time when he was Professor of Physiology at the University of Halle. This elegantly written book, beautifully illustrated with engravings, deals with the way in which the brain forms constructs of the internal and external worlds through information referred to it from the various sense organs. In its content, it follows the teachings of Johannes Müller and Hermann Helmholtz. Though Bernstein was an authority on the nerve impulse, not once is the nature of the impulse mentioned, only the fact that the sense organs induce an "irritation" of the nerve fibers.
38. Ranvier, L-A (1878). *Leçons sur l'histologie du système nerveux.* Weber: Paris.

49 Queen Anne Street, London, Possibly February 15, 1890

1. The incident of Gaskell at Burdon Sanderson's dinner was recounted by Sherrington and given in Bynum, WF (1976). A short history of the Physiological Society 1926–1976. *Journal of Physiology*: **268**: 23–72. The date of the dinner is uncertain but has been made to coincide with a meeting of the Physiological Society at University College London at which Sanderson had been in the chair. The earlier meeting at Sanderson's house, at which the formation of a physiological society was discussed, is described in Sharpey-Shafer, Sir Edward (1927). History of the Physiological Society during its first fifty years 1876–1926. *Journal of Physiology, Supplement.*

Chapter 3: Catching Up

1. See, for example, Ferrier, D (1873). Experimental research in cerebral physiology and pathology. *West Riding Lunatic Asylum Medical Reports*; **3**: 30–96.
2. Besterman, E, & Creese, R (1979). Waller: pioneer of electrocardiography. *British Heart Journal*; **42**: 61–4.
3. Sharpey-Schafer, E (1927). History of the Physiological Society during its first fifty years. *Journal of Physiology*, Supplement: 1–181.
4. There are a number of books about T. H. Huxley, including that written by his son, Leonard Huxley—*Life and Letters of Thomas Henry Huxley*. New York: Appleton, 1901. A lively account of T. H. Huxley and his descendants is given in Clark, RW (1968). *The Huxleys.* London: Heinemann.
5. Huxley, TH (1866). *Lessons in Elementary Physiology.* London: Methuen.
6. Goodman, J (2005). *The Rattlesnake.* London: Faber & Faber.
7. Huxley, TH (1893). *Science and Education: Collected Essays*, vol. 3. London: Methuen.
8. Langley, JN (1907). Sir Michael Foster. In memoriam. *Journal of Physiology*; **35**: 234–246.
9. Henry Head, later Sir Henry, would become a leading neurologist, and well known for his experiments on cutaneous sensibility, including one series in which he had one of his own nerves surgically divided. He also developed the concept of two parallel fiber systems, responsible for "protopathic" and "epicritic" aspects of skin sensation respectively. In his youth he had studied physiology in Germany, estimating the duration of the nerve action potential—as it turned out, inaccurately.
10. See Chapter 6.

11. Raverat, G (1952). *Period Piece.* London: Faber & Faber. This delightful book, the often hilarious reminiscences of a Cambridge childhood, has never been out of print since its first appearance and quickly became a classic. Raverat, who was a granddaughter of Charles Darwin and had enjoyed a friendship with Virginia Woolf, became a noted artist, specializing in wood engravings, and lived the middle part of her life in Provence. She died in 1957, having returned to Cambridge after her husband's death.
12. See Note 3.

Burlington House, London, March 8, 1884

1. The details of the dinner in Cajal's honor are given on pages 420–421 in the English translation of his autobiography (Note 1 in Chapter 4). The preparation of the lantern slides by Sherrington is mentioned by Sherrington's "last" student, William Gibson: Gibson, W (1986). Pioneers in neurosciences: the Sherrington era. *Canadian Journal of Neurological Sciences*; **13**: 295–300.

Chapter 4: The Anatomist's Eye

1. All the information in this chapter pertaining to Cajal has been taken from his autobiography, *Recuerdos de Mi Vida*, (Recollections of My Life), published in Madrid in 1901. The English translation was made by Dr. E. Horne Craigie, an Associate Professor of Anatomy and Neurology in the University of Toronto, and published by the American Philosophical Society in 1937. There is now a paperback edition, brought out by the MIT Press (Cambridge, MA) in 1989, and this is the one the author has used. A number of articles on Cajal and Golgi appeared on the centenary of their shared award of the 1906 Nobel Prize. A recent reappraisal of their work and an account of their nomination for the prize can be found in the October 2007, volume 55, issue of *Brain Research Reviews*.
2. Cajal, S Ramon y (1952). *Histologie du Système Nerveux de l'Homme et des Vertébrés.* Tome premier. Madrid: Instituto Ramon y Cajal. Translated by Mazzarello, P (1999). *A Scientific Biography of Camillo Golgi.* Oxford: Oxford University Press.
3. Cajal, S Ramon y (1989). *Recollections of My Life.* Translated by Craigie, EH. Cambridge, MA: MIT Press, p. 325.
4. Ibid., p. 332.
5. Ibid., p. 369.
6. Ibid., p. 352.

7–9. Ibid., p. 356.

10. Ibid., p. 358.
11. Golgi's biography is available at numerous sites on the Internet. A good example is: http://nobelprize.org/nobel-prizes/medicine/articles/golgi.
12. Much later it was shown, with the electron microscope, that axons *may* form connections with other axons, an arrangement that allows a special type of inhibition to occur (*presynaptic* inhibition).
13. Cajal, SR (1989), pp. 336–337.
14. Cajal's stay with the Sherringtons and his visit to Cambridge are recounted by Sherrington's former student, William Gibson, in: Eccles, JC, & Gibson, WC (1979). *Sherrington. His Life and Thought.* Berlin: Springer-Verlag.
15. Cajal, SR (1989), pp. 552–553.
16. Worth half a million dollars in today's currency!
17. Cajal, SR (1989), p. 565.
18. Ibid., p. 595.

Chapter 5: Cambridge, 1904: The Engineer

1. Gotch, F, & Horsley, V (1888). Observations upon the electromotive changes in the mammalian spinal cord following electrical excitation of the cortex cerebri. *Proceedings of the Royal Society*; **45**: 18–26.

2. Darwin, H, & Bayliss, W (1917). Keith Lucas, 1879–1916. *Proceedings of the Royal Society*, series B; **XC**: xxxi–xlii. Horace Darwin was one of Charles Darwin's sons and the founder of Cambridge Scientific Instruments, of which Lucas was a director.
3. Lucas, K (1905). On the gradation of activity in a skeletal muscle-fibre. *Journal of Physiology*; **33**: 125–137.
4. Lucas, K (1909). The "all or none" contraction of the amphibian skeletal muscle fibre. *Journal of Physiology*; **38**: 113–133.
5. Sherrington, C (1929). Some functional problems attaching to convergence. *Proceedings of the Royal Society*, series B; **40**: 332–362.
6. See Figure 8–2, page 118.
7. The description of the Cambridge Physiological Laboratory is Adrian's and is taken from *Keith Lucas*, ed. Fisher, WM, & Keith-Lucas, A, chapter VII, p. 87. Cambridge: W. Heffer & Son, 1934. Alys, the widow of Keith Lucas and one of the editors of *Keith Lucas*, had her surname and those of her children changed to Keith-Lucas after her husband's death.
8. Reminiscences by L. A. Orbeli in Katz, B (1978). Archibald Vivian Hill. September 26, 1886–June 3, 1977. *Biographical Memoirs of Fellows of the Royal Society*; **24**: 71–149. See also Burn, JH (1978). Essential pharmacology. *Annual Review of Pharmacology*; **9**: 1–20.
9. Cajal, SR (1989). *Recollections of My Life*, translated from Spanish by Craigie, EH. Cambridge, MA: MIT Press, p. 427.
10. Hodgkin, A (1979). Edgar Douglas Adrian, Baron Adrian of Cambridge. November 30, 1889–August 4, 1977. *Biographical Memoirs of Fellows of the Royal Society*; **25**: 1–73.
11. Lucas, K (1910). Quantitative researches on the summation of inadequate stimuli in muscle and nerve, with observations on the time factor in electric excitation. *Journal of Physiology*; **39**: 461–475.
12. Adrian, ED (1912). On the conduction of subnormal disturbances in normal nerve. *Journal of Physiology*; **45**: 389–412. These experiments on narcotized nerve were later criticized by Japanese and American workers. Although ingeniously designed, the experiments had two major shortcomings, the first of which was that the endpoint was whether or not a muscle twitched at all when the nerve was stimulated—rather than the amplitude of the contraction. The second problem was that the results applied to the whole population of nerve fibers rather than to individual fibers. A diminution in the nerve impulse could theoretically be due to similar reductions in all the fibers or to total failure in some with preservation in others. Adrian was well aware of these difficulties and was correct in his conclusion, namely that the nerve impulse regained its full amplitude on emerging from a narcotized region. However, the definitive proof of this type of phenomenon could only come from recordings of the nerve action potential itself when stimulated on either side of a damaged region (e.g., as in a carpal tunnel syndrome in a human subject), or from animal studies of demyelinated single nerve fibers (e.g., Rasminsky, M, & Sears, TA (1972). Internodal conduction in undissected demyelinated nerve fibres. *Journal of Physiology*; **227**: 323–350).
13. Lucas, K (1917). *The Conduction of the Nervous Impulse*. London: Longmans Green & Co. This book is based on a series of lectures that Lucas had given at University College London, and was edited posthumously by Adrian.
14. Adrian, ED, & Lucas, K (1912). On the summation of propagated disturbances in nerve and muscle. *Journal of Physiology*; **44**: 68–124.
15. See page 67
16. Einthoven, W (1901). Un nouveau galvanometre. *Archives Néerlandais des Sciences Exactes et Naturelles*; **6**: 623–633. A nicely illustrated article on Einthoven can be found on the Internet (http://chem.ch.huji.ac.il/history/einthoven.html), on which Figure 5–6 is based.
17. Lucas, K (1912). On a mechanical method of correcting photographic records obtained from the capillary electrometer. *Journal of Physiology*; **44**: 225–242.
18. Adrian's description of Lucas is given in Note 7.
19. See Note 2.

Upavon, Wiltshire, October 5, 1916

1. It is a fact that Lucas was killed in a mid-air collision by the airscrew (propeller) of the other plane, that both planes were B.E.2cs, and that the other pilot involved was a certain Lieutenant Jacques. It is thought that the crash took place over Jenners Firs, close to Upavon. The cause of the collision is not known, but the "blind areas" formed by the wings of the biplane would have been a possible cause. The B.E.2cs were not liked by their pilots because of their low maximum speed (72 m.p.h.) and poor maneuverability, and they were not designed for combat. The pilot sat in the rear of the two seats.

Chapter 6: The Cathode Ray Oscilloscope

1. Cajal, SR (1989) *Recollections of my Life*, translated from the Spanish by Craigie, EH. Cambridge, MA: MIT Press, p. 583.
2. The bibliographical material available on Gasser includes his own autobiography, written, with characteristic modesty, in the third person: Gasser, HS (1964). Herbert Spencer Gasser. 1888–1963. *Experimental Neurology*, Supplement 1, 1–38. Additional sources have been:

   (a) Chase, MW, & Hunt, CC (1995). Herbert Spencer Gasser. July 5, 1888–May 11, 1963. *Biographical Memoirs. National Academy of Sciences (USA)*; **67**: 147–177.
   (b) Adrian, Lord (1964). Herbert Spencer Gasser. 1888–1963. *Biographical Memoirs of the Royal Society*; **10**: 75–82.
   (c) Marshall, LH (1983). The fecundity of aggregates: the axonologists at Washington University, 1922–1942. *Perspectives in Biology and Medicine*; **26**: 613–636.
   (d) Schmitt, FO (1990). The never-ceasing search. *Memoirs of the American Philosophical Society*; **188**. This entertaining autobiography provides the most intimate portraits of Gasser and Erlanger and describes life in the early years of the physiology department at Washington University, in which Schmitt was the first graduate student. Schmitt maintained a close friendship with Gasser until the latter's death. His descriptions of Erlanger show the latter in a favorable light as head of department (and as a keen participant in games of horseshoes).

3. Most of the biographical material on Erlanger is taken from:

   (a) Davis, H (1970). Joseph Erlanger. January 5, 1874–December 5, 1965. *Biographical Memoirs. National Academy of Sciences (USA)*; **410**: 111–139.
   (b) Erlanger, J (1964). A physiologist reminiscences. *Annual Review of Physiology*; **26**: 93–106.

4. Erlanger, J (1904). A new instrument for determining the minimum and maximum blood-pressures in man. *Johns Hopkins Hospital Reports*; **12**: 53–100.
5. Erlanger, J (1905). On the physiology of heart-block in mammals, with special reference to the causation of Stokes-Adams disease. *Journal of Experimental Medicine*; **7**: 676–724.
6. See page 44.
7. Flexner, A (1910). *Medical Education in the United States and Canada.* New York: Carnegie Foundation.
8. For example: Erlanger, J, & Gasser, HS (1919). Hypertonic gum acacia and glucose in the treatment of secondary shock. *Annals of Surgery*; **69**: 389–421.
9. Gasser, HS, & Newcomer, HS (1921). Physiological action currents in the phrenic nerve. An application of the thermionic vacuum tube to nerve physiology. *American Journal of Physiology*; **57**: 1–26.
10. Lucas, K (1912). On a mechanical method of correcting photographic records obtained from the capillary electrometer. *Journal of Physiology*; **44**: 225–242.
11. There was also a very able departmental "mechanician" whom Erlanger credited with building the initial, motor-driven, commutator that started the sweep on the cathode ray oscillograph.

12. See Note 3 (b), p. 106.
13. Bishop, GH (1969). Washington University School of Medicine, Oral History Program. Interview Number 4. Cited by Marshall, LH, in Note 2 (c).
14. Gasser, HS, & Erlanger, J (1922). A study of the action currents of nerve with the cathode ray oscillograph. *American Journal of Physiology*; **62**: 496–524. The paper was received for publication by the *American Journal of Physiology* on August 1, 1922.
15. See Note 3 (b), p. 106.
16. Bishop, GH (1965). My life among the axons. *Annual Review of Physiology*; **27**: 1–18.
17. See Figure 2–9, page 24.
18. See Note 16, p. 2.
19. See Note 14.
20. The biographical material on Bishop comes from his own reminiscences (Note 16), as well as:

    (a) Landau, WM (1985). George Holman Bishop, 1889–1973. *Biographical Memoirs. National Academy of Sciences (USA)*; **55**: 45–66.
    (b) Marshall, LH (1983), Note 2 (c).
    (c) Discussion with Dr. C. Barber Mueller. See *Preface*, Note 7.

21. Bishop, GH (1923). Body fluid of the honey bee larva. I. Osmotic pressure, specific gravity, pH, O_2 capacity, CO_2 capacity, and buffer value, and their changes with larval activity and metamorphosis. *Journal of Biological Chemistry*; **58**: 543–565.
22. Erlanger, J, & Gasser, HS, with the collaboration, in some of the experiments, of GH Bishop (1924). The compound nature of the action current of nerve as disclosed by the cathode ray oscillograph. *American Journal of Physiology*; **70**: 624–666.
23. Erlanger, J, Bishop, GH, & Gasser, HS (1926). The action potential waves transmitted between the sciatic nerve and its spinal roots. *American Journal of Physiology*; **78**: 574–591.
24. Gasser, HS, & Erlanger, J (1927). The role played by the sizes of the constituent fibers of a nerve trunk in determining the form of its action potential wave. *American Journal of Physiology*; **80**: 522–547.
25. Erlanger, J, Gasser, HS, & Bishop, GH (1927). The absolutely refractory phase of the alpha, beta and gamma fibres in the sciatic nerve of the frog. *American Journal of Physiology*; **81**: 473.
26. Reminiscences of Dr. C. Barber Mueller. See *Preface*, Note 7.
27. Hodgkin, AL (1992). *Chance and Design. Reminiscences of Science in Peace and War.* Cambridge: University of Cambridge.
28. Mueller, CB (2002). *Evarts A. Graham. The Life, Lives, and Times of the Surgical Spirit of St. Louis.* Hamilton: Decker. See also Note 2 (c).
29. Gasser, HS, & Graham, HT (1933). Potentials produced in the spinal cord by stimulation of dorsal roots. *American Journal of Physiology*; **103**: 303–320. The recordings were technically very superior to those of Gotch and Horsley (1888), who had stimulated the motor cortex and had not had the advantage of a valve amplifier or a cathode ray oscilloscope.
30. Bishop, GH. Remarks at the Erlanger Memorial Service, January 14, 1969, at Washington University. Cited by Marshall, LH, in Note 2 (c).
31. See *Preface*, Note 7.
32. Bishop, GH, & Heinbecker, P (1930). Differentiation of axon types in visceral nerves by means of the potential record. *American Journal of Physiology*; **94**: 170–200.

33–35. Erlanger, J, & Gasser, HS (1936). *Electrical Signs of Nervous Activity*. Philadelphia: University of Pennsylvania. xi + 242 pp.

36. Gasser, HS (1960). Effect of the method of leading on the recording of the nerve fiber spectrum. *Journal of General Physiology*; **43**: 927–940.

Chapter 7: The Code

1. Nearly all the biographical information on Adrian has come from: Hodgkin, Sir Alan (1979). Edgar Douglas Adrian, Baron Adrian of Cambridge. *Biographical Memoirs of Fellows of the Royal Society*; **25**: 1–73.

2. Adrian, ED (1916). The electrical reactions of muscles before and after nerve injury. *Brain*; **39**: 1–33.
3. Adrian, ED, & Yealland, LR (1917). The treatment of some common war neuroses. *Lancet* (June): 3–24.
4. Adrian, ED (1921). The recovery process of excitable tissues. II. *Journal of Physiology*; **55**: 193–225.
5. Lillie, RS (1920). The recovery of transmissivity in passive iron wire as a model of recovery processes in irritable living systems. Parts I and II. *Journal of General Physiology*; **3**; 107–128 and 129–143.
6. Adrian, ED (1921). The recovery process of excitable tissues. I. *Journal of Physiology*; **54**: 1–31.
7. See page 109.

8 Adrian, Lord (1954). Memorable experiences in research. *Diabetes*; **3**: 17–18.

9. Adrian, ED, & Zotterman, Y (1926). The impulses produced by sensory nerve endings. Part 2. The response of a single end-organ. *Journal of Physiology*; **61**: 151–171.
10. Adrian, ED, & Zotterman, Y (1926). The impulses produced by sensory nerve-endings. III. Impulses set up by touch and pressure. *Journal of Physiology*; **61**: 465–483.
11. Adrian, ED (1926). The impulses produced by sensory nerve-endings. IV. Impulses from pain receptors. *Journal of Physiology*; **62**: 33–51.
12. Adrian, ED, & Bronk, DW (1928). The discharge of impulses in motor nerve fibres. Part 1. Impulses in single fibres of the phrenic nerve. *Journal of Physiology*; **66**: 81–101.
13. Adrian, ED, & Bronk, DW (1929). The discharge of impulses in motor nerve fibres. II. The frequency of discharge in reflex and voluntary contractions. *Journal of Physiology*; **67**: 119–151.
14. Recordings from single human nerve fibers were made possible by introducing fine tungsten wire microelectrodes through the skin into peripheral nerves. This methodology was introduced by Karl-Erik Hagbarth and his colleagues in 1975 and has since been used by others, with considerable success. See: Hagbarth, KE, Wallin, G, & Löfstedt, L (1975). Muscle spindle activity in man during voluntary fast alternating movements. *Journal of Neurology, Neurosurgery and Psychiatry*; **38**: 625–635.
15. See Figure 8–5, page 126.
16. Ratcliffe, JA (1975). Physics in a university laboratory before and after World War II. *Proceedings of the Royal Society, A*; **342**: 457–462. Sir Alan Hodgkin cites Ratcliffe in his own autobiography, *Chance and Design. Reminiscences of Science in Peace and War.* Cambridge: Cambridge University Press. 1992.

17,18. Letters from Zottermann to Professor Johansson, cited by Hodgkin (Note 1).

19. Forbes, A, & Thacher, C (1920). Amplification of action currents with the electron tube in recording with the string galvanometer. *American Journal of Physiology*; **52**: 409–471.
20. Forbes, A, Campbell, CJ, & Williams, HB (1924). Electrical records of afferent nerve impulses from muscular receptors. *American Journal of Physiology*; **69**: 283–303.
21. Adrian, ED (1965). Alexander Forbes 1882–1965. *Electroencephalography and Clinical Neurophysiology*; **19**: 109–111.

22,23. Eccles, JC (1970). Alexander Forbes and his achievement in electrophysiology. *Perspectives in Biology and Medicine*; **13**: 388–404.

24. Adrian, ED (1928). *The Basis of Sensation. The Action of the Sense Organs.* London: Christophers.
25. Adrian, ED (1932). *The Mechanism of Nervous Action.* Oxford: Oxford University Press.
26. Adrian, ED (1947). *The Physical Background of Perception.* Oxford: Clarendon Press.
27. See page 132.
28. The demonstration was not one of the published proceedings in the *Journal of Physiology.* Such publication would have been unnecessary, since a full paper on the "Berger rhythm" appeared later that year: Adrian, ED, & Matthews, BHC (1934). The Berger rhythm: potential changes from the occipital lobes in man. *Brain*; **57**: 355–385.
29. It was not generally realized at the time that, even before Berger, the EEG had been successfully recorded in humans by the Russian, Pravdich-Neminsky, as is apparent from his publication Pravdich-Neminsky, VV (1912). Eim Versuch der Registrierung der electrischen

Gehirnerscheinungen *Zbl Physiol*; **27**: 951–960. Sadly, Berger, later nominated for a Nobel Prize, committed suicide in 1941, having become intensely depressed over the political situation in Nazi Germany.

30. Matthews, BHC (1928). A new electrical recording system for physiological work. *Journal of Physiology*; **65**: 225–242. This was Matthews' first publication. In Figure 22–2, page 308, he can be seen holding one of his oscillographs.
31. See Figure 1–1, page 5.
32. Matthews, BHC (1934). A special purpose amplifier. *Journal of Physiology*; **81**: 28P.
33,34. Gray, Sir John (1990). Bryan Harold Cabot Matthews, June 14, 1906–July 23, 1986. *Biographical Memoirs of Fellows of the Royal Society*; **35**: 265–279.
35. It did, indeed, become possible to record the electrical potentials in the sense organs (receptors) themselves, as opposed to their nerve fibers. Leaving aside the electroretinogram, the first such report may have been that of Katz on the muscle spindle: Katz, B (1950). Depolarization of sensory terminals and the initiation of impulses in the muscle spindle. *Journal of Physiology*; **111**: 261–282. Even the Pacinian corpuscle, the exquisitely sensitive touch ending that Adrian had found impossible to investigate, eventually yielded up its secrets and was found to generate a receptor potential.
36. See Note 24, also: Adrian, ED (1943). Afferent areas in the brain of ungulates. *Brain*; **66**: 89–103.
37. Adrian, ED (1953). Sensory messages and sensation: The response of the olfactory organ to different smells. *Acta Physiologica Scandinavica*; **29**: 4–14.
38. Adrian, ED (1943). Discharges from vestibular receptors in the cat. *Journal of Physiology*; **107**: 389–407.
39. There were a number of studies on the eye and vision, including some on human subjects. The earliest investigation, however, was on the conger eel with Rachel Eckhard (who later married Bryan Matthews). For a review of research on vision, see Note 24 and: Adrian, ED (1943). The dominance of vision. *Transactions of the Ophthalmological Society of the United Kingdom*; **63**: 194–207.
40. Adrian, ED (1943). Afferent areas in the cerebellum connected with the limbs. *Brain*; **66**: 289–315.

Chapter 8: Excitation and Inhibition

1. Sherrington has been the subject of three full biographies:

    (a) Cohen, Lord (1958). *Sherrington—Physiologist, Philosopher and Poet.* Springfield, IL: Thomas.
    (b) Granit, R (1966). *Charles Scott Sherrington: An Appraisal.* London: Thomas Nelson.
    (c) Eccles, Sir John, & Gibson, WC (1979). *Sherrington. His Life and Thought.* Berlin: Springer-Verlag.

    Of the three biographies, Granit's is perhaps the most satisfying, partly because of its coherence and stylishness, and also because of its deeper analysis of Sherrington's research and of the historical context in which it was carried out. On the other hand, that by Eccles and Gibson contains the most information and is the best organized. Also very worthwhile are:

    (d) Fulton, JF (1952). Charles Scott Sherrington, OM. *Journal of Neurophysiology*; **15**: 167–190.
    (e) Liddell, EGT (1952). Charles Scott Sherrington, 1857–1952. *Obituary Notices of Fellows of the Royal Society*; **8**: 241–270.

2. Note 1 (c), p. 84.
3. Granit, Eccles, and Gibson.
4. See page 4.
5. Howard Florey (for work on penicillin), Granit and Eccles (neurophysiology).

6. "It is a patient."
7. Brown, GT, & Sherrington, CS (1912). On the instability of a cortical point. *Proceedings of the Royal Society, B*; **85**: 250–277.
8. For example:

   (a) Langley, JN, & Sherrington, CS (1894). Secondary degeneration of nerve tracts following removal of the cortex of the cerebrum in the dog. *Journal of Physiology*; **5**: 49–65.
   (b) Hadden, WB, & Sherrington, CS (1888). The pathological anatomy of a case of locomotor ataxy, with special reference to ascending degenerations in the spinal cord and medulla oblongata. *Brain*; **11**: 325–335.

9. Note 1 (c), p. 5.
10. Note 1 (b), p. 19, and Note 1 (c), p. 46.
11. Note 1 (c), pp. 3–4.
12. Note 1 (c), pp. 7–9.
13. Cushing, H (1904). Diary. Cited by Bliss, M (2005). *Harvey Cushing. A Life in Surgery.* Oxford: Oxford University Press.
14. Sherrington, CS (1906). *The Integrative Action of the Nervous System.* New Haven: Yale University Press.
15. See page 65.
16. Sherrington, CS (1928). A mammalian myograph. *Journal of Physiology*; **66**: iii-v.
17. Sherrington, CS (1893). Experiments in examination of the peripheral distribution of the fibres of the posterior roots of some spinal nerves. I. *Philosophical Transactions of the Royal Society, B*; **184**: 641–763.
18. Sherrington, CS (1898). Experiments in examination of the peripheral distribution of the fibres of the posterior roots of some spinal nerves. II. *Philosophical Transactions of the Royal Society, B*; **190**: 45–186, pl 3–6.
19. Sherrington, CS (1906). On the proprio-ceptive system, especially in its reflex aspect. *Brain*; **29**: 467–482.
20. Sherrington did this by cutting the posterior nerve roots above and below the nerve root of interest, and mapping out the area of skin that retained sensation. See Notes 14, 15.
21. Fulton, JF (1938). *The Physiology of the Nervous System.* London: Oxford University Press. Fulton critically assessed the contributions of both Magendie and the Scotsman, Charles Bell, in this matter, and concluded that Magendie's work was superior.
22. Hall, M (1833). On the reflex function of the medulla oblongata and medulla spinalis. *Proceedings of the Royal Society of London, Series B*; **123**: 635–665.
23. Sherrington, CS (1893). Note on the knee-jerk and the correlation of action of antagonistic muscles. *Proceedings of the Royal Society*; **52**: 556–564.
24. Sherrington, Sir Charles (1929). Ferrier Lecture: Some problems attaching to convergence. *Proceedings of the Royal Society, B*; **105**: 332–362.
25. Hoffmann, P (1910). Beitrag zur Kenntnis der menschlichen Reflexe mit besonder Berucksichtigung der elektrischen Erscheinungen. *Archiv fuer Anatomie und Physiologie*; **1**: 223–246.
26. Eccles, JC, & Sherrington, CS (1930). Numbers and contraction-values of individual motor-units examined in some muscles of the limb. *Proceedings of the Royal Society, B*; **106**: 326–357.
27. See Figure 5–3, page 67.
28. Eccles, JC, & Sherrington, CS (1931). Studies on the flexor reflex. III. The reflex response evoked by two centripetal volleys. *Proceedings of the Royal Society, B*; **107**: 535–556.
29. Eccles, JC, & Sherrington, CS (1931). Studies on the flexor reflex. VI. Inhibition. *Proceedings of the Royal Society, B*; **109**: 91–113.
30. Sherrington, Sir Charles. Inhibition as a coordinative factor (Nobel Lecture). Available at: http://www.nobel.se/medicine/laureates/1932/sherrington-lecture.html

31. Sherrington, Sir Charles (1941*). Man on His Nature.* New York: MacMillan.

Cambridge, August 1933

1. It is certain that Hodgkin was at Adrian's demonstration, since he describes it on page 52 of his autobiography: Hodgkin, Sir Alan (1992). *Chance and Design. Reminiscences of Science in Peace and War.* Cambridge: Cambridge University Press. Adrian's remark is, however, fictitious.

Chapter 9: The Messengers

1. Especially helpful in the preparation of this chapter was the monograph Bacq, ZM (1975). *Chemical Transmission of Nerve Impulses. A Historical Sketch.* Oxford: Pergamon Press, 106 pp. Bacq was active in the field himself and knew the main scientists involved.
2. Adrian papers: Trinity College Library, Cambridge.
3. Du Bois Reymond, E (1877). In: *Gesammelte Abhandlungen der allgemeinen Muskel-und Nervenphysik*; **2**: 700.
4. Elliott, TR (1904). Physiological Society note. Cited by Bacq (Note 1).
5. Langley, JN (1906). On nerve endings and on special excitable substances in cells. Croonian Lecture. *Proceedings of the Royal Society, B*; **78**: 170–194.
6. Adrian, E, & Lucas, K (1912). On the summation of propagated disturbances in nerve and muscle. *Journal of Physiology*; **44**: 68–124.
7. Lapicque, L & M (1906). *Comptes rendus hebdomadaires des séances et mémoires. Société de biologie, Paris*; **58**: 991.
8. Loewi, O (1921). Über humorale Übertragbarkeit der Herznervenwirkung. I. Mitteilung. *Pflügers Archiv fur die Gesamte Physiologie des Menschen und der Tiere*; **189**: 239–242.
9. Loewi, O (1940). An autobiographical sketch. In *Perspectives in Biology and Medicine, IV.* University of Chicago Press.
10. Kahn, RH (1926). Über humorale Übertragbarkeit der Herznervenwirkung. *Pflügers Archiv fur die Gesamte Physiologie des Menschen und der Tiere*; **214**: 492–498.
11. Adrenaline is not the only chemical released by the adrenal medulla and sympathetic nerve endings—there is also noradrenaline (norepinephrine).
12. Information about Sir Henry Dale came from:
    Dale, Sir Henry Hallett (1953). *Adventures in Physiology with Excursions into Autopharmacology.* London: Pergamon Press.
    Feldberg, WS (1970). Henry Hallett Dale, 1875–1968. *Biographical Memoirs of the Royal Society*; **16**: 77–174.
13. At the time Dale was a George Henry Lewes Student. Lewes was a prominent Victorian with a large number of interests, initially literature and drama and then physiology and pharmacology. A penetrating thinker and writer, he was also the companion of Mary Ann Evans—the writer George Eliot. In relation to the nervous system, Lewes was the first to suggest that the sensory qualities associated with particular nerves (e.g., vision with the optic nerve) depended on the sense organ to which the nerve fibers were connected rather than to any property of the nerve fibers themselves.
14. The hormone was secretin, which is released into the bloodstream from the duodenum and which stimulates the pancreas to secrete a bicarbonate-rich fluid.
15. Through the generosity of one of its two founders, Sir Henry Wellcome, Burroughs Wellcome was ultimately responsible for the creation of the world's largest medical charity, the Wellcome Trust. Only the Bill and Melinda Gates Foundation is of comparable size.
16. Dale, HH (1963). Pharmacology during the past sixty years. *Annual Review of Pharmacology*; **3**: 75–82.
17. The ergot fungus affects grasses and grains. Ergot poisoning can cause vasoconstriction, progressing to gangrene, insanity, seizures, and death.

18. Dale, HH (1914). The action of certain esters and ethers of choline and their relation to muscarine. *Journal of Pharmacology*; **6**: 147–190.
19. For details of Feldberg's life, see:
Feldberg, W (1977). The early history of synaptic and neuromuscular transmission by acetylcholine: reminiscences of an eye witness. In *The Pursuit of Nature*. Cambridge: Cambridge University Press, pp 65–83.
Bisset, GW and Bliss, TVP (1997). Wilhelm Siegmund Feldberg, CBE. November 19, 1900—October 23, 1993. *Biographical Memoirs of Fellows of the Royal Society*; **43**: 145–170.
20. As recounted by Feldberg in: Feldberg, W (1977). The early history of synaptic and neuromuscular transmission by acetylcholine: reminiscences of an eye witness. In *The Pursuit of Nature*. Cambridge: Cambridge University Press, pp. 65–83.
21. Feldberg, W, & Gaddum, JH (1934). The chemical transmitter at synapses in a sympathetic ganglion. *Journal of Physiology*; **81**: 305–319.
22. Dale, HH, & Feldberg, W (1934). The chemical transmitter of vagus effects to the stomach. *Journal of Physiology*; **81**: 320–34.
23. Dale, HH, Feldberg, W, & Vogt, M (1936). Release of acetylcholine at voluntary motor nerve endings. *Journal of Physiology*; **86**: 353–380.
24. The former graduate student was Derek Denny-Brown, who had left Sherrington's laboratory for the National Hospital for Nervous Diseases in London's Queen Square, where he was pursuing a career as neurologist. Denny-Brown had visited the hospital, St. Alfreges, in which Mary Walker worked, and had been consulted about her patient. Aware of the clinical similarity between myasthenia gravis and curare poisoning, and of physostigmine's ability to ameliorate the latter, Denny-Brown had suggested that physostigmine be tried in the patient. It is curious that Mary Walker chose not to acknowledge Denny-Brown's help in her letter to *The Lancet*. As it was, her discovery made her famous. See: Keesey, JC (1998). Contemporary opinions about Mary Walker: a shy pioneer of therapeutic neurology. *Neurology*; **51**: 1433–1439.
25. For details of Brown's life, see: MacIntosh, FC, & Paton, WDM (1974). George Lindor Brown, 1903–71. *Biographical Memoirs of Fellows of the Royal Society*; **20**: 41–73.
26. Brown, GL, Dale, HH, & Feldberg, W (1936). Reactions of the normal mammalian muscle to acetylcholine and to eserine. *Journal of Physiology*; **87**: 394–424.
27. Eccles, JC (1982). The synapse: from electrical to chemical transmission. *Annual Review of Neuroscience*; **5**: 325–339.
28. Brown, GL, & Eccles, JC (1934). The action of a single vagal volley on the rhythm of the heart beat. *Journal of Physiology*; **82**: 211–40.
29. Katz, Sir Bernard (1996). In Squire, LR, ed. *The History of Neuroscience in Autobiography*; **2**: 350–381. Washington: Society for Neuroscience.
30. Feldberg describes Dale being furious with him for showing little concern over his personal funding. See: Feldberg, W (1977). The early history of synaptic and neuromuscular transmission by acetylcholine: reminiscences of an eye witness. In *The Pursuit of Nature*. Cambridge: Cambridge University Press, pp. 65–83.
31. Dale would have been referring to the type of oratory practiced in London's Hyde Park, where anyone can stand on a soapbox and hold forth on any subject he or she pleases.

Chapter 10: The Squid Giant Axon

1. For a portrait of Gasser in the New York phase of his career, see Chase, MW, & Hunt, CC (1995). Herbert Spencer Gasser. July 5, 1888–May 11, 1963. *Biographical Memoirs*; **67**: 144–177, and Schmitt, FO (1990). The never-ceasing search. *Memoirs of the American Philosophical Society*; **188**: 399 pp.
2. Gasser, HS (1964). Herbert Spencer Gasser. 1888–1963. *Experimental Neurology, Supplement*; **1**: 1–38 (published posthumously).
3. Gasser, HS. First year's report to the Board of Scientific Directors of the Rockefeller Institute for Medical Research, April 18, 1936. *Rockefeller Institute Archives*.

4. Woolsey, TA (1990). Rafael Lorente de Nó. April 8, 1902–April 2, 1990. *Biographical Memoirs.* National Academy of Sciences, pp. 84–102.
5. Lorente de Nó (1935). The synaptic delay of the motoneurones. *American Journal of Physiology*; **111**: 272–282.
6. Lorente de Nó (1939). Transmission of impulses through cranial nerve nuclei. *Journal of Neurophysiology*; **2**: 402–464.
7. Hodgkin, Sir Alan (1992). *Chance and Design. Reminiscences of Science in Peace and War.* Cambridge: Cambridge University Press.
8. Author's recollection. See also:
   Messenger, J (1997). Obituary. John Zachary Young (1907–1997). *Nature*; **388**: 726.
   Boycott, BB (1998). John Zachary Young. March 18, 1907–July 4, 1997. *Biographical Memoirs of Fellows of the Royal Society*; **44**: 487–509.
9. Young, JZ (1936). The giant nerve fibres and epistellar body of cephalopods. *Quarterly Journal of Microscopical Science*; **78**: 367–368.
10. Dr. L. W. Williams was an anatomist at Harvard and published a monograph on the anatomy of the squid in 1912.
11. Details of Cole's life are to be found in:
    Cole, KS (1979). Mostly membranes. *Annual Review of Physiology*; **41**: 1–24.
    Huxley, Sir Andrew (1992). Kenneth Stewart Cole. July 10, 1900–April 18, 1984. *Biographical Memoirs of Fellows of the Royal Society of London*; **38**: 98–110.
12. Bernstein, in his 1912 monograph *Elektrobiologie*, draws attention to this property of plants and provides an illustration of a traveling signal recorded in a Venus flytrap with a capillary electrometer.
13. *Impedance* is a measure of the opposition to the passage of an alternating current. In a biological tissue the impedance is determined by both *resistance* and *capacitance*, being inversely proportional to the latter. In a membrane the lipid, acting as an insulator, accounts for the capacitance, while the resistance is determined by the number of channels available for ions to flow through—the fewer the channels, the higher the resistance. The opposite of resistance (or impedance) is *conductance*.
14. Cole, KS, & Curtis, HJ (1938). Electric impedance of *Nitella* during activity. *Journal of General Physiology*; **22**: 37–64.
15. Cole, KS, & Curtis, HJ (1939). Electric impedance of the squid giant axon during activity. *Journal of General Physiology*; **22**: 649–670.
16. Later on, Oxford became a leader in the field, largely through Wilfrid LeGros Clark and, later, Tom Powell, Max Cowan, and Edward Jones.

Chapter 11: The Neuromuscular Junction

1. Eccles, JC, & Sherrington, CS (1931). Studies on the flexor reflex. VI. Inhibition. *Proceedings of the Royal Society of London, B*; **109**: 91–113.
2. Eccles, JC, & Sherrington, CS (1930). Numbers and contraction-values of individual motor units examined in some muscles of the limb. *Proceedings of the Royal Society of London, B*; **106**: 326–357.
3. Brown, GL, & Eccles, JC (1934). The action of a single vagal volley on the rhythm of the heart beat. *Journal of Physiology*; **82**: 211–240.
4. Eccles, JC (1937). Synaptic and neuro-muscular transmission. *Physiological Reviews*; **17**: 538–555.
5. See page 145.
6. The chapter relies heavily on Eccles' own account of his years in Australia: Eccles, JC (1977). My scientific odyssey. *Annual Review of Physiology*; **39**: 1–18.
7. Kuffler's life story was put together after his death by Sir Bernard Katz: Katz, Sir Bernard (1990). In McMahan, UJ, ed. *Steve. Remembrances of Stephen W. Kuffler.* Sunderland, MA: Sinauer, pp. 107–141.

8. Katz, Sir Bernard (1996). Sir Bernard Katz. In Squire, LR, ed. *The History of Neuroscience in Autobiography*; **1**: 350–381.
9. See page 69.
10. Each thermopile comprised hundreds of constantan–silver junctions. When heated, such dissimilar metals develop small voltage differences across their ends, and these summate with those of other junctions. Hill's technician was Mr. A. C. Downing.
11. Hill's experiences in research are summarized in Hill, AV (1965). *Trails and Trials in Physiology*, while his obituary was written, most appropriately, by Katz—Katz, Sir Bernard (1978). Archibald Vivian Hill. September 26, 1886–June 3, 1977. *Biographical Memoirs of Fellows of the Royal Society*; **24**: 7. London: Edward Arnold.
12. See Note 8 for an account of Katz's life and career.
13. Hill, AV (1926). The heat production of nerve. *Journal of Pharmacology*; **29**: 161–165.
14. Hill, AV (1933). The physical nature of the nerve impulse. *Nature*; **81**: 501–508. Hill had, in fact, been interested in the basis of the impulse before this, publishing a mathematical treatment in 1910.
15. Katz, B (1939). *Electric Excitation of Nerve*. Oxford: Oxford University Press.
16. Now Sri Lanka.
17. Eccles replaced the electric mower with a gas-powered one, allowing Katz to claim credit for persuading him to switch from an electric mechanism to a chemical one—as in synaptic transmission!
18. See Note 3.
19. Eccles, JC, Katz, B, & Kuffler, SW (1941). Nature of the "endplate potential" in curarized muscle. *Journal of Neurophysiology*; **4**: 362–387.
20. Interestingly, a bundle of South American arrows, tipped with curare, lay undetected on top of a cupboard in the Cambridge Physiological Laboratory for many years and were discovered only during a clean-up in the early 1950s. They had probably been put there by Langley in the mid-1880s. (Personal communication by Dr. Rose Mason)
21. Eccles, JC, & Kuffler, SW (1941). Initiation of muscle impulses at neuro-muscular junction. *Journal of Neurophysiology*; **4**: 402–417.
22. Eccles, JC, Katz, B, & Kuffler, SW (1942). Effect of eserine on neuromuscular transmission. *Journal of Neurophysiology*; **5**: 211–230. There is a footnote from the journal editor to the effect that "Dr Eccles' paper was posted from Australia on December 17, 1941. Owing to conditions prevailing in the South Pacific he has not been able to examine proofs." The "conditions" were those of the 1939–45 war!

Plymouth, August 1939

1. As stated in his autobiography, *Chance and Design,* Hodgkin had arrived in Plymouth before Huxley. According to Huxley, in an interview with the author, they had stayed in the same boarding house, one situated in a street behind the main hotels. It is not known in which of the Marine Biological Laboratory rooms their work was carried out in 1939.

Chapter 12: The Giant Axon Impaled

1. Hodgkin's life and family history are recounted in his autobiography: Hodgkin, Sir Alan (1992). *Chance and Design. Reminiscences of Science in Peace and War.* Cambridge: Cambridge University Press. Included in the book are some of the letters he wrote to his mother as a young man.
2. See: Barlow, HB (1986). William Rushton. December 8, 1901–June 21, 1980. *Biographical Memoirs of Fellows of the Royal Society*; **32**: 423–459.
3. The attack had come from Kato in Japan, who had instigated some simple experiments in which the different lengths of narcotized nerve were compared with the times taken for impulse block to occur. Other experiments, by an American group that included Alexander Forbes, also cast doubt on Adrian's conclusions. Much later studies, on single nerve fibers that had been partially demyelinated, showed that impulses could be conducted over one or more

nodes of Ranvier with variable amplitudes and with conduction times varying from node to node, depending on the degree of demyelination (see Rasminsky, M, & Sears, TA (1952). Internodal conduction in undissected single nerve fibres. *Journal of Physiology*; **227**: 323–350).

4. See page 33.
5. Rushton's early thinking about nerve excitation was summarized in Rushton, W (1937). Initiation of the propagated disturbance. *Proceedings of the Royal Society of London, B*; **124**: 210–243. Further important papers on nerve appeared in the *Journal of Physiology* in 1949 and 1951, followed by an analysis of all the factors affecting the excitability of a single fiber within a nerve trunk: Lussier, JJ, & Rushton, W (1952). The excitability of a single fibre in a nerve trunk. *Journal of Physiology*; **117**: 87–108.
6. There are at least two apocryphal stories about Rushton's difficulties. In the first, he was one of a group of students at the bedside of an unconscious patient, the victim of a head injury. "What is wrong with this patient, Rushton?" the consulting neurologist is said to have asked, whereupon Rushton, the neuroscientist, launched into a complex explanation involving nerve excitability, a topic in which he could, with justification, claim to be the world expert. "No, no, Rushton," interrupted the consultant. Then, pointing to the next student, "You tell him." Whereupon came the reply, "Commotio cerebri" and the further comment, "There, you see, Rushton. Remember that!"
   In the second story, Rushton was having difficulties in the oral part of the final examination in medicine. One of the examiners, observing this, prepared a note and slipped it to Rushton's interrogator. The note said, "I think you should know that this man may be examining your son in Cambridge next year." Rushton passed.
7. Lucas, K (1910). Quantitative researches on the summation of inadequate stimuli in muscle and nerve, with observations on the time-factor in electric excitation. *Journal of Physiology*; **39**: 461–475.
8. Hodgkin, AL (1937). Evidence for electrical transmission in nerve. Part I. *Journal of Physiology*; **90**: 183–210.
9. Hodgkin, AL (1937). Evidence for electrical transmission in nerve. Part II. *Journal of Physiology*; **90**: 211–232.
10. Hodgkin, AL (1938). The subthreshold potentials in a crustacean nerve fibre. *Proceedings of the Royal Society of London, B*; **126**: 87–121.
11. Hodgkin, AL (1939). The relation between conduction velocity and the electrical resistance outside a nerve fibre. *Journal of Physiology*; **94**: 560–570.
12. The differing accounts of Hodgkin's visit to Woods Hole in 1938 are given respectively in Hodgkin's autobiography, *Chance and Design* (Note 1), and in Cole's reminiscences: Cole, KS (1982). Squid axon membrane: impedance decrease to voltage clamp. *Annual Review of Neuroscience*; **5**: 305–323.
13. See Note 11.
14. Cole, KS, & Hodgkin, AL (1939). Membrane and protoplasm resistance in the squid giant axon. *Journal of General Physiology*; **22**: 671–687.
15. A number of books have been written about the Huxleys, and about its scientific founder, T. H. Huxley, in particular. A very readable account is Clark, RW (1968). *The Huxleys*. London: Heinemann.
16. Huxley, AF (2004). In Squire, LR, ed. *The History of Neuroscience in Autobiography*; **4**: 282–318.
17. See Note 1.
18. The main reason why the mercury droplet did not fall is that the contents of nerve fibers are semi-solid, due to the presence of neurofilaments and microtubules.
19. Baker, PF, Hodgkin, AL, & Shaw, TI (1962). Replacement of the axoplasm of giant nerve fibres with artificial solutions. *Journal of Physiology*; **164**: 330–354.
20. See Note 1.
21. Hodgkin, AL, & Huxley, AF (1939). Action potentials recorded from inside a nerve fibre. *Nature*; **144**: 710–711.

Chapter 13: The War Years

1. Clamp, AL. *The Blitz of Plymouth*. Plymouth: PDS Printers.
2. Adrian, ED (1965). Alexander Forbes 1882–1965. *Electroencephalography and Clinical Neurophysiology*; **19**: 109–111. See also Fenn, WO (1979). Alexander Forbes. 1882–1965. *Biographical Memoir*; **40**: 113–141. Washington, DC: National Academy of Sciences.
3. Morison, R, & Dempsey, EW (1941). A study of thalamo-cortical relations. *American Journal of Physiology*; **135**: 281–292.
4. Katz, Sir Bernard (1996). In Squire, LR, ed. *The History of Neuroscience in Autobiography,* **2**: 350–381. Washington: Society for Neuroscience.
5. See page 205.
6. Katz, Sir Bernard (1990). In McMahan, UJ, ed. *Steve. Remembrances of Stephen W. Kuffler.* Sunderland, MA: Sinauer.
7. Eccles, JC (1944). The nature of synaptic transmission in a sympathetic ganglion. *Journal of Physiology*; **103**: 27–54.
8. Eccles, JC (1977). My scientific odyssey. *Annual Review of Physiology*; **39**: 1–18.
9. Huxley, AF (2004). In Squire, LR, ed. *The History of Neuroscience in Autobiography*; **4**: 282–318.
10. Hodgkin, Sir Alan (1992). *Chance and Design. Reminiscences of Science in Peace and War.* Cambridge: Cambridge University Press, p. 228.
11. Ibid.
12. Gray, Sir John (1990). Bryan Harold Cabot Matthews. June 14, 1906–July 23, 1986. *Biographical Memoirs of Fellows of the Royal Society*; **35**: 265–279. A more complete account of Matthews' wartime activities is given in: Gibson, TM, & Harrison, MH (1984). *Into Thin Air. A History of Aviation Medicine in the RAF.* London: Robert Hale. One of the Laboratory' s achievements was the design of an economical oxygen breathing system that came to be used by pilots flying at high altitude. After the war, the same system continued to be employed, not only by commercial and military pilots but also by Hillary and Tensing in the first ascent of Everest.
13. Adrian papers, Trinity College Library.
14. Adrian's work on the somatosensory cortex in different species is summarized in Adrian, ED (1947). *The Physical Background of Perception.* Oxford: Clarendon Press.
15. Adrian papers, Trinity College Library.
16. See, for example, Molnar, Z, & Blakemore, C (1991). Lack of regional specificity for connections formed between thalamus and cortex in coculture. *Nature*; **351**: 475–477.
17. Curtis, HJ, & Cole, KS (1942). Membrane resting and action potentials from the squid giant axon. *Journal of Cellular and Comparative Physiology*; **19**: 135–144.
18. Cole's experiences in wartime are included in his memoir: Cole, KS (1979). Mostly membranes. *Annual Review of Physiology*; **41**: 1–24.
19. Lloyd, DPC (1946). Facilitation and inhibition of spinal motoneurons. *Journal of Neurophysiology*; **9**: 421 438.
20. Heinbecker, P, & Bishop, GH (1929). Differentiation between types of fibres in certain components of involuntary nervous system. *Proceedings of the Society for Experimental Biology and Medicine*; **26**: 645–647.
21. Erlanger, J (1947). Some observations on the responses of single nerve fibres. *Nobel Lectures, Physiology or Medicine, 1942–1962*; **1**: 50–73. Amsterdam: Elsevier, 1964.
22. Bishop's life and research accomplishments are summarized in Landau, WM (1985). George Holman Bishop. 1889–1973. *Biographical Memoirs*; **55**: 44–66. Washington: National Academy of Sciences.
23. It was also the opinion of Lorente de Nó, who had worked in St. Louis for several years before transferring to the Rockefeller Institute, where he had been with Gasser. Of the three original axonologists, he considered Bishop the most brilliant (personal communication to Louise H. Marshall, cited in Marshall, LH (1983). The fecundity of aggregates: the axonologists at Washington University, 1922–1942. *Perspectives in Biology and Medicine*; **26**: 613–636.). Bishop also impressed A. V. Hill, a person not given to undue praise, who, after meeting Bishop for the

first time, described him as "a lion in a cage" (letter to Louise H. Marshall from Dr. James O' Leary, cited in Marshall, LH (1983, above)).

24. Gasser, HS (1964). Herbert Spencer Gasser. 1888–1963. *Experimental Neurology, Supplement*; **1**: 1–38 (published posthumously).
25. Erlanger's dismissal of Bishop's achievements was criticized by Gasser, at the time Erlanger and Gasser were preparing the important Johnson Foundation Lectures (letter from Gasser to Erlanger, cited by Marshall, LH (1983)).
26. Gasser, in refusing to submit any material of his own for consideration by the Nobel Committee, could not have corrected Erlanger's omission of Bishop's contributions.
27. Helen Graham's life is recounted in Mueller, CB (2002). *Evarts A. Graham. The Life, Lives, and Times of the Surgical Spirit of St. Louis.* Hamilton: Decker. Her neurophysiological contributions are summarized in Marshall, LH (1983; note 23).
28. Gasser, HS, & Graham, HT (1932). The end of the spike-potential of nerve and its relation to the beginning of the after-potential. *American Journal of Physiology*; **101**: 316–330.
29. Gasser, HS, & Graham, HT (1933). Potentials produced in the spinal cord by stimulation of dorsal roots. *American Journal of Physiology*; **103**: 303–320.
30. Hodgkin, Sir Alan (Note 10).
31. It is unlikely that Evarts Graham, on his own, would have overlooked Bishop. Indeed, he had earlier chosen Bishop, rather than Gasser or Erlanger, to be the scientific mentor for the aspiring young thoracic surgeon Peter Heinbecker.

Rockefeller Institute, New York, May 7, 1945

1. The details of the experiment are accurate and are taken from Figure 10, p. 28, in Lorente de Nó, R (1947). A study of nerve physiology, Part 1, In *Studies from the Rockefeller Institute for Medical Research*; **131.** New York. Lorente's thoughts, however, are fictitious.

Chapter 14: Sodium Unmasked

1. Cole, KS (1979). Mostly membranes. *Annual Review of Physiology*; **41**: 1–24.
2. Hodgkin, AL, & Huxley, AF (1945). Resting and action potentials in single nerve fibres. *Journal of Physiology*; **104**: 176–195.
3. Hodgkin, AL, & Huxley, AF (1947). Potassium leakage from an active nerve fibre. *Journal of Physiology*; **106**: 341–367.
4. Personal communication from Sir Andrew Huxley.
5. Overton, E (1902). Beiträge zur allgemeinen Muskel-und-Nervenphysiologie. *Pflügers Archiv fur die Gesamte Physiologie des Menschen und der Tiere*; **92**: 346–386. The translation of the excerpt is that given in Clarke, E, & O'Malley, CD (1968). *The Human Brain and Spinal Cord.* Berkeley: University of California Press. It should be added that Overton made other contributions to cell membrane theory, including the suggestion that the actions of anesthetics depended on their solubility in the membrane lipids.
6. Curtis, HJ, & Cole, KS (1942). Membrane resting and action potentials from the squid giant axon. *Journal of Cellular and Comparative Physiology*; **19**: 135–144.
7. See page 35.
8. See page 35.
9. Katz, Sir Bernard (1996). Sir Bernard Katz. In Squire, LR (ed). *The History of Neuroscience in Autobiography*; **1**: 350–381. Washington: Society for Neuroscience.
10. Katz, B (1947). The effect of electrolyte deficiency on the rate of conduction in a single nerve fibre. *Journal of Physiology*; **106**: 411–417.
11. Clamp, AL. *The Blitz of Plymouth, 1940–44.* Plymouth: PDS Printers.
12. Hodgkin, AL, & Katz, B (1949). The effect of sodium ions on the electrical activity of the giant axon of the squid. *Journal of Physiology*; **108**: 37–77.
13. Goldman, DE (1943). Potential, impedance and rectification in membranes. *Journal of General Physiology*; **27**: 37–60.

14. See Note 1.
15. Lorente de Nó, R (1947). A study of nerve physiology, Parts 1 and 2. *Studies from the Rockefeller Institute for Medical Research*; **131, 132**. New York.

Chapter 15: The Voltage Clamp

1. Penfield, W (1955). The role of the temporal lobe in certain psychical phenomena. *British Journal of Psychiatry*; **101**: 451–465.
2. Adrian, ED, & Matthews, BHC (1934). The Berger rhythm: potential changes from the occipital lobes in man. *Brain*; **57**: 355–385.
3. Hughes, J, Smith, TW, Kosterlitz, HW, Fothergill, LA, Morgan, BA, & Harris, HR (1975). Identification of two related pentapeptides from the brain with potent opiate agonist activity. *Nature*; **258**: 577–579.
4. See: Osmundsen JA (1965). "Matador" with a radio stops wired bull. Modified behaviour in animals the subject of brain study. *The New York Times*, May 17, 1965.
5. Loewi, O (1921). Über humorale Übertragbarkeit der Herznerwirking. I. Mitteilung. *Pflügers Archiv fur die Gesamte Physiologie des Menschen und der Tiere*; **189**: 239–242.
6. Hodgkin, AL, Huxley, AF, & Katz, B (1952). Measurement of current-voltage relations in the giant axon of *Loligo. Journal of Physiology*; **116**: 442–448.
   Hodgkin, AL, & Huxley, AF (1952a). Currents carried by sodium and potassium ions through the membrane of the giant axon of *Loligo. Journal of Physiology*; **116**: 449–472.
   Hodgkin, AL, & Huxley, AF (1952b). The components of membrane conductance in the giant axon of *Loligo. Journal of Physiology*; **116**: 473–496.
   Hodgkin, AL, & Huxley, AF (1952c). The dual effect of membrane potential on sodium conductance in the giant axon of *Loligo. Journal of Physiology*; **116**: 497–506.
   Hodgkin, AL, & Huxley, AF (1952d). A quantitative description of membrane current and its application to conduction and excitation in nerve. *Journal of Physiology*; **116**: 500–544.
7. Marmont, G (1949). Studies on the axon membrane. *Journal of Cellular and Comparative Physiology*; **34**: 351–382.
8. Cole, KS (1982). Squid axon membrane: impedance decrease to voltage clamp. *Annual Review of Neuroscience*; **5**: 305–323.
9. Hodgkin, Sir Alan (1992). Page 278 in *Chance and Design. Reminiscences of Science in Peace and War.* Cambridge: University of Cambridge Press.
10,11. Cole (1982). Note 8.
12. Personal communication from Sir Bernard Katz to Dr. Jack Diamond.
13. Hodgkin is said to have made the remark during a seminar to the Department of Neurobiology at Harvard (communication to the author by Dr. Jack Diamond).
14. Cole, KS (1949). Dynamic electrical characteristics of the squid giant axon membrane. *Archives des Sciences Physiologiques*; **3**: 253–258.
   Marmont, G (1949). Studies on the axon membrane. *Journal of Cellular and Comparative Physiology*; **34**: 351–382.
15. It is uncertain what Cole meant by this. Hodgkin, Huxley, and Katz had devised and constructed their own intracellular electrodes for their voltage clamp work, while the feedback amplifier had been built to Hodgkin's design.
16. Personal communication to the author by Sir Andrew Huxley.

Chapter 16: Aftermath

1. Nachmanson, D (1977). Nerve excitability: transition from descriptive phenomenology to chemical analysis of mechanisms. *Klinishce Wochenschrift*; **55**: 715–723.
2. Katz, B (1969). *The Release of Neural Transmitter Substances.* Liverpool: Liverpool University Press.
3. Overton, E (1902). Beiträge zur allgemeinen Muskel-und-Nervenphysiologie. *Pflügers Archiv fur die gesamte Physiologie des Menschen und der Tiere*; **92**: 346–386. Overton calculated that

during the life of a 70-year-old person with an average pulse rate of 70/s there would be more than 2.5 *billion* heartbeats. For more information on Overton and on subsequent research on the sodium pump, see Glynn, IM (2002). A hundred years of sodium pumping. *Annual Review of Physiology*; **64**: 1–18.

4. Dean, RB (1941). Theories of electrolyte equilibrium in muscle. *Biol Symp*; **3**: 331–348.
5. Lorente de Nó (1947). A study of nerve physiology, Parts 1 and 2. *Studies from the Rockefeller Institute for Medical Research*; **131, 132**. New York.
6. Sir Geoffrey Keynes' interesting and varied life is beautifully captured in his autobiography: Keynes, Sir Geoffrey (1985). *The Gates of Memory.* Oxford: Oxford University Press.
7. A treatment recently rediscovered and now hailed as a great therapeutic advance (!)
8. Hodgkin, Sir Alan (1992). *Chance and Design. Reminiscences of Science in Peace and War.* Cambridge: University of Cambridge Press.
9. Keynes, RD (1951). The leakage of radioactive potassium from stimulated nerve. *Journal of Physiology*; **113**: 99–114.
10. Keynes, RD (1951). The ionic movements during nervous activity. *Journal of Physiology*; **114**: 119–150.
11. Hodgkin, AL, & Keynes, RD (1955). The potassium permeability of a giant nerve fibre. *Journal of Physiology*; **128**: 61–88.
12. Hodgkin, AL, & Keynes, RD (1955). Active transport of cations in giant axons from *Sepia* and *Loligo. Journal of Physiology*; **128**: 28–60.
13. Caldwell, PC, Hodgkin, AL, Keynes, RD, & Shaw, TI (1960). The effects of injecting "energy-rich" phosphate compounds on the active transport of ions in the giant axons of *Loligo. Journal of Physiology*; **152**: 561–590.
14. Skou, JC (1957). The influence of some cations on an adenosine-triphosphatase from peripheral nerves. *Biochemica and Biophysica Acta*; **23**: 394–401.
15. See page 35.
16. Baker, PF, Hodgkin, AL, & Shaw, TI (1962). Replacement of the axoplasm of giant nerve fibres with artificial solutions. *Journal of Physiology*; **164**: 330–354.
17. Baker, PF, Hodgkin, AL, & Shaw, TI (1962). The effects of changes in internal ionic concentrations on the electrical properties of perfused giant nerve fibres. *Journal of Physiology*; **164**: 355–374.
18. Huxley, AF, & Stämpfli, R (1949). Evidence for saltatory conduction in peripheral myelinated nerve fibres. *Journal of Physiology*; **108**: 315–339. Unbeknownst to the Cambridge group, workers in Japan had already shown the existence of saltatory conduction in single nerve fibers. Since the Japanese work had been published during the 1939–45 war in a German journal, it did not come to light until considerably later. See Tasaki, I, & Takeuchi, T (1942). Weitere Studien über den Aktionstrom der markhaltigen Nervenfaser und über die elektrosaltatorische Übertragung des Nervenimpulses. *Pflügers Archiv fur die Gesamte Physiologie des Menschen und der Tiere*; **245**: 764–782.
19. Huxley, AF, & Stämpfli, R (1951). Direct determination of membrane resting and action potential in single myelinated nerve fibres. *Journal of Physiology*; **112**: 476–495.
20. Curare had been used by natives in the Amazon rain forest for many years, the poison being prepared from certain plants and used to coat arrows. In 1844, Claude Bernard, Professor of Experimental Physiology in Paris, showed that, in a frog paralyzed with curare, electrical stimulation of muscles could evoke twitches, even though nerve stimulation was ineffective.
21. See Kao, CY (1966). Tetrodotoxin, saxitonin and their significance in the study of excitation phenomena. *Pharmacological Reviews*; **18**: 997–1049.
22. Cook, JA (1777). *A Voyage Towards the South Pole and Around the World*; vol. **2**, pp. 112–113. London: Strahan & Cadell.
23. Moore, JW, Narahashi, T, & Shaw, TI (1967). An upper limit to the number of sodium channels in nerve membrane? *Journal of Physiology*; **188**: 89–105.
24. Keynes, RD, Ritchie, JM, & Rojas, E (1971). The binding of tetrodotoxin to nerve membranes. *Journal of Physiology*; **213**: 235–254.

25. Neumcke, B, & Stämpfli, R (1982). Sodium currents and sodium-current fluctuations in rat myelinated nerve fibres. *Journal of Physiology*; **329**: 163–184.
26. Assuming the exposed membrane at a node of Ranvier has a surface area of 30 μm.[2]
27. Values based on patch clamping. See chapter 19.

Chapter 17: Muscle: The New Physiology

1. Ling, G, & Gerard, RW (1949). The normal membrane potential of frog sartorius muscle. *Journal of Cellular and Comparative Physiology*; **34**: 383–396.
2. Nastuk, WL, & Hodgkin, AL (1950). The electrical activity of single muscle fibres. *Journal of Cellular and Comparative Physiology*: **35**: 35–73.
3. The first of a series of papers on cardiac muscle fibers was Draper, MH, & Weidmann, S (1951). Cardiac resting and action potentials recorded with an intracellular electrode. *Journal of Physiology*; **115**: 74–94.
4. Sjöstrand, FS, & Andersson, E (1954). Electronmicroscopy of the intercalated discs of cardiac muscle tissue. *Experientia*; **10**: 369–370.
5. Katz, B (1950). Depolarization of sensory terminals and the initiation of impulses in the muscle spindle. *Journal of Physiology*; **111**: 261–282.
6. Katz, B (1961). The terminations of the afferent fibre in the muscle spindle of the frog. *Philosophical Transactions of the Royal Society, A*; **243**: 221–240.
7. Katz's observations on the satellite cells went unnoticed and, instead, the cells were "discovered" by Mauro at about the same time: Mauro, A (1961). Satellite cell of skeletal muscle fibres. *Journal of Biophysical and Biochemical Cytology*; **9**: 493–495. Their importance for normal muscle fiber growth and for regeneration was first demonstrated by others a little later. See, for example, MacConnachie, HF, Enesco, M, & LeBlond, CP (1964). The mode of increase in the number of skeletal muscle nuclei in the postnatal rat. *American Journal of Anatomy*; **114**: 245–253.
8. See Figure 11–3, page 170.
9. See page 142, and also Brown, GL, Dale, HH, & Feldberg, W. (1936). Reactions of the normal mammalian muscle to acetylcholine and eserine. *Journal of Physiology*; **87**: 394–424.
10. Eccles, JC, Katz, B, & Kuffler, SW (1941). Nature of the "endplate" potential in curarized muscle. *Journal of Neurophysiology*; **4**: 362–387. Also: Eccles, JC, & Kuffler, SW (1941). Initiation of muscle impulses at neuro-muscular junction. *Journal of Neurophysiology*; **4**: 402–417.
11. Katz shared the 1970 Nobel Prize in Medicine or Physiology with the Swedes Ulf von Euler and Julius Axelrod.
12. Fatt, P, & Katz, B (1952). Spontaneous subthreshold activity of motor nerve endings. *Journal of Physiology*; **117**: 109–128.
13. Ernst Ruska won the Nobel Prize in Physics in 1986, some 53 years after he had developed the principle of the electron microscope and built the first machine. His original research had been on the improvement of the electron beam for the cathode ray oscilloscope. As in the case of Peyton Rouse (see p. 185), longevity was sometimes a requisite for the Prize.
14. See, for example, Robertson, JD (1959). The ultrastructure of cell membranes and their derivatives. *Biochemistry Society Symposia*; **16**: 3–43. De Robertis, in Argentina, was another of the early electron microscopists to study cell membranes.
15. Birks, R, Huxley, HE, & Katz, B (1960). The fine structure of the neuromuscular junction of the frog. *Journal of Physiology*; **150**: 134–144.
16. For an overview of excitation at the neuromuscular junction, see Katz, B (1969). *The Release of Neural Transmitter Substances*. Liverpool: Liverpool University Press, 60 pp.
17. Katz, B, & Miledi, R (1965). Propagation of electric activity in motor nerve terminals. *Proceedings of the Royal Society, B*; **161**: 453–482.
18. Katz, B, & Miledi, R (1967). The timing of calcium action during neuromuscular transmission. *Journal of Physiology*; **189**: 535–544.
19. Whittaker, VP, Michaelson, IA, & Kirkland, RJA (1964). The separation of synaptic vesicles from nerve-ending particles ("synaptosomes"). *Biochemical Journal*; **90**: 293–303.
20. Katz, B, & Miledi, R (1963). A study of spontaneous miniature potentials in spinal motoneurones. *Journal of Physiology*; **163**: 389–422.

21. See Figure 12–5, page 181.
22. As described in Huxley, Sir Andrew (1980). *Reflections on Muscle.* Princeton, NJ: Princeton University Press, 111 pp.
23. Huxley, TH (1879). *The Crayfish. An Introduction to the Study of Zoology.* London: Negan Paul.
24. See page 205.
25. Krause, W (1873). Die Contraction der Muskelfasser. *Pflügers Archiv fur die Gesamte Physiologie des Menschen und der Tiere*; **7**: 508–514.
26. See page 163.
27. Fenn, WO (1923). A quantitative comparison between the energy liberated and the work performed by the isolated sartorius muscle of the frog. *Journal of Physiology*; **58**: 175–203.
28. Eggleton, P, & Eggleton, GP (1927). The inorganic phosphate and a labile form of organic phosphate in the gastrocnemius of the frog. *Biochemical Journal*; **21**: 190–195.
29. Lohmann, K (1929). Über die fermentative Kohlenhydrat-Phosphorsäurenveresterung in Gegenwart von Fluorid, Oxalat und Citrat. *Klinische Wochenschrift*; **8**: 2009.
30. Engelhardt, VA, & Lyubimova, MN (1939). Myosine and adenosinetriphosphatase. *Nature*; **144**: 668–669.
31. Huxley, AF, & Niedergerke, R (1954). Structural changes in muscle during contraction. Interference microscopy of living muscle fibres. *Nature*; **173**: 971–973.
32. Huxley, HE (1996). A personal view of muscle and motility mechanisms. *Annual Review of Physiology*; **58**: 1–19.
33. Huxley, HE (1953). X-ray analysis and the problem of muscle. *Proceedings of the Royal Society of London, B*; **141**: 69–62.
34. Huxley, HE, & Hanson J (1954). Changes in the cross-striations of muscle during contraction and stretch and their structural interpretation. *Nature*; **173**: 973–976.
35. Huxley, HE (1957). The double array of filaments in cross-striated muscle. *Journal of Biophysical and Biochemical Cytology*; **3**: 631–648.
36. Gordon, AM, Huxley, AF, & Julian, FJ (1966). The variation in isometric tension with sarcomere length in vertebrate muscle fibres. *Journal of Physiology*; **184**: 170–192.
37. Huxley, AF, & Taylor, RE (1958). Local activation of striated muscle fibres. *Journal of Physiology*; **144**: 426–441.
38. Veratti, E (1902). Richerche sulla fine structtura della fibra muscolare striata. *Memorie Reale Istituto Lombardi*; **19**: 87–133.

Chapter 18: More Triumphs with Microelectrodes

1. Details of Eccles' life are given in Eccles, JC (1977). My scientific odyssey. *Annual Review of Physiology*; **39**: 1–18, and also in Curtis, DR, & Andersen, P (2001). Sir John Eccles, AC. January 27, 1903–May 2, 1997. *Biographical Memoirs of Fellows of the Royal Society*; **47**: 160–187.
2. Sir Karl Popper (1902–1994) was born in Austria, escaping Hitler to pursue his career as a philosopher in New Zealand and then in the United Kingdom. He is best known for his work on the scientific method, insisting that hypotheses can never be verified and can only remain tenable through not having been proven false. A corollary is that a hypothesis must be capable of being tested.
3. Buller, AJ, Eccles, JC, & Eccles, RM (1960). Interactions between motoneurones and muscle in respect of the characteristic speeds of their responses. *Journal of Physiology*; **150**: 417–439.
4. Brock, LG, Coombs, JS, & Eccles, JC (1952). The recording of potentials from motoneurones with an intracellular electrode. *Journal of Physiology*; **117**: 431–460.
5. Florey's life and achievements are the subject of a fine biography by Gwyn Macfarlane (1979). *Howard Florey. The Making of a Great Scientist.* Oxford: Oxford University Press, 369 pp. Florey was an Australian who achieved great things as Professor of Experimental Pathology at Oxford, quite apart from producing penicillin. Yet in spite of his many honors, including the Presidency of the Royal Society, he remained almost unknown to the general public and, for that matter, to the scientific community at large.

6. Eccles, JC (1953). *The Neurophysiological Basis of Mind: The Principles of Neurophysiology*. Oxford: Clarendon Press.
7. See Figure 17–3, page 243.
8. See page 109 and Gray, Sir John (1990). Bryan Harold Cabot Matthews, June 14, 1906–July 23, 1986. *Biographical Memoirs of Fellows of the Royal Society*; **35**: 265–279.
9. Lloyd, DPC (1946). Facilitation and inhibition of spinal motoneurons. *Journal of Neurophysiology*; **9**: 421–438.
10. Woodbury, JW, & Patton, HD (1952). Electrical activity of single spinal cord elements. *Cold Spring Harbor Symposia in Quantitative Biology*; **17**: 185–188.
11. Now Dr. Rose Mason.
12. Eccles, JC, Fatt, P, & Koketsu, K (1954). Cholinergic and inhibitory synapses in a pathway from motor-axon collaterals to motoneurones. *Journal of Physiology*; **126**: 524–562.
13. In recent years, some exceptions have been found to Dale's Principle. For example, neuropeptides may be released, together with a classical transmitter, from the same nerve terminal. See, for example, Schultzberg, M, Hökfelt, T, & Lundberg, JM (1982). Coexistence of classical transmitters and peptides in the central and peripheral nervous systems. *British Medical Bulletin*; **38**: 309–313.
14. Coombs, JS, Eccles, JC, & Fatt, P (1955). The electrical properties of the motoneuronal membrane. *Journal of Physiology*; **130**: 291–325.
    Coombs, JS, Eccles, JC, & Fatt, P (1955). The specific ionic conductances and the ionic movements across the motoneuronal membrane that produce the inhibitory post-synaptic potential. *Journal of Physiology*; **130**: 326–373.
    Coombs, JS, Eccles, JC, & Fatt, P (1955). Excitatory synaptic action in motoneurones. *Journal of Physiology*; **130**: 374–395.
    Coombs, JS, Eccles, JC, & Fatt, P (1955). Inhibitory suppression of reflex discharges from motoneurones. *Journal of Physiology*; **130**: 396–413.
15. Eccles, JC (1962). Central connections of muscle afferent fibres. In *Muscle Receptors* (ed. D. Barker), pp 81–101. Hong Kong University Press.
16. Andersen, P, Eccles, JC, Oshima, T, & Schmidt, RF (1965). Mechanisms of synaptic transmission in the cuneate nucleus. *Journal of Neurophysiology*; **27**: 1096–1116.
17. Andersen, P, Eccles, JC, & Løyning, Y (1964). Pathway of postsynaptic inhibition in the hippocampus. *Journal of Neurophysiology*; **27**: 608–619.
18. Eccles, JC, Ito, M, & Szentàgothai, J (1967). *The Cerebellum as a Neuronal Machine*. New York: Springer-Verlag.

Chapter 19: The Single Ion Channel

1. See page 221.
2. Takeuchi, A, & Takeuchi, N (1959). Active phase of frog's end-plate potential. *Journal of Neurophysiology*; **22**: 395–411.
3. See page 237.
4. Cole, KS (1982). Squid axon membrane: impedance decrease to voltage clamp. *Annual Review of Neuroscience*; **5**: 305–323.
5. See page 236 for further information on this toxin.
6. See page 325 for more on Dawson and signal averaging.
7. Armstrong, CM, & Bezanilla, FM (1974). Charge movement associated with the opening and closing of the activation gates of the Na channels. *Journal of General Physiology*; **63**: 675–689.
8. Keynes, RD, & Rojas, E (1974).Kinetics and steady-state properties of the charged system controlling sodium conductance in the squid giant axon. *Journal of Physiology*; **239**: 393–434.
9. Katz, B, & Miledi, R (1970). Membrane noise produced by acetylcholine. *Nature*; **226**: 962–963.
10. Katz, B, & Miledi, R (1971). Further observations on acetylcholine noise. *Nature*; **232**: 124–126.

11. Neher, E, & Sakmann, B (1976). Noise analysis of drug-induced voltage clamp currents in denervated frog muscle fibres. *Journal of Physiology*; **258**: 705–729.
12. Neher, E, & Sakmann, B (1976). Single-channel currents recorded from membrane of denervated frog muscle fibres. *Nature*; **260**: 779–802.
13. Hamill, OP, Marty, A, Neher, E, Sakmann, B, & Sigworth, FJ (1981). Improved patch-clamp techniques for high-resolution current recording from cells and cell-free membrane patches. *Pflügers Archiv fur die Gesamte Physiologie des Menschen und der Tiere*; **391**: 85–100.
14. See Figure 20–5, page 227, for examples of ion channel opening.
15. Bertil Hille is the author of the definitive book on ion channels: Hille (2001). *Ionic Channels of Excitable Membranes* (3rd ed). Sunderland, MA: Sinauer. He has himself worked extensively in the field, particularly on Na and K channels, and on their functional modifications by other metallic ions and by pharmacological agents.
16. Hille, B (1989). Ionic channels: evolutionary origins and modern roles. *Quarterly Journal of Experimental Physiology*; **74**: 785–804.
17. For example: Bormann, J, Hamill, OP, & Sakmann, B (1987). Mechanism of anion permeation through channels gated by glycine and gamma-aminobutyric acid in mouse cultured spinal neurons. *Journal of Physiology*; **385**: 243–286.
18. See Figure 2–7, page 21, for illustration of the electric organ of the ray.
19. Raftery, MA, Hunkapiller, MW, Strader, CD, & Hood, LE (1980). Acetylcholine receptor: complex of homologous subunits. *Science*; **208**: 1454–1457.
20. Noda, M, Furutani, T, Takahashi, H, Toyosato, M, Tanabe, T, et al. (1983). Cloning and sequence analysis of calf cDNA and human genomic DNA encoding a-subunit precursor of muscle acetylcholine receptor. *Nature*; **305**: 818–823.
21. The same idea, but using a different source of mRNA, had occurred independently to Eric Barnard, both Miledi and Barnard publishing accounts in 1982, and Barnard taking out a patent for this technology.
22. Toyoshima, C, & Unwin, N (1988). Ion channel of acetylcholine receptor reconstructed from images of postsynaptic membranes. *Nature*; **336**: 247–250.
23. Noda, M, Shimizu, S, Tanabe, T, Takai, T, Kayano, T, et al. (1984). Primary structure of *Electrophorus electricus* sodium channel deduced from cDNA sequence. *Nature*; **312**: 121–127.
24. For a review of the approaches used in determining the functions of the different parts of the Na channel, see Hille, B (2001; see Note 15).
25. Armstrong, C, & Bezanilla, F (1977). Inactivation of the sodium channel. II. Gating current experiments. *Journal of General Physiology*; **70**: 567–590.
26. See Note 15.

Chapter 20: Myotonic Goats and Migraines

1. Sir Andrew Huxley, personal communication.
2. In 1974 the author was invited to the farm of Dr. Virgil LeQuire outside Nashville, Tennessee, to inspect the owner's flock of myotonic goats. It was while he was there that Dr. LeQuire kindly gave him a copy of the following letter, which tells the story of the hired hand and his peculiar animals:

> Dear Mr Goode:
>
> I do not recall the exact date, but early in the 1880s, a stranger appeared one day at the home of my neighbour, Mr J M Porter. Besides the clothes on his back, the man's only possession was a sacred cow, three nanny goats, and one billy. Mr Porter invited them all to stay and they all did.
>
> During the months the man was there, the Porters never succeeded in finding out from whence the stranger had come. It was believed by some, however, that he had arrived, rather circuitously, from Nova Scotia. Many things about the man mystified the Porters but they were most impressed I think by the fact that his goats were subject

to strange fits or fainting spells, the like of which had never been seen before in these parts.

The goats were indeed interesting, and one day after I had seen one of these incredible attacks for myself, I offered the man $36.00 for his goats. At that time, he refused to even consider selling them, but he did promise to let me know if he ever decided to part with them.

About a month later, accompanied as usual by his cow and his goats, he appeared at my home and said he would sell the goats. I paid him for them, and both they and their former owner settled down at my place, the stranger putting in his time working on the farm. Not once while he was there did the stranger ever eat at my table. He took all his meals in the barn where the sacred cow was kept.

At the end of about three weeks, the stranger and his sacred cow left, going over to Lick Creek, a little town in Maury County, Tennessee. There he promptly married an old lady by the name of Barnhill. On her farm that summer the stranger raised an excellent corn crop. But one night after the crop was in, without any warning to his wife, he left with the sacred cow and was never heard of again.

From the goats which I bought from this man I raised a number of other goats and sold them in different parts of Kentucky and Tennessee. These goats were, I am sure, the progenitors of all such goats in this section of the country, and from which you say it is quite possible that the entire breed originates from this source.

Sincerely yours,
Dr H H Mayberry

3. Clark, SM, Luton, FH, & Cutler, JT (1939). A form of congenital myotonia in goats. *Journal of Nervous and Mental Disease*; **90**: 297–309.
4. Kolb, LC (1938). Congenital myotonia in goats. Description of the disease. The effect of quinine, various cinchona alkaloids and salts upon the myotonic symptom. *Johns Hopkins Hospital Bulletin*; **63**: 221–237.
5. Kolb, LC, Harvey, AM, & Whitehill, MR (1938). A clinical study of myotonic dystrophy and myotonia congenita with special reference to the therapeutic effect of quinine. *Johns Hopkins Hospital Bulletin*; **62**: 188–215.
6. Brown, GL, & Harvey, AM (1939). Congenital myotonia in the goat. *Brain*; **62**: 341–363.
7. See page 111 for more on the Matthew's oscillograph.
8. See page 105.
9. MacIntosh, FC, & Paton, WDM (1974). George Lindor Brown. 1903–1971. *Biographical Memoirs of Fellows of the Royal Society*; **20**; 41–73.
10. Bryant, SH (1969). Cable properties of external intercostal muscle fibres from myotonic and nonmyotonic goats. *Journal of Physiology*; **294**; 539–550.
11. Bryant, SH (1971). Chloride conductance in normal and myotonic muscle fibres and the action of monocarboxylic aromatic acids. *Journal of Physiology*; **219**: 367–383.
12. Adrian, RH, & Bryant, SH (1974). On the repetitive discharge in myotonic muscle fibres. *Journal of Physiology*; **240**: 505–515.
13. See page 240 for Andrew Huxley's experimental observations.
14. Thomsen, J (1876). Tonische krämpfe in willkürlich beweglichen muskeln in Folge von erebter Physischer-Disposition (Ataxia muscularis?). *Archiv fur Psychiatrie und Nervenkrankheiten*; **6**: 702–718.
15. The author of the story was the same Mr. Goode who had received Dr. Mayberry's letter (see Note 2).
16. Koch, MC, Steinmeyer, K, Lorenz, C, Ricker, K, Wolf, F, et al. (1992). The skeletal muscle chloride channel in dominant and recessive myotonia. *Science*; **257**: 797–800.
17. See Talbott, JH (1941). Periodic paralysis. *Medicine (Baltimore)*; **20**: 85–143.
18. Singer, HD, & Goodbody, FW (1901). A case of family periodic paralysis with a critical digest of the literature. *Brain*; **24**: 257–285.

19. Goldflam was probably the first to try stimulating the muscle fibers directly (actually, through their intramuscular nerve twigs), and to show the loss of excitability. Similar but less marked changes were found between attacks as well, indicating the presence of more permanent changes in the fibers. See Goldflam (1891). Ueber eine eigenthümlicher Form von periodischer, familiärer, wahrschleinlich auto-intoxicatorischer Paralyse. *Zeitschrift fur Klinische Medecin, suppl*; **240**: 242.
20. Biemond and Daniels were the first to detect a fall in serum potassium during attacks, while a rise in potassium was emphasized in a particularly thorough and extensive study by Ingrid Gamstorp in Denmark. See Biemond, A, & Daniels, AP (1934). Familial periodic paralysis and its transition into spinal muscular atrophy. *Brain*; **57**: 91–108. Also: Gamstorp, I (1956). Adynamia episodica hereditaria. *Acta Paediatrica*: **45,** suppl. 108: 1–126.
21. See page 208.
22. Brooks, JE (1969). Hyperkalaemic periodic paralysis. Intracellular EMG studies. *Archives of Neurology*; **20**: 13–18.
23. McComas, AJ, Mrozek, K, & Bradley, WG (1968). The nature of the electrophysiological defect in adynamia episodica. *Journal of Neurology, Neurosurgery & Psychiatry*; **31**: 448–452.
24. Lehmann-Horn, F, Rüdel, R, Ricker, K, Lorkovic, H, Dengler, R, & Hopf, HC (1983). Two cases of adynamia episodica hereditaria: in vitro investigation of muscle membrane and contractile parameters. *Muscle and Nerve*; **6**: 113–121.
25. Barchi, RL (1995). Molecular pathology of the skeletal muscle sodium channel. *Annual Review of Physiology*; **57**: 355–385.
26. Jurkat-Rott, K, Lehmann-Horn, F, Elbz, A, Heine, R, Gregg, RG, et al. (1994). A calcium channel mutation causing hypokalemic period paralysis. *Human Molecular Genetics*; **3**: 1415–1419.
27. The paper in question was Patrick, J, & Lindstrom, J (1973). Autoimmune response to acetylcholine receptor. *Science*; **180**: 871–872, while the neuroimmunologist was Dr. Vanda Lennon.
28. Ophoff, RA, Terwindt, GM, Vergouwe, MN, van Eijk, R, Oefner, PJ, et al. (1996). Familial hemiplegic migraine and episodic ataxia type-2 are caused by mutations in the Ca^2+ channel gene *CACNL1A4*. *Cell*; **87**: 543–552.
29. Zhuchenko, O, Bailey, J, Bonnen, P, Ashizawa, T, Stockton, DW, et al. (1997). Autosomal dominant cerebellar ataxia (SCA6) associated with small polyglutamine expansions in the alpha 1A-voltage-dependent calcium channel. *Nature Genetics*; **15**: 62–69.
30. For a review of the entire field of channelopathies, see Ashworth, FM (2000). *Ion Channels and Disease*. San Diego: Academic Press, 481 pp.

Chapter 21: The Swinging Gate

1. Lorente de Nó, R (1949). On the effect of certain quaternary ammonium ions upon frog nerve. *Journal of Cellular and Comparative Physiology*; **33**, suppl.: 1–231.
2. Tempel, BL, Papazian, DM, Schwarz, TL, Jan, YN, & Jan, LY (1987). Sequence of a probable potassium channel component encoded at *Shaker* locus of *Drosophila*. *Science*; **237**: 770–775.
3. Kamb, A, Iversen, LE, & Tanouye, MA (1987). Molecular characterization of *Shaker*, a Drosophila gene that encodes a potassium channel. *Cell*; **50**; 405–413.
4. MacKinnon, R, & Miller, C (1988). Mechanism of charybdotoxin in block of the high-conductance, Ca^2+- activated K+ channel. *Journal of General Physiology*; **91**: 335–349.
5. MacKinnon, R (1991). Determination of the subunit stoichiometry of a voltage-activated potassium channel. *Nature*; **350**: 232–235.
6. MacKinnon's unusual scientific path is given in his Nobel Lecture: MacKinnon, R (2003). Potassium channels and the atomic basis of selective ion conduction. *Les Prix Nobel. The Nobel Prizes 2003*. Tore Frängsmyr, ed. Stockholm: Nobel Foundation, 2004. Additional information appears in his accompanying autobiography. A video of the lecture can be found on the Internet: http://nobelprize.org/nobel_prizes/chemistry/laureates/2003/mackinnon-lecture.html
7. Related to Alan Hodgkin through marriage to Hodgkin's cousin, Thomas.

8. Bernal, JD, & Crowfoot, D (1934). X-ray photographs of crystalline pepsin. *Nature*; **133**: 795–797.
9. Crowfoot, D, Bunn, CW, Rogers-Low, BW, & Turner-Jones, A (1949). X-ray crystallographic investigation of the structure of penicillin. In *Chemistry of Penicillin*. Princeton: Princeton University Press, 300 pp. An account of these pioneering studies, carried out with meager funding, is given in Dorothy Crowfoot Hodgkin's 1964 Nobel Lecture: The X-ray analysis of complicated molecules. *Nobel Lectures, Chemistry 1963–1970*. Amsterdam: Elsevier, 1972.
10. Perutz, MF (1962). Nobel Lecture. X-ray analysis of haemoglobin. *Nobel Lectures, Chemistry 1942–1962*. Amsterdam: Elsevier. 1964. The lecture includes an excellent account of the principles used in analyzing the diffraction pictures.
11. Watson, JD (1968). *The Double Helix. A Personal Account of the Discovery of the Structure of DNA*. New York: Athenaeum.
12. For a biography of Rosalind Franklin, including her difficult professional dealings with Watson and Crick, see Maddox, B (2002). *Rosalind Franklin. The Dark Lady of DNA*. London: Harper Collins.
13. See Note 6.
14. See page 257 for further information on Popper.
15. Doyle, DA, Cabral, JM, Pfuetzner, RA, et al. (1998). The structure of the potassium channel: molecular basis of K+ conduction and selectivity. *Science*; **280**: 69–77.
16. Miller, C (2001). See potassium run. *Nature*; **414**: 23–24.
17. Roux, D, & MacKinnon, R (1999). The cavity and pore helices in the Kcs K+ channel: electrostatic stabilization of monovalent cations. *Science*; **285**: 100–102.
18. Zhou, M, Morais-Cabral, JH, Mann, S, & MacKinnon, R (2001). Potassium channel receptor site for the inactivation gate and quaternary amine inhibitors. *Nature*; **411**: 657–662.
19. Zhou, Y, Morais-Cabral, JH, Kaufman, A, & MacKinnon, R (2001). Chemistry of ion coordination and hydration revealed by a K+ channel-Fab complex at 2.0 Å resolution. *Nature*; **414**: 43–48.
20. Morais-Cabral, JH, Zhou, Y, & MacKinnon, R (2001). Energetic optimization of ion conduction rate by the K+ selectivity filter. *Nature*; **414**: 37–42.
21. Ruta, V, Jiang, Y, Lee, A, Chen, J, & MacKinnon, R (2003). Functional analysis of an archaebacterial voltage-dependent K+ channel. *Nature*; **422**: 180–185.
22. Jiang, Y, Lee, A, Chen, J, Ruta, V, Cadene, M, Chalt, BT, & MacKinnon, R (2003). X-ray structure of a voltage-dependent K+ channel. *Nature*; **423**: 33–41.
23. Jiang, Y, Ruta, V, Chen, J, Lee, A, & MacKinnon, A (2003). The principle of gating charge movement in a voltage-dependent K+ channel. *Nature*; **423**: 42–48.
24. Watson, JD, & Crick, FHC (1953). A structure for deoxyribonucleic acid. *Nature*; **171**: 737–738.
25. Sigworth, FJ (2003). Life's transistors. *Nature*; **423**: 21–22.
26. See Note 6.

Chapter 22: Departures

1. This is no longer quite true. There is now an oil painting, in the entrance foyer to the Department of Physiology, depicting the elderly Sherrington with a youthful Eccles. The painting is based on a photograph of the two, taken in the back garden of Sherrington's Ipswich home after his retirement (see Figure 9–9, page 144).
2. Tansey, EM (2008). Working with Cambridge physiologists. *Notes and Records of the Royal Society of London*; **62**: 131–137. The article is an interview with Mr. Clive Hood, a long-serving technician in the Physiological Laboratory at Cambridge, and it is Mr. Hood's description of Adrian's laboratory that is quoted.
3. Adrian papers, Trinity College Library. The tap had been left running by a graduate student, whose subsequent career was not adversely affected, at least in the long term—Dr. Jared Diamond, the physiologist and Pulitzer Prize–winning author of *Guns, Germs and Steel* and other books.

4. See *The Photograph*, page 255.
5. Seen by the author, but no reference obtainable.
6. There is now a third generation of physiological Matthews, in the form of Dr. Hugh Matthews, currently at Cambridge as a Fellow of St. John's College and a member of the Department of Physiology, Development and Neuroscience.
7. Krnjevic, K, & Phillis, JW (1963). Iontophoretic studies of neurons in the mammalian cerebral cortex. *Journal of Physiology*; **165**: 274–303.
8. Gasser, HS (1958). Comparison of the structure, as revealed with the electron microscope, and the physiology of the unmedullated fibrers in the skin nerves and in the olfactory nerves. *Experimental Cell Research*; **Suppl 5**: 3–17.
9. Gasser, HS (1960). Effect of the method of leading on the recording of the nerve fiber spectrum. *Journal of General Physiology*; **43**: 927–940.
10. Erlanger, J, & Schoepfle, GM (1951). Observations on the local response in single medullated nerve fibres. *American Journal of Physiology*; **167**: 134–146.
11. Davis, H (1970). Joseph Erlanger. January 5, 1874–December 5, 1965. *Biographical Memoirs. National Academy of Sciences (USA)*; **41**: 111–139.
12. See page 317.
13. Landau, WM (1985). George Holman Bishop. June 27, 1889–October 11, 1973. *Biographical Memoirs. National Academy of Sciences (USA)*; **55**: 45–66.
14. Marshall, LH (1983). The fecundity of aggregates: the axonologists at Washington University. *Perspectives in Biology and Medicine*; **26**: 613–636.
15. Woolsey, TA (2001). Rafael Lorente de Nó. April 8, 1902–April 2, 1990. *Biographical Memoirs. National Academy of Sciences*; **79**: 85–105.
16. See page 37 for an account of the meeting in Burdon Sanderson's house.
17. Cole, KS (1982). Squid axon membrane: impedance decrease to voltage clamp. *Annual Review of Neuroscience*; **5**: 305–323.
18. Huxley, Sir Andrew (1992). Kenneth Stewart Cole. July 10, 1900–April 18, 1984. *Biographical Memoirs of Fellows of the Royal Society*; **38**: 98–110.
19. See pages 218–221 for Hodgkin's later experiments on the squid giant axon.
20. Eccles, JC, & Popper KR (1977). *The Self and its Brain. An Argument for Interactionism.* Berlin: Springer International.
21. Eyzaguirre, C, & Kuffler, SW (1955). Processes of excitation in the dendrites and in the soma of single isolated sensory nerve cells of the lobster and crayfish. *Journal of General Physiology*; **39**: 87–119.
22. Kuffler, SW (1946). Inhibition at the nerve-muscle junction in Crustacea. *Journal of Neurophysiology*; **9**: 337–346.
23. See page 322–324 for an account of Hubel and Wiesel's work.

Chapter 23: Postscript

(The references given are an extremely small sample of the enormous bibliography available and should be viewed as representative examples of the particular investigator's work.)

1. Burke, RE (1967). Motor unit types of cat triceps surae muscle. *Journal of Physiology*; **193**: 141–160.
2. Gillespie, CA, Simpson, DR, & Edgerton, VR (1974). Motor unit recruitment as reflected by muscle fibre glycogen loss in a prosimian (bushbaby) after running and jumping. *Journal of Neurology, Neurosurgery and Psychiatry*; **37**: 817–824.
3. Henneman, E, Somjen, G, & Carpenter, DO (1965). Functional significance of cell size in spinal motoneurones. *Journal of Neurophysiology*; **28**: 560–580.
4. Granit, R, Kernell, D, & Smith, RS (1963). Delayed depolarization and the repetitive response to intracellular stimulation of mammalian motoneurones. *Journal of Physiology*; **168**: 890–910.
5. Jankowska, E, & McCrea, DA (1983). Shared reflex pathways from Ib tendon organ afferents and Ia muscle spindles afferents in the cat. *Journal of Physiology*; **338**: 99–112.

6. Phillips, CG (1956). Intracellular records from Betz cells in the cat. *Quarterly Journal of Experimental Physiology*; **41**: 58–69.
7. Krnjevic, K, & Phillis, JW (1963). Iontophoretic studies of neurones in the mammalian cerebral cortex. *Journal of Physiology*; **165**: 274–303.
8. Dahlström, A, & Fuxe, K (1964). Evidence for the existence of monoamine containing neurons in the central nervous system. I: Demonstration of monoamines in the cell bodies of brain stem neurons. *Acta Physiologica Scandinavica*; **62**, suppl. 232: 1–55.
9. Woolsey, CN, Marshall, WH, & Bard, P (1942). Representation of cutaneous tactile sensibility in the cerebral cortex of the monkey as indicated by evoked potentials. *Bulletin of the Johns Hopkins Hospital*; **70**: 399–441.
10. Mountcastle, VB (1957). Modality and topographic properties of single neurons of cat's somatic sensory cortex. *Journal of Neurophysiology*; **20**: 408–434.
11. Lorente de Nó, R (1938). The cerebral cortex: architecture, intracortical connections and motor projections. In Fulton JF. *Physiology of the Nervous System*. London: Oxford University Press, pp. 291–325.
12. See page 327 for this period of Kuffler's life.
13. Kuffler, SW (1953). Discharge patterns and functional organization of the mammalian retina. *Journal of Neurophysiology*; **16**: 37–68.
14. Li, C-L, & Jasper, H (1953). Microelectrode studies of the electrical activity of the cerebral cortex in the cat. *Journal of Physiology*; **121**: 117–140.
15. Or was it a "fine dark line caused by a crack on one of their slides"? See: Barlow, HB (1982). David Hubel and Torsten Wiesel. Their contributions towards understanding the primary visual cortex. *Trends in Neurosciences*; **5**: 145–152.
16. Hubel, DH, & Wiesel, TN (1959). Receptive fields of single neurons in the cat's striate cortex. *Journal of Physiology*; **148**: 574–591. See also Hubel, DH. Evolution of ideas on the primary visual cortex, 1955–1978: a biased historical account. *Les Prix Nobel. The Nobel Prizes 1981*, ed. W Oldenberg. Stockholm: Nobel Foundation, 1982.
17. Hubel, DH, & Wiesel, TN (1970). The period of susceptibility to the physiological effects of unilateral eye closure in kittens. *Journal of Physiology*; **206**: 419–436.
18. Evarts, EV, & Tanji, J (1976). Reflex and intended responses in motor cortex pyramidal tract neurons of monkeys. *Journal of Neurophysiology*; **39**: 169–180.
19. McIlwain, H, Buchel, L, & Cheshire, JX (1951). The inorganic phosphate and phosphocreatine of brain especially during metabolism in vitro. *Biochemistry Journal*; **48**: 12–20.
20. Hillman, HH, & McIlwain, H (1961). Membrane potentials in mammalian cerebral tissue in vitro; dependence on ionic environment. *Journal of Physiology*; **157**: 263–278.
21. In fact, Choh-Luh Li had already demonstrated the electrophysiological viability of brain slices while working in McIlwain's laboratory. See Li, C-L, & McIlwain, H (1957). Maintenance of resting potentials in slices of mammalian cerebral cortex and other tissues in vitro. *Journal of Physiology*; **139**: 178–190.
22. Hoffmann, P (1918). Uber die Beziehungen der Sehnenreflexe sur willkürlichen bewegung und zum Tonus. *Zeitschrift fur Biologie*; **68**: 351–370.
23. Vallbo, ÅB, Hagbarth, K-E, Torebjörk, HE, & Wallin, BG (1979). Somatosensory, proprioceptive and sympathetic activity in human peripheral nerves. *Physiological Reviews*; **59**: 919–957.
24. Dawson, GD (1947). Cerebral responses to electrical stimulation of peripheral nerve in man. *Journal of Neurology, Neurosurgery and Psychiatry*; **10**: 137–140.
25. Dawson, GD (1954). Cerebral responses to electrical stimulation of peripheral nerve in man. *Electroencephalography and Clinical Neurophysiology*; **6**: 65–84.
26. Flexner, A (1910). *Medical Education in the United States and Canada*. New York: Carnegie Foundation.

BIBLIOGRAPHY

Adrian, ED (1912). On the conduction of subnormal disturbances in normal nerve. *Journal of Physiology*; **45**: 389–412.

Adrian, ED (1914). The relation between the size of the propagated disturbance and the rate of conduction in nerve. *Journal of Physiology*; **48**: 53–72.

Adrian, ED (1916). The recovery of conductivity and of excitability in nerve. *Journal of Physiology*; **50**: 345–363.

Adrian, ED (1916). The electrical reactions of muscles before and after nerve injury. *Brain*; **39**: 1–33.

Adrian, ED (1921). The recovery process of excitable tissues. I. *Journal of Physiology*; **54**: 1–31.

Adrian, ED (1921). The recovery process of excitable tissues. II. *Journal of Physiology*; **55**: 193–225.

Adrian, ED (1926). The impulses produced by sensory nerve-endings. IV. Impulses from pain receptors. *Journal of Physiology*; **62**: 33–51.

Adrian, ED (1928). *The Basis of Sensation. The Action of the Sense Organs.* London: Christophers.

Adrian, ED (1932). *The Mechanism of Nervous Action.* Oxford: Oxford University Press.

Adrian, ED (1934). In Fisher, WM, & Keith-Lucas, A, eds. *Keith Lucas*; chapter 7, p. 87. Cambridge: Heffer.

Adrian, ED (1943). Afferent areas in the brains of ungulates. *Brain*; **66**: 89–103.

Adrian, ED (1943). Afferent areas in the cerebellum connected with the limbs. *Brain*; **66**: 289–315.

Adrian, ED (1943). Discharges from vestibular receptors in the cat. *Journal of Physiology*; **107**: 389–407.

Adrian, ED (1943). Sensory areas of the brain. *Lancet*; July: 33–40.

Adrian, ED (1943). The dominance of vision. *Transactions of the Ophthalmological Society of the United Kingdom*; **63**: 194–207.

Adrian, ED (1947). *The Physical Background of Perception.* Oxford: Clarendon Press.

Adrian, ED (1953). Sensory messages and sensation: The response of the olfactory organ to different smells. *Acta Physiologica Scandinavica*; **29**: 4–14.

Adrian, Lord (1954). Memorable experiences in research. *Diabetes*; **3**: 17–18.

Adrian, Lord (1964). Herbert Spencer Gasser, 1888–1963. *Biographical Memoirs of the Royal Society*; **10**: 75–82.

Adrian, ED (1965). Alexander Forbes 1882–1965. *Electroencephalography and Clinical Neurophysiology*; **19**: 109–111.

Adrian, ED, & Bronk, DW (1928). The discharge of impulses in motor nerve fibres. I. Impulses in single fibres of the phrenic nerve. *Journal of Physiology*; **66**: 81–101.

Adrian, ED, & Bronk, DW (1929). The discharge of impulses in motor nerve fibres. II. The frequency of discharge in reflex and voluntary contractions. *Journal of Physiology*; **67**: 119–151.

Adrian, E, & Lucas, K (1912). On the summation of propagated disturbances in nerve and muscle. *Journal of Physiology*; **44**: 68–124.

Adrian, ED, & Matthews, BHC (1934). The Berger rhythm: potential changes from the occipital lobes in man. *Brain*; **57**: 355–385.

Adrian, ED, & Yealland, LR (1917). The treatment of some common war neuroses. *Lancet* (June): 3–24.

Adrian, ED, & Zotterman, Y (1926). The impulses produced by sensory nerve endings. Part 2. The response of a single end-organ. *Journal of Physiology*; **61**: 151–171.

Adrian, ED, & Zotterman, Y (1926). The impulses produced by sensory nerve-endings. III. Impulses set up by touch and pressure. *Journal of Physiology*; **61**: 465–483.

Adrian, RH, & Bryant, SH (1974). On the repetitive discharge in myotonic muscle fibres. *Journal of Physiology*; **240**: 505–515.

Andersen, P, Eccles, JC, & Løyning, Y (1964). Pathway of postsynaptic inhibition in the hippocampus. *Journal of Neurophysiology*; **27**: 608–619.

Andersen, P, Eccles, JC, Oshima, T, & Schmidt, RF (1965). Mechanisms of synaptic transmission in the cuneate nucleus. *Journal of Neurophysiology*; **27**: 1096–1116.

Armstrong, C, & Bezanilla, F (1977). Inactivation of the sodium channel. II. Gating current experiments. *Journal of General Physiology*; **70**: 567–590.

Armstrong, CM, & Bezanilla, FM (1974). Charge movement associated with the opening and closing of the activation gates of the Na channels. *Journal of General Physiology*; **63**: 675–689.

Ashcroft, FM (2000). Ion Channels and Disease. San Diego: Academic Press.

Avery, OT, MacLeod, CM, & McCarty, M (1944). Studies on the chemical nature of the substance inducing transformation on pneumococcal type I: Induction of transformation by a desoxyribonucleic acid fraction isolated from pneumococcus type III. *Journal of Experimental Medicine;* **79**: 137–158.

Bacq, ZM (1975). *Chemical Transmission of Nerve Impulses. A Historical Sketch.* Oxford: Pergamon Press, 106 pp.

Baker, PF, Hodgkin, AL, & Shaw, TI (1962). The effects of changes in internal ionic concentrations on the electrical properties of perfused giant nerve fibres. *Journal of Physiology*; **164**: 355–374.

Barchi, RL (1995). Molecular pathology of the skeletal muscle sodium channel. *Annual Review of Physiology*; **57**: 355–385.

Barlow, HB (1986). William Rushton. December 8, 1901–June 21, 1980. *Biographical Memoirs of Fellows of the Royal Society*; **32**: 423–459.

Bennett, MR (1999). The early history of the synapse: From Plato to Sherrington. *Brain Research Bulletin*; **50**: 95–118.

Bentivoglio, M (1996). 1898–1996: The centennial of the axon. *Brain Research Bulletin;* **41**: 319–325.

Bernal, JD, & Crowfoot, D (1934). X-ray photographs of crystalline pepsin. *Nature*; **133**: 795–797.

Bernstein, J (1868). Ueber den zeitlichen Verlauf der negativen Schwankung des Nervenstroms. *Pflügers Archiv fur die Gesamte Physiologie des Menschen und der Tiere*; **1**: 173–207.

Bernstein, J (1902). Untersuchungen zur thermodynamik der bioelektrischen ströme. *Archiv fur die Gesamte Physiologie des Menschen und der Tiere*; **92**: 521–562.

Bernstein, J (1912). *Electrobiologie. Die Lehre von den Elektrischen Vorgängen im Organizmus auf Moderner Grundlage Dargestellt.* Braunschweig: Fr Vieweg. ix + 215 pp.

Besterman, E, & Creese, R (1979). Waller: pioneer of electrocardiography. *British Heart Journal*; **42**: 61–64.

Biemond, A, & Daniels, AP (1934). Familial periodic paralysis and its transition into spinal muscular atrophy. *Brain*; **57**: 91–108.

Birks, R, Huxley, HE, & Katz, B (1960). The fine structure of the neuromuscular junction of the frog. *Journal of Physiology*; **150**: 134–144.

Bishop, GH (1923). Body fluid of the honey bee larva. I. Osmotic pressure, specific gravity, pH, O2 capacity, and buffer value, and their changes with larval activity and metamorphosis. *Journal of Biological Chemistry*; **58**: 543–565.

Bishop, GH (1965). My life among the axons. *Annual Review of Physiology*; **27**: 1–18.

Bishop, GH (1969). Washington University School of Medicine, Oral History Program. Interview Number 4. Cited by Marshall, LH (1983).

Bishop, GH, & Heinbecker, P (1930). Differentiation of axon types in visceral nerves by means of the potential record. *American Journal of Physiology*; **94**: 170–200.

Bisset, GW, & Bliss, TVP (1997). Wilhelm Siegmund Feldberg, CBE. November 19, 1900–October 23, 1993. *Biographical Memoirs of Fellows of the Royal Society*; **43**: 145–170.

Bliss, M (2005). *Harvey Cushing. A Life in Surgery*. Oxford: Oxford University Press.

Bormann, J, Hamill, OP, & Sakmann, B (1987). Mechanism of anion permeation through channels gated by glycine and gamma-aminobutyric acid in mouse cultured spinal neurons. *Journal of Physiology*; **385**: 243–286.

Boycott, BB (1998). John Zachary Young. March 18, 1907–July 4, 1997. *Biographical Memoirs of Fellows of the Royal Society*; **44**: 487–509.

Brock, LG, Coombs, JS, & Eccles, JC (1952). The recording of potentials from motoneurones with an intracellular electrode. *Journal of Physiology*; **117**: 431–460.

Brooks, JE (1969). Hyperkalaemic periodic paralysis. Intracellular EMG studies. *Archives of Neurology*; **20**: 13–18.

Brown, GL, Dale, HH, & Feldberg, W (1936). Reactions of the normal mammalian muscle to acetylcholine and to eserine. *Journal of Physiology*; **87**: 394–424.

Brown, GL, & Eccles, JC (1934). The action of a single vagal volley on the rhythm of the heart beat. *Journal of Physiology*; **82**: 211–240.

Brown, GL, & Harvey, AM (1939). Congenital myotonia in the goat. *Brain*; **62**: 341–363.

Brown, GT, & Sherrington, CS (1912). On the instability of a cortical point. *Proceedings of the Royal Society, B*; **85**: 250–277.

Bryant, SH (1969). Cable properties of external intercostal muscle fibres from myotonic and non-myotonic goats. *Journal of Physiology*; **294**: 539–550.

Bryant, SH (1971). Chloride conductance in normal and myotonic muscle fibres and the action of monocarboxylic aromatic acids. *Journal of Physiology*; **219**: 367–383.

Buchthal, F, & Rosenfalck, A (1966). Evoked action potentials and conduction velocity in human sensory nerves. *Brain Research*; **3**: 1–122.

Buller, AJ, Eccles, JC, & Eccles, RM (1960). Interactions between motoneurones and muscle in respect of the characteristic speeds of their responses. *Journal of Physiology*; **150**: 417–439.

Burke, RE (1967). Motor unit types of cat triceps surae muscle. *Journal of Physiology*; **193**: 141–160.

Burn, JH (1978). Essential pharmacology. *Annual Review of Pharmacology*; **9**: 1–20.

Bynum, WF (1976). A short history of the Physiological Society, 1926–1976. *Journal of Physiology*; **268**: 23–72.

Cahan, D, ed. (1993). *Hermann von Helmholtz and the Foundation of Nineteenth Century Science*. Berkeley: University of California Press.

Cajal, R (1901). *Recuerdos de mi Vida*. English translation by Craigie, EH. Cambridge, MA: MIT Press, 1989.

Cajal, S Ramon y (1952). *Histologie du Système Nerveux de l'Homme et des Vertébrés*. Tome premier. Madrid: Instituto Ramon y Cajal. Translated by Mazzarello, P (1999). *A Scientific Biography of Camillo Golgi*. Oxford: Oxford University Press.

Caldwell, PC, Hodgkin, AL, Keynes, RD, & Shaw, TI (1960). The effects of injecting "energy-rich" phosphate compounds on the active transport of ions in the giant axons of *Loligo*. *Journal of Physiology*; **152**: 561–590.

Cannon, SC, Brown, RH, & Corey, DP (1991). A sodium channel defect in hyperkalemic periodic paralysis: potassium-induced failure of inactivation. *Neuron*; **6**: 619–626.

Cannon, WB (1945). *The Way of an Investigator*. New York: Norton.

Chase, MW, & Hunt, CC (1995). Herbert Spencer Gasser, July 5, 1888–May 11, 1963. *Biographical Memoirs, National Academy of Sciences (USA)*; **67**: 147–177.

Clamp, AL (1980). *The Blitz of Plymouth*. Plymouth: PDS Printers.

Clark, RW (1968). *The Huxleys*. London: Heinemann, xvi + 399 pp.

Clark, SM, Luton, FH, & Cutler, JT (1939). A form of congenital myotonia in goats. *Journal of Nervous and Mental Disease*; **90**: 297–309.

Clarke, E, & O'Malley, CD (1968). *The Human Brain and the Spinal Cord. A Historical Study Illustrated by Writings from Antiquity to the Twentieth Century*. Berkeley: University of California Press.

Cohen, Lord (1958). *Sherrington—Physiologist, Philosopher and Poet.* Springfield, IL: Thomas.

Cole, KS (1979). Mostly membranes. *Annual Review of Physiology*; **41**: 1–24.

Cole, KS (1982). Squid axon membrane: impedance decrease to voltage clamp. *Annual Review of Neuroscience*; **5**: 305–323.

Cole, KS, & Curtis, HJ (1938). Electric impedance of *Nitella* during activity. *Journal of General Physiology*; **22**: 37–64.

Cole, KS, & Curtis, HJ (1939). Electric impedance of the squid giant axon during activity. *Journal of General Physiology*; **22**: 649–670.

Cole, KS, & Hodgkin, AL (1939). Membrane and protoplasm resistance in the squid giant axon. *Journal of General Physiology*; **22**: 671–687.

Cook, JA (1777). *A Voyage Towards the South Pole and Around the World*; vol 2: pp 112–113. London: Strahan & Cadell.

Coombs, JS, Eccles, JC, & Fatt, P (1955). The electrical properties of the motoneuronal membrane. *Journal of Physiology*; **130**: 291–325.

Coombs, JS, Eccles, JC, & Fatt, P (1955). The specific ionic conductances and the ionic movements across the motoneuronal membrane that produce the inhibitory post-synaptic potential. *Journal of Physiology*; **130**: 326–373.

Coombs, JS, Eccles, JC, & Fatt, P (1955). Excitatory synaptic action in motoneurones. *Journal of Physiology*; **130**: 374–395.

Coombs, JS, Eccles, JC, & Fatt, P (1955). Inhibitory suppression of reflex discharges from motoneurones. *Journal of Physiology*; **130**: 396–413.

Crowfoot, D, Bunn, CW, Rogers-Low, BW, & Turner-Jones, A (1949). X-ray crystallographic investigation of the structure of penicillin. In *Chemistry of Penicillin*. Princeton: Princeton University Press, 300 pp.

Curtis, DR, & Andersen, P (2001). Sir John Eccles, AC. January 27, 1903–May 2, 1997. *Biographical Memoirs of Fellows of the Royal Society*; **47**: 160–187.

Curtis, HJ, & Cole, KS (1942). Membrane resting and action potentials from the squid giant axon. *Journal of Cellular and Comparative Physiology*; **19**: 135–144.

Cushing, H (1904). Cited by Bliss, M (2005).

Dahlström, A, & Fuxe, K (1964). Evidence for the existence of monoamine containing neurons in the central nervous system. I: Demonstration of monoamines in the cell bodies of brain stem neurons. *Acta Physiologica Scandinavica*; **62**, Suppl. 232: 1–55.

Dale, HH (1914). The action of certain esters and ethers of choline and their relation to muscarine. *Journal of Pharmacology*; **6**: 147–190.

Dale, Sir Henry Hallett (1953). *Adventures in Physiology with Excursions into Autopharmacology*. London: Pergamon Press.

Dale, HH (1963). Pharmacology during the past sixty years. *Annual Review of Pharmacology*; **3**: 75–82.

Dale, HH, & Feldberg, W (1934). The chemical transmitter of vagus effects to the stomach. *Journal of Physiology*; **81**: 320–334.

Dale, HH, Feldberg, W, & Vogt, M (1936). Release of acetylcholine at voluntary motor nerve endings. *Journal of Physiology*; **86**: 353–380.

Darwin, H, & Bayliss, W (1917). Keith Lucas, 1879–1916. *Proceedings of the Royal Society, B*; **90**: 31–42.

Davis, H (1970). Joseph Erlanger. January 5, 1874–December 5, 1965. *Biographical Memoirs. National Academy of Sciences (USA)*; **410**: 53–100.

Dawson, GD (1947). Cerebral responses to electrical stimulation of peripheral nerve in man. *Journal of Neurology, Neurosurgery and Psychiatry*; **10**: 137–140.

Dawson, GD (1954). Cerebral responses to electrical stimulation of peripheral nerve in man. *Electroencephalography and Clinical Neurophysiology*; **6**: 65–84.

Dean, RB (1941). Theories of electrolyte equilibrium in muscle. *Biological Symposia*; **3**: 331–348.

Doyle, DA, Cabral, JM, Pfuetzner, RA, et al (1998). The structure of the potassium channel: molecular basis of K+ conduction and selectivity. *Science*; **280**: 69–77.

Draper, MH, & Weidmann, S (1951). Cardiac resting and action potentials recorded with an intracellular electrode. *Journal of Physiology*; **115**: 74–94.

Du Bois-Reymond, E (1877). *Gesammelte abhandlungen der allgemeinen Muskel-und Nervenphysik*; volume 2. Leipzig:Veit. 753 pp.

Eccles, JC (1937). Synaptic and neuro-muscular transmission. *Physiological Reviews*; **17**: 538–555.

Eccles, JC (1944). The nature of synaptic transmission in a sympathetic ganglion. *Journal of Physiology*; **103**: 27–54.

Eccles, JC (1953). *The Neurophysiological Basis of Mind: The Principles of Neurophysiology*. Oxford: Clarendon Press.

Eccles, JC (1957). *The Physiology of Nerve Cells*. London: Oxford University Press.

Eccles, JC (1962). Central connections of muscle afferent fibres. In *Muscle receptors* (ed. D Barker), pp 81–101. Hong Kong University Press.

Eccles, JC (1977). Alexander Forbes and his achievement in electrophysiology. *Perspectives in Biology and Medicine*; **13**: 388–404.

Eccles, JC (1977). My scientific odyssey. *Annual Review of Physiology*; **39**: 1–18.

Eccles, JC (1982). The synapse: from electrical to chemical transmission. *Annual Review of Neuroscience*; **5**: 325–339.

Eccles, JC, & Gibson, WC (1979). *Sherrington. His Life and Thought*. Berlin: Springer-Verlag, 269 pp.

Eccles, JC, & Granit, R (1929). Crossed extensor reflexes and their interaction. *Journal of Physiology*; **67**: 97–118.

Eccles, JC, & Popper KR (1977). *The Self and its Brain. An Argument for Interactionism*. Berlin: Springer International.

Eccles, JC, & Sherrington, CS (1930). Numbers and contraction-values of individual motor-units examined in some muscles of the limb. *Proceedings of the Royal Society, B*; **106**: 326–357.

Eccles, JC, & Sherrington, CS (1931). Studies on the flexor reflex. III. The reflex response evoked by two centripetal volleys. *Proceedings of the Royal Society, B*; **107**: 535–556.

Eccles, JC, & Sherrington, CS (1931). Studies on the flexor reflex. VI. Inhibition. *Proceedings of the Royal Society, B*; **109**: 91–113.

Eccles, JC, Fatt, P, & Koketsu, K (1954). Cholinergic and inhibitory synapses in a pathway from motor-axon collaterals to motoneurones. *Journal of Physiology*; **126**: 524–562.

Eccles, JC, Ito, M, & Szentàgothai, J (1967). *The Cerebellum as a Neuronal Machine*. New York: Springer-Verlag.

Eccles, JC, Katz, B, & Kuffler, SW (1941). Nature of the "endplate potential" in curarized muscle. *Journal of Neurophysiology*; **4**: 362–387.

Eccles, JC, Katz, B, & Kuffler, SW (1941). Initiation of muscle impulses at neuro-muscular junction. *Journal of Neurophysiology*; **4**: 402–417.

Eccles, JC, Katz, B, & Kuffler, SW (1942). Effect of eserine on neuromuscular transmission. *Journal of Neurophysiology*; **5**: 211–230.

Edström, L, & Kugelberg, E (1968). Histochemical composition, distribution of fibres and fatiguability of single motor units. Anterior tibial muscle of the rat. *Journal of Neurology, Neurosurgery and Psychiatry*; **31**: 424–433.

Eggleton, P, & Eggleton, GP (1927). The inorganic phosphate and a labile form of organic phosphate in the gastrocnemius of the frog. *Biochemical Journal*; **21**: 190–195.

Einthoven, W (1901). Un nouveau galvanometre. *Archives Néerlandais des Sciences Exactes et Naturelles*; **6**: 623–633.

Elliott, TR (1904). Physiological Society note, cited by Bacq (1975).

Encyclopaedia Britannica. 15th ed., **2**: 1035, 1979.

Engelhardt, VA, & Lyubimova, MN (1939). Myosine and adenosinetriphosphatase. *Nature*; **144**: 668–669.

Erlanger, J (1904). A new instrument for determining the minimum and maximum blood-pressures in man. *Johns Hopkins Hospital Reports*; **12**: 53–100.

Erlanger, J (1905). On the physiology of heart-block in mammals, with special reference to the causation of Stokes-Adams disease. *Journal of Experimental Medicine*; **7**: 676–724.

Erlanger, J (1964). A physiologist reminiscences. *Annual Review of Physiology*; **26**: 93–106.

Erlanger, J, Bishop, GH, & Gasser, HS (1926). The action potential waves transmitted between the sciatic nerve and its spinal roots. *American Journal of Physiology*; **78**: 574–591.

Erlanger, J, & Gasser, HS (1919). Hypertonic gum acacia and glucose in the treatment of secondary shock. *Annals of Surgery*; **69**: 389–421.

Erlanger, J, & Gasser, HS (1936). *Electrical Signs of Nervous Activity.* Philadelphia: University of Philadelphia, xi + 242 pp.

Erlanger, J, & Gasser HS, with the collaboration, in some of the experiments, of GH Bishop (1924). The compound nature of the action current of nerve as disclosed by the cathode ray oscillograph. *American Journal of Physiology*; **70**: 624–666.

Erlanger, J, Gasser, HS, & Bishop, GH (1927). The absolutely refractory phase of the alpha, beta and gamma fibres in the sciatic nerve of the frog. *American Journal of Physiology*; **81**: 473.

Erlanger, J, & Schoepfle, GM (1951). Observations on the local response in single medullated nerve fibres. *American Journal of Physiology*; **167**: 134–146.

Evans, EL (1949). *Principles of Human Physiology*, 10th ed. Philadelphia: Lea & Febiger.

Evarts, EV, & Tanji, J (1976). Reflex and intended responses in motor cortex pyramidal tract neurons of monkeys. *Journal of Neurophysiology*; **39**: 169–180.

Eyzaguirre, C, & Kuffler, SW (1955). Processes of excitation in the dendrites and in the soma of single isolated sensory nerve cells of the lobster and crayfish. *Journal of General Physiology*; **39**: 87–119.

Fatt, P, & Katz, B (1952). Spontaneous subthreshold activity of motor nerve endings. *Journal of Physiology*; **117**: 109–128.

Feldberg, W (1977). The early history of synaptic and neuromuscular transmission by acetylcholine: reminiscences of an eye witness. In *The Pursuit of Nature.* Cambridge: Cambridge University Press, pp. 65–83.

Feldberg, W, & Gaddum, JH (1934). The chemical transmitter at synapses in a sympathetic ganglion. *Journal of Physiology*; **81**: 305–319.

Feldberg, WS (1970). Henry Hallett Dale, 1875–1968. *Biographical Memoirs of the Royal Society*; **16**: 77–174.

Fenn, WO (1923). A quantitative comparison between the energy liberated and the work performed by the isolated sartorius muscle of the frog. *Journal of Physiology*; **58**: 175–203.

Fenn, WO (1979). Alexander Forbes. 1882–1965. *Biographical Memoir*; **40**: 113–141. Washington, DC: National Academy of Sciences.

Ferrier, D (1873). Experimental research in cerebral physiology and pathology. *West Riding Lunatic Asylum Medical Reports*; **22**: 229–232.

Finger, S (2000). *Minds Behind the Brain. A History of the Pioneers and their Discoveries*. Oxford: Oxford University Press, xii + 364 pp.

Finkelstein, G (2006). Emil du Bois-Reymond vs Ludimar Hermann. *Comptes Rendus Biologies*; **329**: 340–347.

Fisher, WM, & Keith-Lucas, A (1934). *Keith Lucas.* Cambridge: Heffer.

Flexner, A (1910). *Medical Education in the United States and Canada.* New York: Carnegie Foundation.

Fontana, F (1781). *Traité sur le venim de la vipère, sur les poisons americains, sur le laurier-cerise et sur quelques autres poisons végétaux. On y a joint des observations sur la structure primitive du corps animal. Différentes experiences sur la reproduction des nerfs et la description d'un nouveau canal de l'oeil.* Florence. Cited by Bentivoglio, M (1996).

Forbes, A, & Thacher, C (1920). Amplification of action currents with the electron tube in recording with the string galvanometer. *American Journal of Physiology*; **52**: 409–471.

Forbes, A, Campbell, CJ, & Williams, HB (1924). Electrical records of afferent nerve impulses from muscular receptors. *American Journal of Physiology*; **69**: 283–303.

Fulton, JF (1938). *The Physiology of the Nervous System.* London: Oxford University Press.

Fulton, JF (1952). Charles Scott Sherrington, OM. *Journal of Neurophysiology*; **15**: 167–190.

Galvani, L (1791). De viribus electricitatis in motu musculari. Commentarius. *De Bononiensi Scientarum et Artium Instituto atque Academia Commentarii*; **7**: 363–418.

Gamstorp, I (1956). Adynamia episodica hereditaria. *Acta Paediatrica*; **45**, suppl 108: 1–126.

Gasser, HS (1936). First year's report to the Board of Scientific Directors of the Rockefeller Institute for Medical Research, April 18, 1936. Rockefeller University Archives.

Gasser, HS (1958). Comparison of the structure, as revealed with the electron microscope, and the physiology of the unmedullated fibrers in the skin nerves and in the olfactory nerves. *Experimental Cell Research*; Suppl 5: 3–17.

Gasser, HS (1960). Effect of the method of leading on the recording of the nerve fiber spectrum. *Journal of General Physiology*; **43**: 927–940.

Gasser, HS (1964). Herbert Spencer Gasser, 1888–1963. *Experimental Neurology,* Supplement 1: 1–38.

Gasser, HS, & Erlanger, J (1922). A study of the action currents of nerve with the cathode ray oscillograph. *American Journal of Physiology*; **62**: 496–524.

Gasser, HS, & Erlanger, J (1927). The role played by the sizes of the constituent fibers of a nerve trunk in determining the form of its action potential wave. *American Journal of Physiology*; **80**: 522–547.

Gasser, HS, & Graham, HT (1932). The end of the spike-potential of nerve and its relation to the beginning of the after-potential. *American Journal of Physiology*; **101**: 316–330.

Gasser, HS, & Graham, HT (1933). Potentials produced in the spinal cord by stimulation of dorsal roots. *American Journal of Physiology*; **103**: 303–320.

Gasser, HS, & Newcomer, HS (1921). Physiological action currents in the phrenic nerve. An application of the termionic vacuum tube to nerve physiology. *American Journal of Physiology*; **57**: 1–26.

Gibson, TM, & Harrison, MH (1984). *Into Thin Air. A History of Aviation Medicine in the RAF*. London: Robert Hale.

Gibson, W (1986). Pioneers in neurosciences: the Sherrington era. *Canadian Journal of Neurological Sciences*; **13**: 295–300.

Gillespie, CA, Simpson, DR, & Edgerton, VR (1974). Motor unit recruitment as reflected by muscle fibre glycogen loss in a prosimian (bushbaby) after running and jumping. *Journal of Neurology, Neurosurgery and Psychiatry*; **37**: 817–824.

Glynn, IM (2002). A hundred years of sodium pumping. *Annual Review of Physiology*; **64**: 1–18.

Goldflam, S (1891). Ueber eine eigenthümliche Form von periodischer, familiärer, wahrscheinlich auto-intoxicatorischer Paralyse. *Zeitschrift fur Klinische Medecin, suppl*; **240**: 242.

Goldman, DE (1943). Potential, impedance and rectification in membranes. *Journal of General Physiology*; **27**: 37–60.

Goodman, J (2005). *The Rattlesnake.* London: Faber & Faber, x + 357 pp.

Gordon, AM, Huxley, AF, & Julian, FJ (1966). The variation in isometric tension with sarcomere length in vertebrate muscle fibres. *Journal of Physiology*; **184**: 170–192.

Gotch, F, & Horsley, V (1888). Observations upon the electromotive changes in the mammalian spinal cord following electrical excitation of the cortex cerebri. *Proceedings of the Royal Society*; **45**: 18–26.

Granit, R (1966). *Charles Scott Sherrington. An Appraisal.* London: Thomas Nelson, 188 pp.

Granit, R, Kernell, D, & Smith, RS (1963). Delayed depolarization and the repetitive response to intracellular stimulation of mammalian motoneurones. *Journal of Physiology*; **168**: 890–910.

Gray, Sir John (1990). Bryan Harold Cabot Matthews, June 14, 1906–July 23, 1986. *Biographical Memoirs of Fellows of the Royal Society*; **35**: 265–279.

Grundfest, H (1963). Julius Bernstein, Ludimar Hermann and the discovery of the overshoot of the axon spike. *Archives Italiennes de Biologie*; **103**: 483–490.

Hadden, WB, & Sherrington, CS (1888). The pathological anatomy of a case of locomotor ataxy, with special reference to ascending degenerations in the spinal cord and medulla oblongata. *Brain*; **11**: 325–335.

Hagbarth, KE, Wallin, G, & Löfstedt, L (1975). Muscle spindle activity in man during voluntary fast alternating movements. *Journal of Neurology, Neurosurgery and Psychiatry*; **38**: 625–635.

Hall, M (1833). On the reflex function of the medulla oblongata and medulla spinalis. *Proceedings of the Royal Society of London, Series B*; **123**: 635–665.

Hamill, OP, Marty, A, Neher, E, Sakmann, B, & Sigworth, FJ (1981). Improved patch-clamp techniques for high-resolution current recording from cells and cell-free membrane patches. *Pflügers Archiv*; **391**: 85–100.

Hartline, HK (1934). Intensity and duration in the excitation of single photoreceptor units. *Journal of Cellular and Comparative Physiology*; **5**: 229–247.

Heinbecker, P, & Bishop, GH (1929). Differentiation between types of fibres in certain components of involuntary nervous system. *Proceedings of the Society for Experimental Biology and Medicine*; **26**: 645–647.

Helmholtz, H von (1850). Messunger über den zeitlichen Verlauf der Zuckung animalischer Muskeln und die fortpflanzungsgeschwindigkeit der Reizung in den Nerven. *Archiv fur Anatomie, Physiologie und Wissenschaftliche Medicin*; 276–367.

Helmholtz, H von (1852). Messungen über Fortpflanzungsgeschwindigkeit der Reizung in den Nerven zweite Reihe. *Archiv fur Anamomie, Physiologie und Wissenschaftliche Medicin*; 199–216.

Helmholtz, H von (1878). The facts of perception. In Warren, RM, & Warren, RP, eds. *Helmholtz on Perception: Its Physiology and Development.* New York: John Wiley, 1968, pp. 207–231.

Henneman, E, Somjen, G, & Carpenter, DO (1965). Functional significance of cell size in spinal motoneurones. *Journal of Neurophysiology*; **28**: 560–580.

Hermann, L (1879). Allgemeine Nervenphysiologie. In: *Handbuch der Physiologie*; **2**: 1–196.

Hill, AV (1926). The heat production of nerve. *Journal of Pharmacology*; **29**: 161–165.

Hill, AV (1933). The physical nature of the nerve impulse. *Nature*; **81**: 501–508.

Hill, AV (1965). *Trails and Trials in Physiology.* London: Edward Arnold.

Hille, B (1989). Ionic channels: evolutionary origins and modern roles. *Quarterly Journal of Experimental Physiology*; **74**: 785–804.

Hille, B (2001). *Ionic Channels of Excitable Membranes* (3rd ed). Sunderland, MA: Sinauer.

Hillman, HH, & McIlwain, H (1961). Membrane potentials in mammalian cerebral tissue in vitro; dependence on ionic environment. *Journal of Physiology*; **157**: 263–278.

Höber, R (1905). Über den Einfluss der Salze auf den Ruhestrom des Froschmuskels. *Pflügers Archiv fur die Gesamte Physiologie des Menschen und der Tiere*; **106**: 599–635.

Hodgkin, Sir Alan (1979). Edgar Douglas Adrian, Baron Adrian of Cambridge, 30 November, 1914–August 4, 1977. *Biographical Memoirs of Fellows of the Royal Society*; **25**: 1–73.

Hodgkin, AL (1937). Evidence for electrical transmission in nerve. Part I. *Journal of Physiology*; **90**: 183–210.

Hodgkin, AL (1937). Evidence for electrical transmission in nerve. Part II. *Journal of Physiology*; **90**: 211–232.

Hodgkin, AL (1938). The subthreshold potentials in a crustacean nerve fibre. *Proceedings of the Royal Society of London, B*; **126**: 87–121.

Hodgkin, AL (1939). The relation between conduction velocity and the electrical resistance outside a nerve fibre. *Journal of Physiology*; **94**: 560–570.

Hodgkin, AL (1992). *Chance and Design. Reminiscences of Science in Peace and War.* Cambridge: University of Cambridge.

Hodgkin, AL, & Huxley, AF (1939). Action potentials recorded from inside a nerve fibre. *Nature*; **144**: 710–711.

Hodgkin, AL, & Huxley, AF (1945). Resting and action potentials in single nerve fibres. *Journal of Physiology*; **104**: 176–195.

Hodgkin, AL, & Huxley, AF (1947). Potassium leakage from an active nerve fibre. *Journal of Physiology*; **106**: 341–367.

Hodgkin, AL, & Huxley, AF (1952a). Currents carried by sodium and potassium ions through the membrane of the giant axon of *Loligo*. *Journal of Physiology*; **116**: 449–472.

Hodgkin, AL, & Huxley, AF (1952b). The components of membrane conductance in the giant axon of *Loligo*. *Journal of Physiology*; **116**: 473–496.

Hodgkin, AL, & Huxley, AF (1952c). The dual effect of membrane potential on sodium conductance in the giant axon of *Loligo*. *Journal of Physiology*; **116**: 497–506.

Hodgkin, AL, & Huxley, AF (1952d). A quantitative description of membrane current and its application to conduction and excitation in nerve. *Journal of Physiology*; **116**: 500–544.

Hodgkin, AL, Huxley, AF, & Katz, B (1952). Measurement of current-voltage relations in the giant axon of *Loligo*. *Journal of Physiology*; **116**: 442–448.

Hodgkin, AL, & Katz, B (1949). The effect of sodium ions on the electrical activity of the giant axon of the squid. *Journal of Physiology*; **108**: 37–77.

Hodgkin, AL, & Keynes, RD (1955). Active transport of cations in giant axons from *Sepia* and Loligo. *Journal of Physiology*; **128**: 28–60.

Hodgkin, AL, & Keynes, RD (1955). The potassium permeability of a giant nerve fibre. *Journal of Physiology*; **128**: 61–88.

Hoffmann, P (1910). Beitrag zur Kenntnis der menschlichen Reflexe mit besonder Berucksichtigung der elektrischen Erscheinungen. *Archiv fur Anatomie und Physiologie*; **1**: 223–246.

Hoffmann, P (1918). Über die Beziehungen der Sehnenreflexe sur willkürlichen Bewegung und sum Tonus. *Zeitschrift fur Biologie*; **68**: 351–370.

Hubel, DH, & Wiesel, TN (1959). Receptive fields of single neurons in the cat's striate cortex. *Journal of Physiology*; **148**: 574–591.

Hubel, DH, & Wiesel, TN (1970). The period of susceptibility to the physiological effects of unilateral eye closure in kittens. *Journal of Physiology*; **206**: 419–436.

Hughes, J, Smith, TW, Kosterlitz, HW, Fothergill, LA, Morgan, BA, & Harris, HR (1975). Identification of two related pentapeptides from the brain with potent opiate agonist activity. *Nature*; **258**: 577–579.

Huxley, Sir Andrew (1980). *Reflections on Muscle*. Princeton, NJ: Princeton University Press, 111 pp.

Huxley, Sir Andrew (1992). Kenneth Stewart Cole. July 10, 1900–April 18, 1984. *Biographical Memoirs of Fellows of the Royal Society of London*; **38**: 98–110.

Huxley, AF (2002). From overshoot to voltage clamp. *Trends in Neurosciences*; **25**: 553–558.

Huxley, AF (2004). In Squire, LR, ed. *The History of Neuroscience in Autobiography*; **4**: 282–318.

Huxley, AF, & Niedergerke, R (1954). Structural changes in muscle during contraction. Interference microscopy of living muscle fibres. *Nature*; **173**: 971–973.

Huxley, AF, & Stämpfli, R (1949). Evidence for saltatory conduction in peripheral myelinated nerve fibres. *Journal of Physiology*; **108**: 315–339.

Huxley, AF, & Stämpfli, R (1951). Direct determination of membrane resting and action potential in single myelinated nerve fibres. *Journal of Physiology*; **112**: 476–495.

Huxley, AF, & Taylor, RE (1958). Local activation of striated muscle fibres. *Journal of Physiology*; **144**: 426–441.

Huxley, HE (1953). X-ray analysis and the problem of muscle. *Proceedings of the Royal Society of London, B*; **141**: 69–62.

Huxley, HE (1957). The double array of filaments in cross-striated muscle. *Journal of Biophysical and Biochemical Cytology*; **3**: 631–648.

Huxley, HE (1996). A personal view of muscle and motility mechanisms. *Annual Review of Physiology*; **58**: 1–19.

Huxley, HE, & Hanson J (1954). Changes in the cross-striations of muscle during contraction and stretch and their structural interpretation. *Nature*; **173**: 973–976.

Huxley, L (1901). *Life and Letters of Thomas Henry Huxley*. New York: Appleton.

Huxley, TH (1866). *Lessons in Elementary Physiology*. London: Methuen.

Huxley, TH (1893). *Science and Education: Collected Essays*, vol 3. London: Methuen.

Jankowska, E, & McCrea, DA (1983). Shared reflex pathways from Ib tendon organ afferents and Ia muscle spindles afferents in the cat. *Journal of Physiology*; **338**: 99–112.

Jiang, Y, Ruta, V, Chen, J, Lee, A, & MacKinnon, R (2003). The principle of gating charge movement in a voltage-dependent K+ channel. *Nature*; **423**: 42–48.

Jiang, Y, Lee, A, Chen, J, Ruta, V, Cadene, M, Chalt, BT, & MacKinnon, R (2003). X-ray structure of a voltage-dependent K+ channel. *Nature*; **423**: 33–41.

Jurkat-Rott, K, Lehmann-Horn, F, Elbaz, A, Heine, R, Gregg, RG, et al (1994). A calcium channel mutation causing hypokalemic period paralysis. *Human Molecular Genetics*; **3**: 1415–1419.

Kahn, RH (1926). Über humorale Übertragbarkeit der Herznervenwirkung. *Pflügers Archiv fur die Gesamte Physiologie des Menschen und der Tiere*; **214**: 482–498.

Kamb, A, Iversen, LE, & Tanouye, MA (1987). Molecular characterization of *Shaker*, a Drosophila gene that encodes a potassium channel. *Cell*; **50**: 405–413.

Kao, CY (1966). Tetrodotoxin, saxitonin and their significance in the study of excitation phenomena. *Pharmacological Reviews*; **18**: 997–1049.

Katz, B (1947). The effect of electrolyte deficiency on the rate of conduction in a single nerve fibre. *Journal of Physiology*; **106**: 411–417.

Katz, B (1950). Depolarization of sensory terminals and the initiation of impulses in the muscle spindle. *Journal of Physiology*; **111**: 261–282.

Katz, B (1961). The terminations of the afferent fibre in the muscle spindle of the frog. *Philosophical Transactions of the Royal Society, A*; **243**: 221–240.

Katz, B (1969). *The Release of Neural Transmitter Substances*. Liverpool: Liverpool University Press, pp. 45–46.

Katz, B (1978). Archibald Vivian Hill, September 26, 1886–June 3, 1977. *Biographical Memoirs of Fellows of the Royal Society*; **24**: 71–149.

Katz, Sir Bernard (1990). In McMahan, UJ, ed. *Steve. Remembrances of Stephen W. Kuffler*. Sunderland, MA: Sinauer, pp. 107–141.

Katz, Sir Bernard (1996). In Squire, LR, ed. *The History of Neuroscience in Autobiography*; **2**: 350–381. Washington: Society for Neuroscience.

Katz, B, & Kuffler, SW (1946). Inhibition at the nerve-muscle junction in Crustacea. *Journal of Neurophysiology*; **9**: 337–346.

Katz, B, & Miledi, R (1963). A study of spontaneous miniature potentials in spinal motoneurones. *Journal of Physiology*; **163**: 389–422.

Katz, B, & Miledi, R (1965). Propagation of electric activity in motor nerve terminals. *Proceedings of the Royal Society, B*; **161**: 453–482.

Katz, B, & Miledi, R (1967). The timing of calcium action during neuromuscular transmission. *Journal of Physiology*; **189**: 535–544.

Katz, B, & Miledi, R (1970). Membrane noise produced by acetylcholine. *Nature*; **226**: 962–963.

Katz, B, & Miledi, R (1971). Further observations on acetylcholine noise. *Nature*; **232**: 124–126.

Keesey, JC (1998). Contemporary opinions about Mary Walker: a shy pioneer of therapeutic neurology. *Neurology*; **51**: 1433–1439.

Keynes, G (1985). *The Gates of Memory*. Oxford: Oxford University Press, 312 pp.

Keynes, RD (1951). The leakage of radioactive potassium from stimulated nerve. *Journal of Physiology*; **113**: 99–114.

Keynes, RD (1951). The ionic movements during nervous activity. *Journal of Physiology*; **114**: 119–150.

Keynes, RD, Ritchie, JM & Rojas, E (1971). The binding of tetrodotoxin to nerve membranes. *Journal of Physiology*; **213**: 235–254.

Keynes, RD, & Rojas, E (1974). Kinetics and steady-state properties of the charged system controlling sodium conductance in the squid giant axon. *Journal of Physiology*; **239**: 393–434.

Koch, MC, Steinmeyer, K, Lorenz, C, Ricker, K, Wolf, F, et al (1992). The skeletal muscle chloride channel in dominant and recessive myotonia. *Science*; **257**: 797–800.

Kolb, LC (1938). Congenital myotonia in goats. Description of the disease. The effect of quinine, various cinchona alkaloids and salts upon the myotonic symptom. *Johns Hopkins Hospital Bulletin*; **63**: 221–237.

Kolb, LC, Harvey, AM, & Whitehill, MR (1938). A clinical study of myotonic dystrophy and myotonia congenita with special reference to the therapeutic effect of quinine. *Johns Hopkins Hospital Bulletin*; **62**: 188–215.

Krause, W (1873). Die Contraction der Muskelfasser. *Pflugers Archiv fur di Gesamte Physiologie des Menschen und der Tiere*; **7**: 508–514.

Krnjevic, K, & Phillis, JW (1963). Iontophoretic studies of neurons in the mammalian cerebral cortex. *Journal of Physiology*; **165**: 274–303.

Kuffler, SW (1953). Discharge patterns and functional organization of the mammalian retina. *Journal of Neurophysiology*; **16**: 37–68.

Landau, WM (1985). George Holman Bishop, 1889–1973. *Biographical Memoirs, National Academy of Sciences (USA)*; **55**: 45–66.

Langley, JN (1906). On nerve endings and on special excitable substances in cells. Croonian Lecture. *Proceedings of the Royal Society, B;* **78**: 170–194.

Langley, JN (1907). Sir Michael Foster. In memoriam. *Journal of Physiology*; **35**: 234–246.

Langley, JN, & Sherrington, CS (1894). Secondary degeneration of nerve tracts following removal of the cortex of the cerebrum in the dog. *Journal of Physiology*; **5**: 49–65.

Lapicque, L & M (1926). *L'excitabilité en fonction du temps.* Paris: Presses Universitaires de France.

Leewenhoek, AV (1719). *Epistolae physiologicae super pluribus naturae arcanis.* Delft: Berman. Cited by Clarke, E, & O'Malley, CD (1968).

Lehmann-Horn, F, Rüdel, R, Ricker, K, Lorkovic, H, Dengler, R, & Hopf, HC (1983). Two cases of adynamia episodica hereditaria: in vitro investigation of muscle membrane and contractile parameters. *Muscle and Nerve*; **6**: 113–121.

Lenoir, T (1986). Models and instruments in the development of electrophysiology, 1845–1012. *Historical Studies in the Physical and Biological Sciences*; **17**: 1–54.

Li, C-L, & Jasper, H (1953). Microelectrode studies of the electrical activity of the cerebral cortex in the cat. *Journal of Physiology*; **121**: 117–140.

Li, C-L, & McIlwain, H (1957). Maintenance of resting potentials in slices of mammalian cerebral cortex and other tissues in vitro. *Journal of Physiology*; **139**: 178–190.

Liddell, EGT (1952). Charles Scott Sherrington, 1857–1952. *Obituary Notices of Fellows of the Royal Society*; **8**: 241–270.

Lillie, RS (1920). The recovery of transmissivity in passive iron wire as a model of recovery processes in irritable living systems. Parts I and II. *Journal of General Physiology;* **3**: 107–128 and 129–143.

Ling, G, & Gerard, RW (1949). The normal membrane potential of frog sartorius muscle. *Journal of Cellular and Comparative Physiology*; **34**: 383–396.

Lloyd, DPC (1946). Facilitation and inhibition of spinal motoneurons. *Journal of Neurophysiology*; **9**: 421–438.

Loewi, O (1921). Über humorale Übertragbarkeit der Herznervenwirkung. I. Mitteilung. *Pflügers Archiv fur die Gesamte Physiologie des Menschen und der Tiere*; **189**: 239–242.

Loewi, O (1940). An autobiographical sketch. In *Perspectives in Biology and Medicine, IV.* University of Chicago Press.

Lohmann, K (1929). Über die fermentative Kohlenhydrat-Phosphorsäurenveresterung in Gegenwart von Fluorid, Oxalat und Citrat. *Klinische Wohnschrift*; **8**: 2009.

Lorente de Nó, R (1935). The synaptic delay of the motoneurones. *American Journal of Physiology*; **111**: 272–282.

Lorente de Nó, R (1938). The cerebral cortex: architecture, intracortical connections and motor projections. In Fulton JF, *Physiology of the Nervous System.* London: Oxford University Press, pp. 291–325.

Lorente de Nó, R (1939). Transmission of impulses through cranial nerve nuclei. *Journal of Neurophysiology*; **2**: 402–464.

Lorente de Nó, R (1947). A study of nerve physiology. *Studies from the Rockefeller Institute for Medical Research*; **131–2**: 496 & 548 pp. New York: Rockefeller Institute.

Lorente de Nó, R (1949). On the effect of certain quaternary ammonium ions upon frog nerve. *Journal of Cellular and Comparative Physiology*; **33**, suppl: 1–231.

Lorenzini, S (1678). *Observazioni interne alle Torpedini.* Florence: Per l'Ononfri. Cited by Finger, S (2000). *Minds Behind the Brain. A History of the Pioneers and their Discoveries.* Oxford: Oxford University Press.

Lucas, K (1905). On the gradation of activity in a skeletal muscle-fibre. *Journal of Physiology*; **33**: 125–137.

Lucas, K (1909). The "all or none" contraction of the amphibian skeletal muscle fibre. *Journal of Physiology*; **38**: 113–133.

Lucas, K (1910). Quantitative researches on the summation of inadequate stimuli in muscle and nerve, with observations on the time factor in electric excitation. *Journal of Physiology*; **39**: 461–475.

Lucas, K (1912). On a mechanical method of correcting photographic records obtained from the capillary electrometer. *Journal of Physiology*; **44**: 225–242.

Lucas, K (1917). *The Conduction of the Nervous Impulse.* London: Longmans Green.

Lussier, JJ, & Rushton, W (1952). The excitability of a single fibre in a nerve trunk. *Journal of Physiology*; **117**: 87–108.

MacConnachie, HF, Enesco, M, & LeBlond, CP (1964). The mode of increase in the number of skeletal muscle nuclei in the postnatal rat. *American Journal of Anatomy*; **114**: 245–253.

Macfarlane, G (1979). *Howard Florey. The Making of a Great Scientist.* Oxford: Oxford University Press, 369 pp.

MacIntosh, FC, & Paton, WDM (1974). George Lindor Brown, 1903–71. *Biographical Memoirs of Fellows of the Royal Society*; **20**: 41–73.

MacKinnon, R (1991). Determination of the subunit stoichiometry of a voltage-activated potassium channel. *Nature*; **350**: 232–235.

MacKinnon, R (2004). Nobel Lecture. Potassium channels and the atomic basis of selective ion conduction. *Bioscience Reports*; **24**: 75–100.

MacKinnon, R, & Miller, C (1988). Mechanism of charybdotoxin in block of the high-conductance, Ca^2+- activated K+ channel. *Journal of General Physiology*; **91**: 335–349.

Maddox, B (2002). *Rosalind Franklin. The Dark Lady of DNA.* London: Harper Collins.

Marmont, G (1949). Studies on the axon membrane. *Journal of Cellular and Comparative Physiology*; **34**: 351–382.

Marshall, LH (1983). The fecundity of aggregates: the axonologists at Washington University, 1922–1942. *Perspectives in Biology and Medicine*; **26**: 613–636.

Matteucci, C (1844). *Traité des phénomenes électro-physiologiques des animaux suivi d'études anatomiques sur le système nerveux et sur l'organe électrique de la torpille par Paul Savi.* Paris: Fortin, Masson.

Matthews, BHC (1928). A new electrical recording system for physiological work. *Journal of Physiology*; **65**: 225–242.

Matthews, BHC (1931). The response of a muscle spindle during active contraction of a muscle. *Journal of Physiology*; **72**: 153–174.

Matthews, BHC (1934). A special purpose amplifier. *Journal of Physiology*; **81**: 28*P*.

Mauro, A (1961). Satellite cell of skeletal muscle fibres. *Journal of Biophysical and Biochemical Cytology*; **9**: 493–495.

McComas, AJ, Mrozek, K, & Bradley, WG (1968). The nature of the electrophysiological defect in adynamia episodica. *Journal of Neurology, Neurosurgery & Psychiatry*; **31**: 448–452.

McComas, AJ (1996). *Skeletal Muscle. Form and Function.* Champaign, IL: Human Kinetics.

McIntosh BR, Gardiner PF, & McComas AJ (2006). Skeletal Muscle: Form and Function, 2nd ed. Champaign, IL: Human Kinetics.

McIlwain, H, Buchel, L, & Cheshire, JX (1951). The inorganic phosphate and phosphocreatine of brain especially during metabolism in vitro. *Biochemistry Journal*; **48**: 12–20.

Messenger, J (1997). Obituary. John Zachary Young (1907–1997). *Nature*; **388**: 726.

Miller, C (2001). See potassium run. *Nature*; **414**: 23–24.

Molnar, Z, & Blakemore, C (1991). Lack of regional specificity for connections formed between thalamus and cortex in coculture. *Nature*; **351**: 475–477.

Moore, JW, Narahashi, T, & Shaw, TI (1967). An upper limit to the number of sodium channels in nerve membrane? *Journal of Physiology*; **188**: 89–105.

Morais-Cabral, JH, Zhou, Y, & MacKinnon, R (2001). Energetic optimization of ion conduction rate by the K+ selectivity filter. *Nature*; **414**: 37–42.

Morison, R, & Dempsey, EW (1941). A study of thalamo-cortical relations. *American Journal of Physiology*; **135**: 281–292.

Moruzzi, G (1996). The electrophysiological work of Carlo Matteucci. *Brain Research Bulletin*; **40**: 69–91.

Mountcastle, VB (1957). Modality and topographic properties of single neurons of cat's somatic sensory cortex. *Journal of Neurophysiology*; **20**: 408–434.

Mueller, CB (2002). *Evarts A Graham. The Life, Lives, and Times of the Surgical Spirit of St. Louis.* Hamilton: BC Decker.

Müller, J (1838). *Handbuch der Physiologie des Menschen für Vorlesungen.* Coblenz: Hölscher.
Nachmanson, D (1977). Nerve excitability: transition from descriptive phenomenology to chemical analysis of mechanisms. *Klinishce Wochenschrift*; **55**: 715–723.
Nastuk, WL, & Hodgkin, AL (1950). The electrical activity of single muscle fibres. *Journal of Cellular and Comparative Physiology*; **35**: 35–73.
Neher, E, & Sakmann, B (1976). Noise analysis of drug-induced voltage clamp currents in denervated frog muscle fibres. *Journal of Physiology*; **258**: 705–729.
Neher, E, & Sakmann, B (1976). Single-channel currents recorded from membrane of denervated frog muscle fibres. *Nature*; **260**: 779–802.
Neumcke, B, & Stämpfli, R (1982). Sodium currents and sodium-current fluctuations in rat myelinated nerve fibres. *Journal of Physiology*; **329**: 163–184.
Noda, M, Furutani, T, Takahashi, H, Toyosato, M, Tanabe, T, et al (1983). Cloning and sequence analysis of calf cDNA and human genomic DNA encoding a-subunit precursor of muscle acetylcholine receptor. *Nature*; **305**: 818–823.
Noda, M, Shimigu, S, Tanabe, T, Takai, T, Kayano, T, et al. (1984). Primary structure of *Electrophorus electricus* sodium channel deduced from cDNA sequence. *Nature*; **312**: 121–127.
Ophoff, RA, Terwindt, GM, Vergouwe, MN, van Eijk, R, Oefner, PJ, et al (1996). Familial hemiplegic migraine and episodic ataxia type-2 are caused by mutations in the Ca^2+ channel gene *CACNL1A4. Cell*; **87**: 543–552.
Osmundsen, JA (1965). "Matador" with a radio stops wired bull. Modified behaviour in animals the subject of brain study. *The New York Times*, May 17, 1965.
Overton, E (1902). Beiträge zur allgemeinen Muskel-und-Nervenphysiologie. *Pflügers Archiv fur die Gesamte Physiologie des Menschen und der Tiere*; **92**: 346–386.
Patrick, J, & Lindstrom, J (1973). Autoimmune response to acetylcholine receptor. *Science*; **180**: 871–872.
Penfield, W (1955). The role of the temporal lobe in certain psychical phenomena. *British Journal of Psychiatry*; **101**: 451–465.
Perutz, MF (1964). Nobel Lecture. X-ray analysis of haemoglobin. *Nobel Lectures, Chemistry 1942–1962.* Amsterdam: Elsevier.
Phillips, CG (1956). Intracellular records from Betz cells in the cat. *Quarterly Journal of Experimental Physiology*; **41**: 58–69.
Piccolini, M (1998). Animal electricity and the birth of electrophysiology. The legacy of Luigi Galvani. *Brain Research Bulletin*; **46**: 381–407.
Pravdich-Neminsky, VV (1912). Eim Versuch der Registrierung der electrischen Gehirnerscheinungen. *Zentralblatt fur Physiologie*; **27**: 951–960.
Raftery, MA, Hunkapiller, MW, Strader, CD, & Hood, LE (1980). Acetylcholine receptor: complex of homologous subunits. *Science*; **208**: 1454–1457.
Ranvier, L-A (1872). Recherches sur l'histologie des nerfs periphériques. *Archives de Physiologie Normale et Pathologique*; 4: 129–149.
Rasminsky, M, & Sears, TA (1972). Internodal conduction in undissected demyelinated nerve fibres. *Journal of Physiology*; **227**: 323–350.
Ratcliffe, JA (1975). Physics in a university laboratory before and after World War II. *Proceedings of the Royal Society, A*; **342**: 457–462.
Robertson, JD (1959). The ultrastructure of cell membranes and their derivatives. *Biochemical Society Symposium*; **16**: 3–43.
Roux, D, & MacKinnon, R (1999). The cavity and pore helices in the Kcs K+ channel: electrostatic stabilization of monovalent cations. *Science*; **285**: 100–102.
Rushton, W (1933). Lapicque's theory of curarization. *Journal of Physiology*; **77**: 337–364.
Rushton, W (1937). Initiation of the propagated disturbance. *Proceedings of the Royal Society of London, B*; **124**: 210–243.
Ruta, V, Jiang, Y, Lee, A, Chen, J, & MacKinnon, R (2003). Functional analysis of an archaebacterial voltage-dependent K+ channel. *Nature*; **422**: 180–185.
Schmitt, FO (1990). The never-ceasing search. *Memoirs of the American Philosophical Society*; 1–406.
Schultzberg, M, Hökfelt, T, & Lundberg, JM (1982). Coexistence of classical transmitters and peptides in the central and peripheral nervous systems. *British Medical Bulletin*; **38**: 309–313.

Scribonius Largus (1529). *De Compositionibus Medicamentorum. Liber unus.* Ed, J Ruel. Paris: Wechel. Cited in Finger, S (2000). *Minds Behind the Brain. A History of the Pioneers and their Discoveries.* Oxford: Oxford University Press.

Sharpey-Shafer, E (1927). History of the Physiological Society during its first fifty years, 1876–1926. *Journal of Physiology, Supplement*; 1–181.

Sherrrington, C (1929). Some functional problems attaching to convergence. *Proceedings of the Royal Society, B*; **105**: 332–362.

Sherrington, CS (1893). Experiments in examination of the peripheral distribution of the fibres of the posterior roots of some spinal nerves. I. *Philosophical Transactions of the Royal Society, B*; **190**: 641–763.

Sherrington, CS (1893). Note on the knee-jerk and the correlation of action of antagonistic muscles. *Proceedings of the Royal Society*; **52**: 556–564.

Sherrington, CS (1898). Experiments in examination of the peripheral distribution of the fibres of the posterior roots of some spinal nerves. II. *Philosophical Transactions of the Royal Society, B*; **190**: 45–186, pl 3–6.

Sherrington, CS (1906). On the proprio-ceptive system, especially in its reflex aspect. *Brain*; **29**: 467–482.

Sherrington, CS (1906). *The Integrative Action of the Nervous System.* New Haven: Yale University Press, xvi + 411 pp.

Sherrington, CS (1913). Reflex inhibition as a factor in the co-ordination of movements and postures. *Quarterly Journal of Experimental Medicine*; **6**: 251.

Sherrington, CS (1928). A mammalian myograph. *Journal of Physiology*; **66**: iii–v.

Sherrington, CS (1940). *Man on His Nature.* The Gifford Lectures, Edinburgh, 1937–38. New York: MacMillan; Cambridge, England: University Press, 1941. viii, 413 pp.

Sherrington, Sir Charles (1932). Inhibition as a coordinative factor (Nobel Lecture). http://www.nobel.se/medicine/laureates/1932/sherrington-lecture.html.

Sherrington, Sir Charles (1941). *Man on His Nature.* New York: MacMillan, viii + 413 pp.

Sigworth, FJ (2003). Life's transistors. *Nature*; **423**: 21–22.

Singer, HD, & Goodbody, FW (1901). A case of family periodic paralysis with a critical digest of the literature. *Brain*; **24**: 257–285.

Sjöstrand, FS, & Andersson, E (1954). Electron microscopy ot the intercalated discs of cardiac muscle tissue. *Experientia*; **10**: 369–370.

Skou, JC (1957). The influence of some cations on an adenosine-triphosphatase from peripheral nerves. *Biochimica et Biophysica Acta*; **23**: 394–401.

Takeuchi, A, & Takeuchi, N (1959). Active phase of frog's end-plate potential. *Journal of Neurophysiology*; **22**: 395–411.

Talbott, JH (1941). Periodic paralysis. *Medicine (Baltimore)*; **20**: 85–143.

Tansey, EM (2008). Working with Cambridge physiologists. *Notes and Records of the Royal Society*; **62**: 131–137.

Tasaki, I, & Takeuchi, T (1942). Weitere Studien über den Aktionstrom der markhaltigen Nervenfaser und über die elektrosaltatorische Übertragung des Nervenimpulses. *Pflügers Archiv fur die Gesamte Physiologie des Menschen und der Tiere*; **245**: 764–782.

Tempel, BL, Papazian, DM, Schwarz, TL, Jan, YN, & Jan, LY (1987). Sequence of a probable potassium channel component encoded at *Shaker* locus of Drosophila. *Science*; **237**: 770–775.

Thomsen, J (1876). Tonische krämpfe in willkürlich beweglichen muskeln in Folge von erebter physischer Disposition (Ataxia muscularis?). *Archiv fur Psychiatrie und Nervenkrankheiten*; **6**: 720–718.

Toyoshima, C, & Unwin, N (1988). Ion channel of acetylcholine receptor reconstructed from images of postsynaptic membranes. *Nature*; **336**: 247–250.

Vallbo, ÅB, Hagbarth, K-E, Torebjörk, HE, & Wallin, BG (1979). Somatosensory, proprioceptive and sympathetic activity in human peripheral nerves. *Physiological Reviews*; **59**: 919–957.

Veratti, E (1902). Richerche sulla fine structtura della fibra muscolare striata. *Memorie Reale Istituto Lombardi*; **19**: 87–133.

Watson, JD (1968). *The Double Helix. A Personal Account of the Discovery of the Structure of DNA.* New York: Athenaeum.

Watson, JD, & Crick, FHC (1953). A structure for deoxyribonucleic acid. *Nature*; **171**: 737–738.

Whittaker, VP, Michaelson, IA, & Kirkland, RJA (1964). The separation of synaptic vesicles from nerve-ending particles ("synaptosomes"). *Biochemical Journal*; **90**: 293–303.

Woodbury, JW, & Patton, HD (1952). Electrical activity of single spinal cord elements. *Cold Spring Harbor Symposia in Quantitative Biology*; **17**: 185–188.

Woolsey, CN, Marshall, WH, & Bard, P (1942). Representation of cutaneous tactile sensibility in the cerebral cortex of the monkey as indicated by evoked potentials. *Bulletin of the Johns Hopkins Hospital*; **70**: 399–441.

Woolsey, TA (2001). Rafael Lorente de Nó. April 8, 1902–April 2, 1990. *Biographical Memoirs*. National Academy of Sciences; **79**: 84–105.

Young, JZ (1936). The giant nerve fibres and epistellar body of cephalopods. *Quarterly Journal of Microscopical Science*; **78**: 367–368.

Young, JZ (1951). *Doubt and Certainty in Science*. London: Clarendon Press.

Zhou, Y, Morais-Cabral, JH, Kaufman, A, & MacKinnon, R (2001). Chemistry of ion coordination and hydration revealed by a K+ channel-Fab complex at 2.0 Å resolution. *Nature*; **414**: 43–48.

Zhou, M, Morais-Cabral, JH, Mann, S, & MacKinnon, R (2001). Potassium channel receptor site for the inactivation gate and quaternary amine inhibitors. *Nature*; **411**: 657–662.

Zhuchenko, O, Bailey, J, Bonnen, P, Ashizawa, T, Stockton, DW, et al (1997). Autosomal dominant cerebellar ataxia (SCA6) associated with small polyglutamine expansions in the alpha 1A-voltage-dependent calcium channel. *Nature Genetics*; **15**: 62–69.

INDEX

Printed in the USA/Agawam, MA
January 4, 2013

571644.090